ADVANCED PRACTICAL ZOOLOGY

(For B.Sc. and M.Sc. Students of Zoology)

P.S. VERMA
M.Sc., Ph.D., F.E.S.I., F.A.Z.
Reader
Department of Zoology
Meerut College, Meerut

P.C. SRIVASTAVA
M.Sc., Ph.D.
Reader
Department of Zoology
K.K. (P.G.) College, Etawah

S. CHAND
PUBLISHING

S Chand And Company Limited

(ISO 9001 Certified Company)

S Chand And Company Limited

(ISO 9001 Certified Company)

Head Office: D-92, Sector–2, Noida – 201301, U.P. (India), Ph. 91-120-4682700

Registered Office: A-27, 2nd Floor, Mohan Co-operative Industrial Estate, New Delhi – 110 044, Phone: 011-49731800

www.**schandpublishing.com**; e-mail: **info@schandpublishing.com**

Marketing Offices:

Chennai	:	Ph: 23632120; chennai@schandpublishing.com
Guwahati	:	Ph: 2738811, 2735640; guwahati@schandpublishing.com
Hyderabad	:	Ph: 40186018; hyderabad@schandpublishing.com
Jalandhar	:	Ph: 4645630; jalandhar@schandpublishing.com
Kolkata	:	Ph: 23357458, 23353914; kolkata@schandpublishing.com
Lucknow	:	Ph: 4003633; lucknow@schandpublishing.com
Mumbai	:	Ph: 25000297; mumbai@schandpublishing.com
Patna	:	Ph: 2260011; patna@schandpublishing.com

First Edition 1994

Subsequent Editions and Reprints 1995, 96, 98, 99, 2000, 2002, 2003, 2004, 2006, 2007, 2008, 2009, 2010, 2012, 2013, 2014 (Twice), 2015, 2016 (Twice), 2017, 2018, 2019, 2020 (Twice), 2021, 2022 (Twice)

Reprint 2024

ISBN : 978-81-219-0518-3 **Product Code** : H6ZOP68ZOOL10ENAC0XR

PRINTED IN INDIA

By Vikas Publishing House Private Limited, Plot 20/4, Site-IV, Industrial Area Sahibabad, Ghaziabad – 201 010 and Published by S Chand And Company Limited, A-27, 2nd Floor, Mohan Co-operative Industrial Estate, New Delhi – 110 044.

PREFACE TO THE THIRD EDITION

With the introduction of +3 pattern in Indian universities at Degree Level, courses have been widely re-oriented and restructured. Several new topics have been added to the syllabi of under-graduate and post-graduate classes, in theory as well as practical curricula. This has created a need for a new book to cover the changed syllabi.

The present book "**Advanced Practical Zoology**" has been written to meet the requirements of the new curriculum in the practical work of B.Sc., B.Sc. (Hons), B.Sc. (Ag) and M.Sc. students of various universities.

The book incorporates practical exercises relating to cytology, microbiology, histochemistry, histology, embryology, economic (applied) zoology, environmental biology, animal behaviour, physiology, toxicology, genetics and some dissections. The chapters on cytology, microbiology, histochemistry, environmental biology, animal behaviour, physiology and toxicology will enable the students to understand some basic and advanced experimental procedures in the field concerned and provide a better understanding of specialised higher practical work. The chapter on Economic (applied) Zoology furnishes an account of some economically important animals from various groups of the Kingdom Animalia. The chapter on genetics deals with some genetical exercises. A chapter on Microtome and Microtomy has been given to enable the students to know the technique of preparing histological slides. A new chapter on Biostatistics has been added in the present edition as was desired by several colleagues and students from different universities and colleges.

We believe that the book will definitely serve the purpose of those for whom it has been written. Constructive suggestions from the readers for further improvement of the book will be highly appreciated and thankfully acknowledged, and incorporated in its future editions.

We sincerely wish to acknowledge our thanks to our colleagues who have inspired us to bring out this book.

AUTHORS

CONTENTS

CHAPTER 1 **General Directions** 1 – 3

Students' equipments 1, Directions to the students 1, Dissections 2 Flag-labelling 2, Mountings 2, Study of museum specimens 2, Study of slides 3, Practical note book 3, Spotting 3.

CHAPTER 2 **Preparation of Laboratory Reagents** 4 – 10

Fixatives or fixing agents 4, Grades of alcohol 6, Stains 6, Other reagents 8, Chemicals 9, Some important physiological reagents 9, Glass apparatus 10.

CHAPTER 3 **Preparation of Permanent Slides** 11 – 14

Temporary mounting 11, Permanent mounting 11, Summary of Permanent mounting 14.

CHAPTER 4 **Culture Methods** 15 – 16

Culture of *Amoeba* 15, Culture of *Paramecium* 15, Culture of *Euglena* 15, Culture of *Vorticella* 16, Culture of *Hydra* 16, Culture of *Planaria* 16, Culture of *Daphnia* 16.

CHAPTER 5 **Important Mountings** 17 – 29

Amoeba 17, *Paramecium* 17, *Euglena* 18, *Vorticella* 18, Rectal ciliates of frog 18, Preparation of stool smear to examine different protozoan parasites 20, Rotifers 21, Ctenophores 22, Freshwater annelids 22, Freshwater arthropods 22, Preparation of permanent slides of *Cimex* and *Pediculus* 23, Preparation of permanent slides of wings, legs and mouth parts of housefly and mosquito 23, Preparation of permanent slides of the developmental stages of mosquito and housefly 23, Preparation of permanent slides of ticks 24, Mites 24, Mouth parts and sting of honeybee 25, *Salpa* 25, *Doliolum* 26, *Amphioxus* 26, Oral hood of *Amphioxus* 26, Velum of *Amphioxus* 26; Squamous epithelium 26, Ciliated epithelium 28, Striated muscle 28, Unstriated muscle 28, T.S. of cartilage 29, Blood of frog 29, Medullated and non-medullated nerve fibres 29, Filoplume of pigeon 29.

CHAPTER 6 **Cytological Preparations** 30 – 42

Mitosis 30, Meiosis 30, Polytene chromosomes 31, Mitochondria 34, Micrographs of: Mycoplasma cell 35, *Escherichia coli* 35, Blue-green alga 36, Plant cell 37, Animal cell 38, Endoplasmic reticulum 38, Mitochondria 39, Golgi body 40, Nucleus 40, Chloroplast 41.

CHAPTER 7 **Microbiology** 43 – 44

Isolation and cultivation of soil micro-organisms 43, Isolate and cultivate bacteria from leaf sample 44, Staining of bacterial smear 44.

CHAPTER 8 **Histochemical Tests** 45 – 48

Histochemical determination of proteins 45, Histochemical determination of lipids 46, Histochemical determination of carbohydrates 46, Histochemical determination of nucleic acids 47, Histochemical determination of acid and alkaline phosphatases 48.

CHAPTER 9 **Histological Slides** 49 – 67

Simple squamous epithelium 49, Unstriated muscle fibres 49, Striated muscle fibres 49, Cardiac muscle fibres 50, Hyaline cartilage 50, Elastic cartilage 51, Fibro-cartilage 51, T.S. bone 51, L.S. bone 51, Blood of frog 52, Blood of man 53, T.S. artery 53, T.S. vein 54, Medullated nerve fibres 54, Non-medullated nerve fibres 55, V.S. skin mammal 55, T.S. oesophagus mammal 56, T.S. stomach mammal 56, T.S. duodenum mammal 57, T.S. liver mammal 59, T.S. pancreas mammal 60, T.S. spleen mammal 60, T.S. lung mammal 61, L.S. kidney mammal 62, T.S. cortex part kidney mammal 62, T.S. medulla part kidney mammal 64, T.S. testis mammal 64, T.S. ovary mammal 65, T.S. spinal cord mammal 66.

CHAPTER 10 **Embryological Slides** 68 – 94

Development of frog; Unfertilized egg 68, Sperm of frog 68, Fertilized egg 69, Cleavage 2-celled stage 69, Cleavage 4-celled stage 69, Cleavage 8-celled stage 70, Cleavage 16-celled stage 70, Cleavage 32-celled stage 70, L.S, early blastula 70, V.S. late blastula 71, V.S. early gastrula 71, V.S. late gastrula 72, V.S. complete gastrula 72, L.V.S. embryo 5 mm 72, Tadpole larva 73, Frontal section tadpole 73, T.S. tadpole through eye region 74, T.S. tadpole through auditory vesicles 74, T.S. tadpole through head and gills 75, Development of chick; Hen's egg 76, Chick embryo - 16 hours (w.m.) 76, Chick embryo - 18 hours (w.m.) 78, Chick embryo - 24 hours (w.m.) 78, Chick embryo - 28 hours (w.m.) 79, Chick embryo - 33 hours (w.m.) 79, Chick embryo - 36 hours (w.m.) 81, Chick embryo - 43 hours (w.m.) 81, Chick embryo - 55 hours (w.m.) 82, Chick embryo - 72 hours (w.m.) 83, Chick embryo - 96 hours (w.m.) 84, T.S. 18 - hours chick embryo 85, T.S. 24 - hours chick embryo 85, T.S. 28 - hours chick embryo 86, T.S. 33 - hours chick embryo 86, T.S. 48 - hours chick embryo 87, 88, 89, T.S. 72 - hours chick embryo through brain and eye 89, T.S. 72 - hours chick embryo through pharynx 90, T.S. 72 - hours chick embryo through auditory vesicles 90, Six-day chick embryo (w.m.) 91, Eight-day chick embryo (w.m.) 92, Twelve-day chick embryo (w.m.) 92, Eighteen-day chick embryo (w.m.) 92, Twenty one-day chick embryo (w.m.) 93, T.S. Chorio-allantoic placenta 93.

CHAPTER 11 **Economic (Applied) Zoology: Specimens and Slides** 95 – 183

Protozoa; *Entamoeba histolytica* 95, *Entamoeba coli* 96, *Entamoeba gingivalis* 96, *Trypanosoma* 97, *Leishmania* 98, *Trichomonas* 99, *Giardia* 100, *Eimeria* 100, *Plasmodium* 102, *Sarcocystis* 103, **Platyhelmenthes;** *Polystoma* 104, *Fasciola* 104, *Schistosoma* 105, *Paramphistomum* 106, *Diphyllobothrium* 107, *Taenia solium* 108, *T. saginata* 108, *Echinococcus granulosus* 109, Hydatid cyst 110, *Hymenolepis nana* 111, *Dipylidium caninum* 112, *Raillietina* 112, *Cotugnia* 113, Miracidium of *Fasciola hepatica* 114, Sporocyst of *F. hepatica* 114, Redia of *F. hepatica* 114, Cercaria of *F. hepatica* 115, Cercaria of *Schistosoma* 116, Mature proglottid *Taenia solium* 116, Scolex *T. solium* 116, Cysticercus (bladderworm) *T. solium* 118, **Aschelminthes;** *Enterobius* 118, *Ascaris* 119, *Ancylostoma* 120, *Dracunculus mediensis* 121, *Wuchereria (Filaria)* 122, *Trichinella*

123, Egg of *Ascaris* 124, **Arthropoda;** *Palaemon*126, *Schistocerca gregaria* 127, *Hieroglyphus* 128, *Odontotermes* 129, *Pediculus* 130, *Cimex* 131, Aphid 132, *Dysdercus koenigii* 132, *Leptocorisa varicornis* 133, *Bagrada* 134, *Pyrilla perpusilla* 134, *Tachardia lacca* 135, A twig encrusted with lac 136, *Aulacophora* 136, *Tribolium* 137, *Apis* 138, Honey bee - stages of life cycle 138, Legs of honey bee 139, Mouth parts of honey bee 141, Sting of honey bee 141, Honey 142, Bees-wax 142, *Corcyra cephalonica* 143, *Chilo* 144, *Earias* 144, *Bombyx mori* 145, Silkmoth – stages of life cycle 146, Caterpillar of silkmoth 146, Silk thread 147, *Xenopsylla* 147, *Musca* 148, Head and mouth parts housefly 149, *Culex* 149, Identification of *Culex and Anopheles* 150, Head and mouth parts female *Culex* and *Anopheles* 151, Head and mouth parts male *Culex* and *Anopheles* 151, **Mollusca;** *Ostrea* 152, *Pinctada* 153, T.S. shell and mantle *Unio* 153, **Pisces;** *Hilsa* 154, *Notopterus* 155, *Labeo* 156, *Catla* 157, *Cyprinus* 157, *Cirrhina* 158, *Heteropneustes* 159, *Clarius* 160, *Wallago* 160, *Mystus* 161, *Rita* 162, *Anabas* 163, **Reptilia;** Poisonous and non-poisonous snakes 164, Key for identification of poisonous and non-poisonous snakes 165, *Typhlops* 166, *Python* 167, *Eryx* 168, *Bungarus* 169, *Naja* 170, *Elaps* 172, *Viper* 173, *Ancistrodon* 173, *Crotalus* 174, *Hydrophis* 175, *Natrix* 176, *Zamenis* 177, *Dendrophis* 178, **Aves;** *Gallus* 179, **Mammalia;** *Lepus* 180, *Rattus* 181.

CHAPTER 12 **Experimental Physiology** **184 – 224**

Carbohydrates; General tests 184, Colour test for Fructose 185, Glucose 185, Reducing monosaccharides 185, Sucrose 186, Starch 187, Glycogen 187, **Proteins;** Colour tests for proteins 188, Colour tests individual amino acids; Arginine 189, Tyrosine 189, Tryptophan 189, Precipitation test for proteins 189, **Lipids** (fats and oils); Colour test 190, Emulsification test 190, Formation of acrolein 190, Solubility test 190, **Enzymes;** Test for salivary amylase 191, Invertase 191, Pepsin 192, Trypsin 192, **Haematological;** Estimation of haemoglobin 193, Enumeration of RBCs 194, Enumeration of WBCs 197, Differential count of WBCs 198, Coagulation of blood 199, Bleeding time 199, Clotting time 199, Preparation of haemin crystals 200, Effect of osmolarity on blood corpuscles 200, Determination of E.S.R. 201, Blood pressure and its measurement 203, Circulation of blood in capillaries of web of frog 204, Recording of heart beat of frog 205, Effect of acetylcholine, adrenaline, hot and cold Ringers solutions, pH change, Ca, Na and K ions on heart beat 207, Recording of muscular contraction 207, Recording of tetanus 209, **Respiration;** To determine oxygen consumption of rat 209, Rate of oxygen consumption of rat 210, **Nervous system;** Reflex action of frog 211, Wiping reflex in frog 211, Knee-jerk reflex in man 211, **Excretory system;** Test of urine 212, **Osmosis;** Its demonstration 212, **Chromatography** 214.

CHAPTER 13 **Ecological Experiments** **225 – 254**

Study of physical factors; recording of atmospheric temperature 225, recording of atmospheric pressure 226, recording of R.H. 227, recording of light intensity 230, Study of abiotic factors; study of physical characteristic of soil 231, study of chemical characteristics of soil 232, Estimation of pH of water 233, To record temperature of water 234, To estimate total hardness of water 234, To estimate the alkalinity of water sample 235, To esitmate total dissolved solids in water sample by gravimatric method 235, To estimate D.O. content of water sample by Winkler method 236, To estimate the amount of free carbon dioxide in water sample 236, Quaiitative study of planktons 237, Quantitative study (total count and differential count) of planktons 238; Study of community; Its study by quadrat method by determining of frequency, density and abundance 241, Study of pyramids and food chain 243, Pyramid of numbers 244, Pyramid of biomass 245, Pyramid of energy 245, Food chain 246, Effect of light on colour changes in frog 246, Adaptive and structural modifications of *Physalia* 248, *Fasciola* 248, *Schistosoma* 249, *Taenia* 249, *Chaetopterus* 249, *Carausius* 249, *Labeo* 250, *Exocoetus* 250, *Anabas* 250, *Synaptura* 251, *Rhacophorus* 252, *Draco* 252, *Phrynosoma* 252, *Corvus* 253, *Cynopterus* 253.

CHAPTER 14 **Toxicological Experiments** **255 — 257**

Bahavioural response of a fish under different doses of DDT stress 255, Determination of LC_{50} 256, Ocular effect of organophosphorus insecticide in rat 256.

CHAPTER 15 **Animal Behaviour** **258 — 260**

Orientation behaviour; Phototaxis in earthworm 258, Phototaxis in housefly 259, Geotaxis in earthworm 259, Chaemotaxis in *Paramecium* 259.

CHAPTER 16 **Endocrine Glands** **261 — 270**

To locate various endocrine glands in a given mammal 261, To locate pituitary gland and pineal gland 263, Comment upon the slides of endocrine glands; V.L.S. anterior lobe of pituitary 264, V.S. thyroid gland 264, T.S. parathyroid gland 265, T.S. adrenal gland 266, Experiments relating to surgical removal of endocrine glands; Thyroidectomy 267, Adrenalectomy 268, Ovariectomy 269.

CHAPTER 17 **Genetical Exercises** **271 — 283**

Based on dominance and recessive 271; Based on incomplete dominance 273; Based on lethal genes 274; Based on law of independent assortment 276; Based on interaction of genes 278; Based on multiple alleles 279; Based on sex-linked inheritance 282.

CHAPTER 18 **Dissections** **284 — 296**

Instructions for dissection 284; **Grasshopper**: General anatomy 284, Digestive system 285; Reproductive system 285, Nervous system 286; ***Loligo***: Nervous system 287, ***Labeo***: General anatomy 288, Cranial nerves 290, Internal ear 291, Weberian ossicles 291, ***Wallago***: Cranial nerves 293; **Accessory respiratory organs**: *Anabas* 294, *Clarias* 295, *Heteropneustes or Saccobranchus* 295; Electric organs 295.

CHAPTER 19 **Microtome and Microtomy** **297 — 303**

Know your microtome 297, Rocking microtome 297, Rotary microtome 298, **Microtomy**: Material 300, Fixation of material, Washing of fixative from material, Dehydration of material, Clearing or dealcoholization of material; Preparation of the material for embedding, Paraffin infiltration and embedding, Block making, Trimming of block, Section cutting, Spreading of ribbons, Staining, dehydration, clearing and mounting, Microscopic study.

CHAPTER 20 **Biostatistics** **304 — 320**

Presentation of Data in the form of Frequency Tables 304, Discrete variable 305, Continuous variable 305, Presentation of Data in the form of Graphs; Pie chart 307, Line charts and bar graphs 308, Histograms, Frequency Polygons and Ogives 308, Measures of Central Tendencies; Mean 309, Median 310, Mode 312, Measures of Dispersion; Range 312, Mean deviation 312, Variance 313, Standard deviation 314, Standard error 316, Coefficient of Variation 316, Tests of Significance; t-test 317, Chi-square test 318.

CHAPTER 1

GENERAL DIRECTIONS

STUDENTS' EQUIPMENTS

Each student, while coming to the laboratory for the practical work, is required to bring certain necessary equipments. However, it is not possible to list all what is required but the following list is considered necessary.

1. Practical note book.
2. A Manual of Practical Zoology.
3. Drawing pencil (preferably H, B and HB).
4. Pencil sharpner (sand paper).
5. Pencil eraser.
6. Measuring scale.
7. Dissecting box containing the following instruments :
   (i) **Scalpels** two or three, sharp and of different sizes.
   (ii) **Scissors** two pairs, a fine pointed pair for fine dissection and other large for general work.
   (iii) **Forceps** two, one fine pointed and other blunt.
   (iv) **Dissecting needles** two, mounted in handles.
   (v) **Blow pipe** one.
   (vi) One edged safety razor or blade.
8. Brush one.
9. Dropper one.
10. Hand lens one.
11. A piece of clean cloth.
12. Alpins.

Do not mind bringing all these things on every turn, you do not know when you may need them.

DIRECTIONS TO THE STUDENTS

1. While coming to the laboratory for practical class check that you possess the **practical note book, pencil, pencil eraser, sharpner, scale, brush** and complete set of **dissecting instruments.**
2. Instruments should be sharp and according to the requirements.
3. Come prepared with the work you are supposed to do in the laboratory.
4. Keep your instruments, practical note book and seat well arranged and tidy.
5. Never encourage the habit of lending either to or from your class fellows. Bring all the requirements of the day.
6. Before starting the work listen carefully the instructions given by your teacher.
7. Remove your all difficulties from your teacher. Do not consult your class fellows for any help.
8. Never rub your pencil on the floor or on the top of the table, use the sharpner.
9. Clean and arrange your seat before leaving the laboratory.

10. Maintain complete silence in the laboratory.

Dissections

1. Before starting the dissection listen the instructions given by the teacher and follow them carefully throughout the dissection.
2. Draw a diagram of the dissection to be done.
3. Dissection should be done under water and should be left completely submerged in water.
4. Always keep the water clean in the dissecting dish during dissection by changing it whenever stained with blood, etc., so that visibility remains the same.
5. Put every waste of dissection in the petridish or in the sank slowly. Do not throw it carelessly on the top of the table or everywhere.
6. Dissect **along** and not **across** the blood vessels and nerves.
7. While doing dissection, the required parts should not only be exposed but these should be properly displayed by cleaning and black papering.
8. A well-labelled diagram of the dissection should be left along with dissection only if it is required.
9. Do tthe dissection patiently and confidently.
10. During the course of dissection, concentrate only on your own dissection:

Flag-labelling

Frequently, the students are asked to dissect some organs, blood vessels or nerves, etc., and flag-label them. In flag-labelling following instructions are to be followed:

1. Small pieces of paper (2.5 x 0.7 cm) are prepared and a needle is passed through each close to one end.
2. The names of organs, blood vessels or nerves are written (in pencil) on these flags.
3. Then each flag is inserted in the dissecting dish close to the organ, blood vessel or nerve bearing the name on the flag.

Mountings

While making the permanent slides (mountings) the following instructions should be followed strictly :

1. Never keep your mounting material for less or more time than the desired time in a alcoholic grade or stain.
2. Keep the material for slightly more time in 90% alcohol and absolute alcohol for complete dehydration.
3. Ensure the complete dehydration after absolute alcohol by putting the material in xylol or benzene. If it gives turbidity with xylol then dehydrate it again.
4. Do the dehydration in closed specimen tubes or in covered cavity blocks.
5. Always use a brush and never forceps for holding the mounting material.
6. Put the required amount of **Canada balsam** or **D.P.X.** on the slides for mounting. Excess of Canada balsam or D.P.X. makes the slides dirty.

Study of Museum Specimens

1. First study the characteristics of the specimen to be drawn.
2. Try to find out those characteristics in the specimens.
3. Draw only that view (e.g., dorsal, ventral or lateral) which exhibits maximum details.
4. Diagrams should be drawn from the original specimen although help of the book can be taken for comparing the details and labelling of various parts.
5. While drawing, special care should be taken in maintaining the proportion of length and width of the animal body.
6. Draw only line diagrams, shading should be avoided.
7. Always write down the classification of the entire specimens.

Study of Slides

1. While studying microscopical preparations (permanent mounts) under the microscope, do not disturb the slide focussed by the teacher.

2. Draw from the slide. Compare and label it with your book.

3. Whenever there is a mount of an entire animal do not forget to write its classification.

Practical Note Book (Practical Record)

1. Practical note book should be neat and clean and uptodate.

2. Put date on the left hand corner of the page of the note book and details of the work on the top in centre.

3. Diagrams should be correctly drawn and well-labelled.

4. Diagrams of all dissections, museum specimens and slides should be drawn and comments of all should be written.

Spotting

While doing spotting during the practical examination, be careful about the following points :

1. First of all write the spot number.

2. Draw a rough but well-labelled diagram of the spot.

3. Write down the classification if it is an entire animal, whether a museum specimen or entire animal mounted on the slide.

4. Give characteristic features of the phylum, subphylum, class, subclass, order, suborder, etc.

5. Only interesting and important features should be written in comments.

CHAPTER 2

PREPARATION OF LABORATORY REAGENTS

Various chemicals, reagents and glass apparatuses, etc., are needed in a laboratory to conduct different practical works, exercises and experiments. However, some important common reagents and the method of their preparation are described herein to enable the laboratory staff to prepare these reagents accurately.

FIXATIVES OR FIXING AGENTS

1. Formalin solution

1. Formalin water ... 10 ml.
2. Distilled water ... 90 ml.

This fixative is used for the fixation of animal material.

2. Alcohol

70%, 90% or absolute alcohols are used as fixatives chiefly for protozoans and other micro-organisms. A few drops of the alcohol are poured on the object.

3. Formalin acetic alcohol (F. A. A.)

1. Formalin 40% ... 10 ml.
2. Glacial acetic acid ... 5 ml.
3. Ethyl alcohol 70% ... 85 ml.

This fixative is generally used for fixing the plant material and nematodes.

4. Bouin's fluid (Aqueous)

1. Saturated aqueous solution of picric acid ... 75 ml.
2. Formalin 40% ... 25 ml.
3. Glacial acetic acid ... 5 ml.

The tissue fixed in aqueous Bouin's fluid must be thoroughly washed in water to remove the traces of picric acid so that it may not interfere in staining and the process of dehydration.

5. Bouin's fluid (Alcoholic)

1. Saturated solution of picric acid in 70% alcohol ... 75 ml.
2. Formalin 40% ... 25 ml.
3. Glacial acetic acid ... 5 ml.

The tissue fixed in alcoholic Bouin's fluid should be thoroughly washed in 70% alcohol after fixation to remove the traces of picric acid.

6. Chromo-acetic acid (Flemming solution)

1. Chromic acid ... 0.2 to 0.25 ml.
2. Acetic acid ... 0.1 ml.
3. Distilled water ... 100 ml.

7. Formalin-picric-trichloracetic fixative

1. Formalin 40% ... 15 ml.
2. Trichloracetic acid ... 0.50 gm.
3. 2% picric acid solution in 50% alcohol ... 85 ml.

8. Carnoy's fluid

(a) Formula No. 1

1. Glacial acetic acid ... 25 ml.
2. Absolute alcohol ... 75 ml

(b) Formula No. 2

1. Glacial acetic acid ... 10 ml.
2. Chloroform ... 30 ml.
3. Absolute alcohol ... 60 ml.

9. Zenker's fluid

1. Potassium dichromate ... 2.5 gm.
2. Mercuric chloride ... 5.0 gm.
3. Sodium sulphate ... 1.0 gm.
4. Glacial acetic acid ... 5.0 ml.
5. Water ... 100.0 ml.

The common practice of making a stock solution of first two ingredients and adding the acetic acid just before use has been found to be unnecessary when chemically pure acetic acid is used.

10. Chromo-aceto-osmic acid

(a) Weak formula

1. Chromic acid ... 0.25 ml.
2. Osmic acid ... 0.1 ml.
3. Glacial acetic acid ... 0.1 ml.

This fluid is good for fixing the smaller animals.

(b) Strong formula

1. Chromic acid 1% ... 15 ml.
2. Osmic acid 2% ... 4 ml.
3. Glacial acetic acid ... 1 ml.

The stock solution of the above constituents is prepared and mixed at the time of use. This fluid is good for fixing the large animals or objects.

11. Corrosive sublimate acetic fixative

1. 5% or saturated solution of mercuric chloride ... 100 ml.
2. Glacial acetic acid ... 5.0 ml.

12. Formol-sublimate fixative

1. Saturated solution of mercuric chloride in 10% formalin.

13. Henking fluid

1. Osmic acid ... 1 ml.
2. Formic acid ... 3 ml.
3. Glycerine ... 16 ml.
4. Dahlia ... 0.04 gm.
5. Distilled water ... 80 ml.

14. Mayer's albumen

1. White of egg (Albumen) ... 50 ml.
2. Glycerine ... 50 ml.
3. Sodium salicylate ... 1.0 gm.

Shake them well together and filter into a clean bottle for use.

Mayer's albumen is used to fix the small organisms and microtome sections on the slide.

15. Normal saline

1. Sodium chloride (analytical) ... 0.78 gm.
2. Distilled water ... 100 ml.

Dissolve sodium chloride in water and mix it thoroughly before use.

GRADES OF ALCOHOL

The alcoholic grades are usually prepared from **rectified spirit** which contains approximately 95% alcohol.

(i) **30% Alcohol**

Rectified spirit ... 31.5 ml.
Distilled water ... 68.5 ml.

(ii) **50% Alcohol**

Rectified spirit ... 52.6 ml.
Distilled water ... 47.4 ml.

(iii) **70% Alcohol**

Rectified spirit ... 73.6 ml.
Distilled water ... 26.4 ml.

(iv) **90% Alcohol**

Rectified spirit ... 94.7 ml.
Distilled water ... 5.3 ml.

Acid alcohol

1. Alcohol 70% ... 100 ml.
2. Hydrochloric acid concentrated ... 1 ml.

STAINS

1. Borax carmine

1. Borax ... 4 gm.
2. Carmine ... 2 gm.
3. Distilled water ... 100 ml.

The solution of borax and carmine is boiled for about half an hour so that it changes its colour. Now add equal volume of 70% alcohol, allow it to stand and filter. After 2–3 days it becomes mature enough to be used in the laboratory.

2. Picro-indigo-carmine

1. Saturated solution of picric acid to 90% ... 1 part
2. Saturated solution of indigo carmine in 70% ... 2 parts

This stain is very conveniently used for the staining of chitinous parts such as mouth parts of insects. It is used after dehydration of material in 70% alcohol.

3. Aceto-carmine

1. Glacial acetic acid ... 45 ml.
2. Distilled water ... 55 ml.
3. Dry carmine stain ... 0.5-1.0 gm.

Mix 45 ml of glacial acetic acid in 55 ml of distilled water and heat to boiling and then add 0.5–1.0 gm. of carmine. Shake well, cool and filter.

This is one of the most commonly used stain for chromosomal studies.

4. Aceto-orcein

1. Glacial acetic acid ... 45 ml.
2. Distilled water ... 55 ml.
3. Orcein stain ... 1 gm.

Mix the above three things together. It is an ideal stain for the study of chromosomes.

5. Haematoxylin

(I) Delafield's haematoxylin

1. Haematoxylin crystals ... 4 gm.
2. Absolute alcohol ... 25 ml.
3. Saturated solution of ammonia alum ... 400 ml.
4. Glycerine ... 100 ml.
5. Methyl alcohol ... 100 ml.

Dissolve 4 gm of haematoxylin crystals in 25 ml of absolute alcohol and add to it 400 ml of saturated solution of ammonia alum. Leave it exposed to light and air in unstoppered bottle for about a week. Filter and add 100 ml of glycerine and 100 ml of methyl alcohol. Allow the solution to stand (uncorked) until the colour is sufficiently dark, then filter.

(II) Ehrlich's acid haematoxylin

1. Haematoxylin ... 2 gm.
2. Absolute alcohol ... 100 ml.
3. Glycerine ... 100 ml.
4. Glacial acetic acid ... 10 ml.
5. Distilled water ... 100 ml.
6. Alum in excess

Dissolve the haematoxylin in absolute alcohol, add the acetic acid and then the glycerine and water. Let the mixture ripen in the light until it acquires a dark red colour. It will then keep with constant power for years, if kept in a well stoppered bottle. Haematoxylin is used as a nuclear stain.

6. Eosin

(I) Aqueous eosin

1. Eosin powder ... 1.0 gm.
2. Distilled water ... 100 ml.

(II) Alcoholic eosin

1. Eosin powder ... 1.0 gm.
2. Alcohol 70% or 90% ... 100 ml.

7. Safranin

1. Safranin ... 1.0 gm.
2. Absolute alcohol ... 100 ml.

Safranin is popular for its brilliancy and permanency in balsam.

8. Light green

1. Light green stain ... 0.82 gm.
2. Alcohol 905 ... 100 ml.

This is a very good plasma stain.

9. Fast green

1. Fast green stain ... 1 gm.
2. Alcohol 90% ... 100 ml.

Mix these together. It is a very good cytoplasmic stain.

10. Acid fuchsin

1. Acid fuchsin ... 0.5 gm.
2. Distilled water ... 100 ml.

This stain is mostly used for the staining of chitinous parts such as mouth parts and other sclerites of arthropods.

11. Leishman stain

1. Leishman stain ... 0.15 gm.
2. Absolute alcohol ... 100 ml

Dissolve the stain in absolute alcohol and leave it for ripening for 48 hours. Leishman stain is very good for the staining of protozoans.

12. Mallory's triple stain

Sol. A.	1.	Acid fuchsin	... 1.0 gm.
	2.	Distilled water	... 100 ml.
Sol. B.	1.	Aniline blue	... 0.5 gm.
	2.	Orange G	... 2.0 gm.
	3.	Phosphomolybdic acid	... 2.0 gm.
	4.	Distilled water	... 100 ml.
Sol. C.	1.	Phosphomolybdic acid	... 1 gm.
	2.	Distilled water	... 100 ml.

For Mallory's triple stain three solutions are prepared. In solution A sections are stained for 30 seconds and after it washed in distilled water for 2 minutes and then stained in solution B for 1–5 minutes and subsequently washed in distilled water for 2–5 minutes. After staining the sections in A and B solutions, sections are kept in solution C for 30 seconds and are washed again in distilled water for 1–2 minutes and then dehydrated and mounted in Canada balsam or D.P.X.

Mallory's triple is a very good stain for staining the sections of animal tissues. It gives very good results.

13. Congo red vital stain

1. Congo red ... 0.5 gm.
2. Distilled water ... 100 ml.

Stir the congo red into the water until all dye is dissolved. Filter to remove any large particles.

14. Crystal violet stain

1. Crystal violet (Gentian violet) ... 2.0 gm.
2. Ethyl alcohol 90% ... 20 ml.
3. Distilled water ... 180 ml.

Dissolve the crystal violet in the alcohol and add the water. This is a simple substitute for Gram's stain.

15. Giema's stain

1. Giema's powder ... 0.5 gm.
2. Glycerine ... 33 ml.
3. Absolute methyl alcohol ... 33 ml.

It takes 1 to 2 hours to dissolve the powder in the glycerine. Then add the acetone free absolute methyl alcohol. This stock solution is diluted 1 : 10 with distilled water before using.

16. Methylene blue

1. Methylene blue ... 1.5 gm.
2. Ethyl alcohol 95% ... 100 ml.

This stain is used for staining the small animals in life.

17. Methyl green

1. Methyl green ... 8.5 gms.
2. Distilled water ... 100 ml.
3. Glacial acetic acid ... 1 ml.

OTHER REAGENTS

1. Acid water

1. Conc. HCl ..... 1 drop
2. Distilled water .... 100 ml.

The following chemicals and glass apparatus are required for performing the colour reactions (tests of carbohydrates, proteins, fats and enzymes, etc.).

CHEMICALS

1. Absolute alcohol.
2. Acetic acid (glacial).
3. Ammonium sulphate.
4. Benedict's solution.
5. Barfoed's reagent.
6. Barium chloride.
7. Bismuth subnitrate.
8. Chloroform.
9. Caustic soda.
10. Caustic soda (alcoholic).
11. Calcium chloride.
12. Copper sulphate 0.5 %.
13. Copper sulphate 1.0 %.
14. Chlorophenol red.
15. Ether.
16. Ethyl acetate.
17. Egg white 5 %.
18. Esbach's reagent.
19. Fehling's solution.
20. Fructose 0.2 %.
21. Fructose 0.5 %.
22. Ferric chloride 0.5 %
23. Ferric chloride 1.0 %.
24. Glucose 1 %.
25. Glucose 2 %.
26. Glucose 0.2 %.
27. Glucose 0.5 %.
28. Glyoxalic acid reagent.
29. Glycerine.
30. Hydrochloric acid (Conc)
31. Iodine solution.
32. Lactose 2 %.
33. Lead acetate.
34. Methylene blue.
35. Maltose 2 %.
36. Mercuric chloride (Saturated).
37. Mercuric sulphate.
38. Mercuric nitrate.
39. Nylander's reagent.
40. Ninhydrin 1 %.
41. Olive oil.
42. Oleic acid.
43. Potassium ferricyanide.
44. Potassium tartarate.
45. Picric acid (Saturated solution).
46. Phenyl hydrazine hydrochloride.
47. Potassium ferrocyanide solution.
48. Peptone.
49. Phenolphthalein.
50. Potassium dihydrogen phosphate.
51. Potassium hydrogen sulphate.
52. Sodium hydrogen sulphate.
53. Sodium chloride.
54. Sodium acetate.
55. Seliwanoff's reagent.
56. Silver nitrate.
57. Sulpho salicylic acid.
58. Sodium nitrate.
59. Sodium hypobromide.
60. Sodium hydroxide.
61. Sucrose 2 %.
62. Sulphuric acid (Conc).
63. Tannic acid.
64. Thymol.
65. Thymol blue.
66. α-naphthol 1 %

SOME IMPORTANT (PHYSIOLOGICAL) REAGENTS

1. **Alpha naphthol**. 1 gm of α-naphthol is dissolved in 95 % alcohol.

2. **Barfoed's reagent**. 13.3 gms of neutral copper acetate crystals are dissolved in 200 ml of distilled water. After filtering it add 1.8 ml of glacial acetic acid.

3. **Benedict's reagent**. Two solutions are prepared :

Solution A. 173 gms of sodium citrate and 100 gms of anhydrous sodium carbonate is dissolved in 800 ml of distilled water by heating.

Solution B. 17.3 gms of copper sulphate crystals dissolved in 100 ml of water.

Both the solutions are to be cooled. Solution B is added to solution A with constant stirring in 1000 ml titration flask and diluted with distilled water up to the mark.

4. **Esbach's reagent**. 1 gm of picric acid and 2 gms of citric acid are dissolved in 100 ml titration flask in distilled water. Dilute with distilled water up to the mark.

5. **Fehling's solution.** There are two types of Fehling's solution, *viz.*, A and B.

   (i) **Fehling's solution A**. 7.93 gms of copper sulphate crystals are dissolved in 100 ml of distilled water.

(ii) **Fehling's solution B.** 250 gms of potassium hydroxide and 320 gms of sodium or potassium tartarate are dissolved in 1000 ml of distilled water.

6. Nylander's reagent. 2 gms of bismuth subnitrate and 4 gms of sodium or potassium tartarate in 100 ml of 10 % potassium hydroxide solution. It is filtered after cooling it.

7. Leishman's stain. It is prepared by dissolving 0.15 gm of Leishman's powder in 100 ml of methyl alcohol.

8. Wright's stain. Take 25 gm of Wright's stain powder and dissolve in 500 ml of methyl alcohol. The stain matures in four weeks.

9. Fearson's reagent. 100% methylamine hydrochloride solution.

10. Seliwanoff's reagent. Dissolve 0.5 gm of resorcinol in 33 ml of conc. HCl (36%). Dilute it with distilled water to make it 100 ml.

11. Foulger's reagent. Dissolve 40 gm of urea in 80 ml of the H_2SO_4 (40%) and add 2 gm of stannous chloride in it. Boil till the mixture appears clear. Cool, then add 20 ml of 40% H_2SO_4 more.

12. Millon's reagent. Dissolve 10 gm of mercuric sulphate in 100 ml of dilute sulphuric acid (40%), add more sulphuric acid to make it 200 ml. Add 20 ml of 0.25% of $NaNO_2$ solution in it.

13. Molisch's reagent. Make it by dissolving 2 gm of α–Naphthol in 100 ml of 95% alcohol.

14. Sudan III reagent. Make it by dissolving 0.1 gm of Sudan III powder in 50 ml of 95% alcohol and add 50 ml glycerol in it.

15. Saturated picric acid solution. Dissolve excess of picric acid crystals in 100 ml of distilled water and filter.

16. Ninhydrin solution. Make 0.1% solution of ninhydrin in distilled water.

17. Iodine solution. Take 2% potassium iodide solution in distilled water and add excess of iodine crystals till the solution turns deep yellow. If essential, filter.

18. Physiological salines. Physiological salines differ for different group of animals, eg.,

(i) For invertebrates 0.60% analytical NaCl solution.

(ii) For frogs 0.64% analytical NaCl solution.

(iii) For birds 0.75% analytical NaCl solution.

(iv) For mammalian tissues except blood 0.9% analytical NaCl solution.

(**Note**. All these salines are made in distilled water)

19. Ringer's solution. Dissolve 0.85 gm of analytical NaCl, 0.025 gm of KCl and 0.03 gm of $CaCl_2$ in distilled water.

GLASS APPARATUS

1. Test-tubes.
2. Funnels.
3. Beakers.
4. Pipettes.
5. Measuring cylinders.
6. Glass stoppered bottles.
7. Spirit lamp.
8. Flasks.
9. Reagent bottles.

Besides these, test-tube holders, test-tube stands, funnel stands, cotton and filter papers are also required.

CHAPTER 3

PREPARATION OF PERMANENT SLIDES

There are two methods of preparation of slides, *viz.*,

1. Temporary mounting ;
2. Permanent mounting.

1. TEMPORARY MOUNTING

Temporary mounts are prepared in water, normal saline and glycerine.

(i) When the mount is prepared in water, a drop of water containing the material to be examined, *e.g.*, *Amoeba, Euglena* and *Paramecium*, is kept on the slide and a cover slip is placed up on the slide and then it is examined under the microscope.

(ii) When the mount is prepared in normal saline solution, the material to be examined is put in the normal saline solution upon the slide which cover slip is placed and then studied under the microscope.

(iii) In case of glycerine preparation, the material is first of all washed in water and stained in aqueous eosin, then mounted in glycerine.

2. PERMANENT MOUNTING

The process of permanent mounting involves a series of treatments. The utility of permanent mountings is that the organism which is to be mounted can be stored for a long a period without any change in its structure. The **Canada balsam** or **D.P.X.** is employed as mounting medium. The process of permanent mounting includes the following steps :

1. Narcotization and killing;
2. Fixing;
3. Washing;
4. Dehydration;
5. Staining;
6. Destaining or differentiation;
7. Clearing or dealcoholization;
8. Mounting;
9. Labelling.

1. Narcotization and killing

Narcotization and killing are the initial and most important stages in the preparation of permanent mounts. The organisms are narcotized or killed with the use of some chemicals in order to prevent the changes in the form of the organisms as they had in the living condition during their life.

The narcotics used are absolute alcohol, chloroform, menthol, ether, cocaine hydrochloride, chloroform water 1 or 2% solution of urethane or by passing a current of CO_2 through the water containing the organism.

For killing the organisms two methods are employed depending upon the purpose of study :

1. If the specimen is to be dissected after killing chloroform is generally used to kill the organism.

2. If the specimen is to be converted into permanent microscopical preparation where only small organisms or small pieces of larger animals are concerned-a reagent is employed which not only kills but penetrates and fixes the tissue so that they are but slightly affected by subsequent

treatment. In some cases smaller organisms are killed by heating. This method of killing also brings about the simultaneous fixation of the organisms. There are some of the chemicals which perform both the functions of killing and fixing, *e.g.*, formol sublimate mixture.

2. Fixing

Fixing is most essential for all types of microscopical preparations, *e.g.*, whole mounts and sections of parts of the body. The fixation of tissues and entire organisms is done by various types of fixatives. The important functions of the fixatives or fixing fluids are as follow :

1. The fixing fluids penetrate the tissues by inward diffusion so that the tissues can be preserved as nearly as possible in natural shape and check the further post-mortem changes.

2. The fixatives harden the tissues and set intra-cellular elements.

3. The fixing fluids render insoluble the various cell constituents and bring out differences in the refractive indices of different portions.

Some of the fixing fluids or fixatives or fixing agents are alcohol, formalin, corrosive sublimate, acetic acid, picric acid and osmic acid, etc., used in various combinations and strengths. The time of fixation may very in different cases.

3. Washing

Washing of the tissue or organism, after fixation, is an essential process to remove the excess of fixative. If the fixative is not removed or washed completely from the tissue it may interfere in the subsequent processes of dehydration and staining. Washing is done by different liquids depending upon the type of fixative used. The aqueous Bouin's fluid is washed in running tap-water, while alcoholic Bouin's fluid is washed with the alcohol of the same strength which is present in the fixative. Formalin and other fixatives containing large amount of water are mostly removed by distilled water.

4. Dehydration

Dehydration means the removal of water from the tissues. Dehydration is an essiental process due to the following reasons:

1. It prevents the tissues from decomposition by removing the water and by killing the germs harboured by tissues.

2. It is necessary because the mounting medium Canada balsam or D.P.X. is not miscible with water and the dehydration renders the mounting medium to penetrate the tissue.

Dehydration is done by processing the tissues in various grades of alcohol. The need of passing the tissues in different grades of alcohol is to avoid the shrinkage and distortion of tissues by abrupt use of strong alcohol. Therefore, dehydration is carried out gradually by passing the tissues from lower grade of alcohol to subsequent higher grade of alcohol such as 30, 50, 70, 90 and 100 per cent. The dehydration is done in corked specimen or staining tubes. The time in each grade of alcohol is to be adjusted according to the thickness of tissue.

5. Staining

By staining, the parts or components of cell or of organism can be clearly differentiated. In staining the tissues or organisms are coloured by particular deys. The deys which are commonly used for staining in the laboratory are: borax carmine, picro-indigo carmine, methylene blue, aniline blue, haematoxylin, eosin, orange G, Victoria blue, Sudan black, Congo red and alizarin, etc. The stains can be classified into the following types :

(i) Nuclear stains are used to stain the nucleus of the cell, *e.g.*, **heamatoxylin**.

(ii) Cytoplasmic stains are used to stain the cytoplasm, *e.g.*, **eosin**.

(iii) General stains are those which stain the entire cell or organism, *e.g.*, **borax carmine, picro-indigo carmine, methylene blue,** and **orange G,** etc.

The stains are prepared in aqueous, alcoholic and acid media.

There are usually two types of staining, *viz.*, single staining and double staining.

Single staining. When only one stain, such as **borax carmine** or **picro-indigo carmine,** is used, the process is called single staining.

Double staining. When two stains, one nuclear stain such as **haematoxylin** and another cytoplasmic stain such as **eosin** are used, the process is called double staining.

6. Destaining or differentiation

Sometimes the tissues or organisms are overstained and render difficulty during observations under the microscope. To avoid this difficulty the excess of stain is removed. This process of removing excess of stain is caled **destaining** or **differentiation**. Destaining is done by acid water or acid alcohol. **Acid water** is used with the **aqueous stains**, while **acid alcohol** is used with the **alcoholic stains**.

7. Clearing or dealcoholization

After dehydration, clearing or dealcoholization of tissues is necessary due to the following reasons :

1. To remove the alcohol from the tissue or organism and to facilitate the penetration of the mounting medium which is mostly Canada balsam or D.P.X.

2. To make the tissue or organism more clear and transparent so that very minute details can be observed under the microscope.

The clearing agents which are commonly used in the laboratories are xylol, benzene, clove oil and cedar wood oil, etc. Sometimes the turbidity is observed just after the use of clearing agent which indicates the presence of moisture, *i.e.*, incomplete dehydration. In such cases tissue or organism should be again dehydrated in 90% and 100% alcohol for complete dehydration and then transferred in clearing agent.

Clove and cedar wood oil are the best clearing agents used for clearing the soft tissues and organisms.

8. Mounting

It is the last step in the permanent mounting and is essential for the study and preservation of the material for longer time. The commonest mounting medium is **Canada balsam** or **D.P.X.** The use of Canada balsam or D.P.X. in the permanent mounting is due to the fact that the refractive index of Canada balsam or D.P.X. is the same as that of the glass of slide.

Procedure of mounting. (i) Clean the slide and cover-slip with a neat cloth.

(ii) Determine the centre of the slide by drawing a diagram of the slide and sketching its diagonals as in Fig. 3.1A.

(iii) Put a drop of mounting medium (Canada balsam or D.P.X.) in the centre of the slide with a glass rod.

(iv) Transfer the tissue or organism from the clearing agent (xylol or clove oil) on the slide by a brush or forceps.

(v) Take a cover slip and place its one edge on the slide touching the drop of Canada balsam or D.P.X. and hold its other edge obliquely either with needle or forceps.

(vi) Now gently lower down the needle or forceps till the coverslip completely touches the Canada balsam or D.P.X. and covers the material.

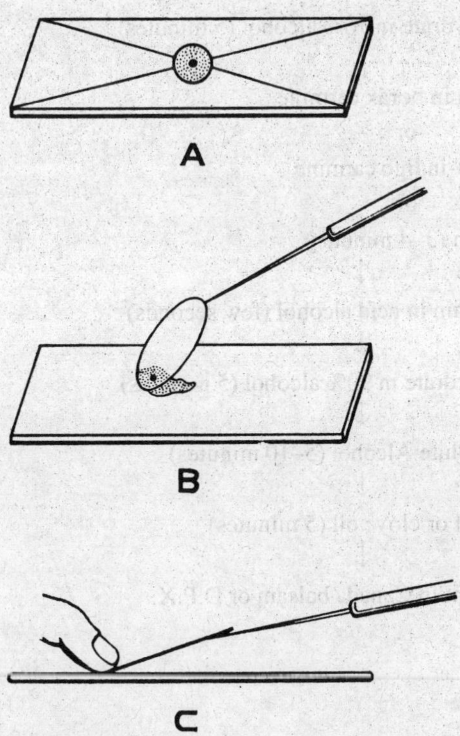

Fig. 3.1 Diagram showing the procedure of mounting.

(vii) After mounting, the slide may be dried in hot chamber at 40–50°C.

(viii) The excess of Canada balsam or D.P.X. which comes out around cover slip, should be washed with cotton soaked in xylol or in a mixture of xylol and absolute alcohol.

Precautions. 1. The slide, cover-slip, needles, brush, etc., should be clean.
2. The cover-slip should be in the centre of slide.
3. The material should be in the centre of cover-slip.
4. There should be no air bubble in the Canada balsam or D.P.X.
5. Excess of Canada balsam or D.P.X. should not come out of the cover-slip.

9. Labelling

After making the permanent mount, the slide should be properly labelled, *i.e.*, identification of material should be pasted on one edge of the slide and name of the student on the other edge of the slide.

SUMMARY OF PERMANENT MOUNTING

Single staining

Fix the material in fixative
↓
Wash in water to remove fixative
↓
Dehydrate in 30% alcohol (5 minutes)
↓
Dehydrate in 50% alcohol (5 minutes)
↓
Dehydrate in 70% alcohol (5 minutes)
↓
Stain in borax carmine
or
Picro-indigo carmine
or
Eosin (3–4 minutes)
↓
Destain in acid alcohol (few seconds)
↓
Dehydrate in 90% alcohol (5 minutes)
↓
Absolute Alcohol (5–10 minutes)
↓
Xylol or clove oil (5 minutes)
↓
Mount in Canada balsam or D.P.X.

Double staining

Fix the material in fixative
↓
Wash in water to remove fixative
↓
Stain in haematoxylin (2–5 minutes)
↓
Wash in water
↓
Destain in acid water
↓
Wash in water
↓
Dehydrate in 30% alcohol (5 minutes)
↓
Dehydrate in 50% alcohol (5 minutes)
↓
Dehydrate in 70% alcohol (5 minutes)
↓
Stain in eosin (5–7 minutes)
↓
Dehydrate in 90% alcohol (5 minutes)
↓
Absolute alcohol (5–10 minutes)
↓
Xylol or clove oil (5 minutes)
↓
Mount in Canada balsam or D.P.X.

CHAPTER 4

CULTURE METHODS

In laboratory different animals (particularly invertebrates) are cultured in artificial media for their use in various experimental purposes. The culture technology has, however, helped the human beings in understanding the different aspects of the life of such animals.

Culture of *Amoeba*

The species of *Amoeba* are found inhabiting the soil, sea and freshwater. *Amoeba proteus*, a freshwater form can be easily cultured by the following methods :

(i) **Hayinfusion method**. For this, dry hay stems or leaves are boiled with freshwater from a pond (tap-water should be avoided) for about 15 minutes. The infusion is then filtered and allowed to cool. Then few drops of amoebae-rich water is added in the infusion. After 2 or 3 days a large number of amoebae can be seen in the culture.

An alternative method of preparing hay infusion is boiling of few grains of wheat with few hay stems or leaves in freshwater for about 10 minutes. Then it is filtered and cooled. Now few drops of amoebae-rich water is added in the filtrate (infusion) so prepared. The container containing infusion is covered and kept for few (8–10) days. A large number of amoebae are developed in the culture.

(ii) With the help of a micropipette collect few *Amoeba proteus* and keep them at the bottom of a clean and sterilized petridish. Add rain water or water from a well and few grains of wheat in the petridish, cover it and keep in dark in a cool room.*Amoeba proteus* adheres the wheat grains and then starts multiplying after sometime. Thus, numerous daughter amoebae are formed in the culture.

Culture of *Paramecium*

Paramecium is found abundantly in ponds, ditches, etc., which are rich in decaying vegetation. For its culture, take decaying submerged weeds from a pond, keep them in a jar containing distilled water, cover the jar and leave to rot. Examine after 5–6 days. Swarms of paramecia are noticed. Now boil hay stems or leaves in water with few grains of wheat for 10 minutes. Cool and filter. Keep the filtrate (infusion) so obtained in dark and cool place. Let it stand for 4–5 days. Then transfer paramecia from first jar into the container containing infusion. The paramecia multiply rapidly and within few days the culture contains numerous paramecia.

Culture of *Euglena*

Euglena is found abundantly in ponds, pools, ditches, lakes, puddles, etc., containing decaying nitrogenous organic matter such as leaves and faeces of animals, etc. *Euglena* can be cultured in laboratory by adopting the following methods :

(i) Boil some cow or horse dung in distilled water in a jar and allow it to cool. After two days put some weeds from a pond containing euglenae into the jar and place the jar near well lighted window. In a few days numerous euglenae will develop in this nitrogenous infusion.

(ii) Boil 25 grains of wheat in distilled water in a small conical flask. Cool and add half-pint milk, plug the mouth of the flask with cotton wool and allow the medium to stand until a rich bacterial scum is formed. Then stir the medium and few euglenae in it by a micro-pipette. Keep the flask near well lighted window. Within a few days the culture will develop several euglenae.

Culture of *Vorticella*

For culturing *Vorticella* in laboratory, same method as mentioned for *Amoeba* can be adopted.

Culture of *Hydra*

Take an aquarium containing pond mud at the bottom and few aquatic plants and leaves etc. Collect few hydras from weeds and decaying leaves of a pond or ditch and put them in the aquarium. Keep on putting some small crustaceans like *Daphnia* and *Cyclops* in the aquarium after regular intervals. Within few days several hydras are developed in the aquarium.

Culture of *Planaria*

For culturing planarians in laboratory first collect few planarians like *Dugesia* from ponds and streams. For this, keep a piece of liver or beef in a jar place it in water near some objects like stones, wooden logs, twigs, etc. Some planarians are collected in the jar surrounding the liver or beef piece. Now transfer these planarians in a clean small trough containing water. Add small pieces of liver or beef twice or thrice a week in the trough. Soon a number of planarians are developed in the trough.

Culture of *Daphnia*

Daphnia can be cultured easily in laboratory in aquarium. For this, first collect some *Daphnia* from ponds or ditches and transfer them in the aquarium and keep regular supply of green algae in the aquarium. Algae serve the purpose of food for *Daphnia.* In place of algae, yeasts may also be added. Within few days a large number of *Daphnia* develop in thhe aquarium.

CHAPTER 5

IMPORTANT MOUNTINGS

Here mounting refers to the placement of biological material, either complete animal or any structure there of after adopting certain procedures, on microscopic glass slides to examine under microscope or with unaided eyes. However, mountings of some important biological materials have been described here under.

1. *Amoeba*

Make a thin film of Mayer's albumen on a slide. Put a few drops from *Amoeba*-culture. Let it spread smoothly and dry. Examine the slide under the microscope for *Amoeba*. If *Amoeba* is there on the slide, then invert it on a bottle containing acetic acid. Keep it for about 15 seconds. The fumes of acetic acid fixes the material. Wash it with distilled water, then follow double staining procedure (dehydration and staining) clear in xylol or benzene in a cavity block and finally mount in Canada balsam or D.P.X. Study the slide under the microscope and draw a neat and well-labelled diagram and note down its characteristic features (Fig. 5.1).

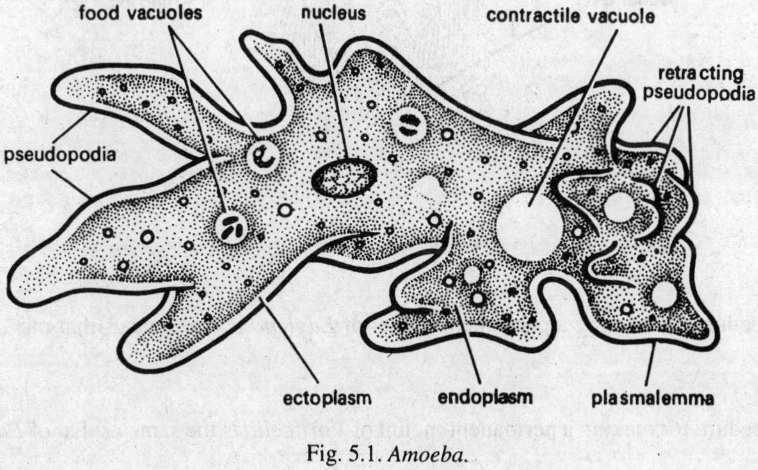

Fig. 5.1. *Amoeba*.

2. *Paramecium*

Make a thin film of Mayer's albumen over the slide. Put a few drops from *Paramecium*-culture. Presence of *Paramecium* can be ascertained by examining the slide under the microscope. If *Paramecium* is there on the slide, fix it with the acetic acid fumes or a drop of absolute alcohol. Let it be dry, wash with distilled water and follow double staining procedure (dehydration, staining), clear in xylol or benzene and mount in Canada balsam or D.P.X. Study the slide under the microscope, draw its diagram, label it and note down its characteristic features (Fig. 5.2).

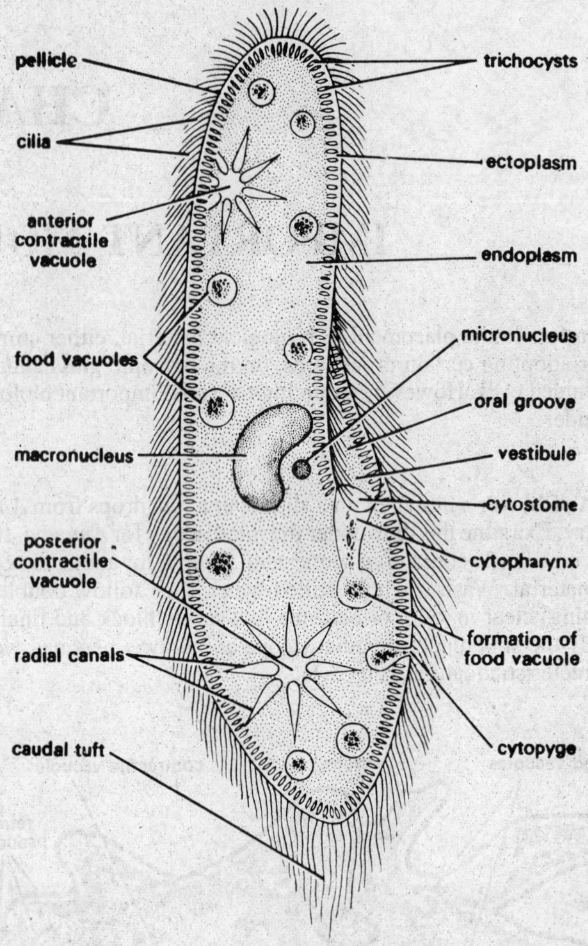

Fig. 5.2. *Paramecium.*

3. *Euglena*

The procedure for making a permanent mount of *Euglena* is the same as that of *Paramecium* (Fig. 5.3).

4. *Vorticella*

The procedure for making a permanent mount of *Vorticella* is the same as that of *Paramecium* (Fig. 5.4).

5. Rectal Ciliates of Frog

The rectal ciliates of frog are *Opalina* (Fig. 5.5) *Balantidium* (Fig. 5.6) and *Nyctotherus* (Fig. 5.7). For the permanent mounting of rectal ciliates following procedure is followed :

Take a freshly-chloroformed frog, cut it open and take out its rectum. Open the rectum by cutting it longitudinally and empty its contents in 0.75 % sodium chloride solution in a small petridish. Take a few drops of solution on a slide and observe it under the microscope to ascertain whether the rectum is infected by rectal ciliates. Make a thin film of Mayer's albumen on the slide and put a few drops of solution containing rectal ciliates. Fix the rectal ciliates with fumes of acetic acid or by pouring a drop of absolute alcohol over the slide. Keep it for drying. Then wash with distilled water and then follow double staining procedure (dehydration and staining), clear in xylol or benzene and

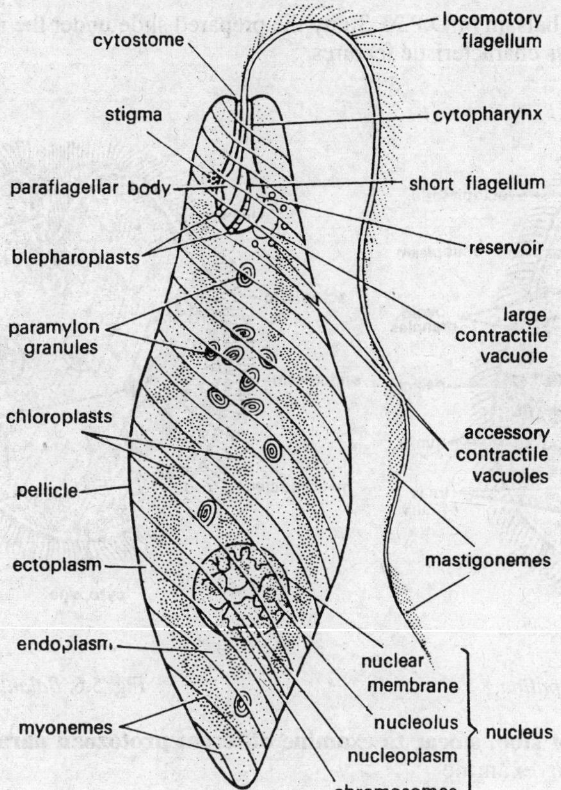

Fig. 5.3. *Euglena.*

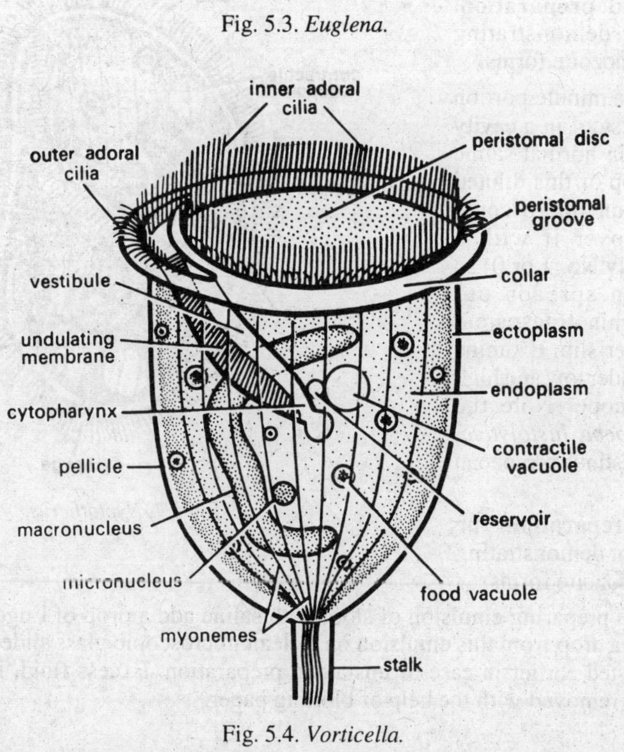

Fig. 5.4. *Vorticella.*

finally mount in Canada balsam or D.P.X. Study the prepared slide under the microscope, draw its diagram and note down its characteristic features.

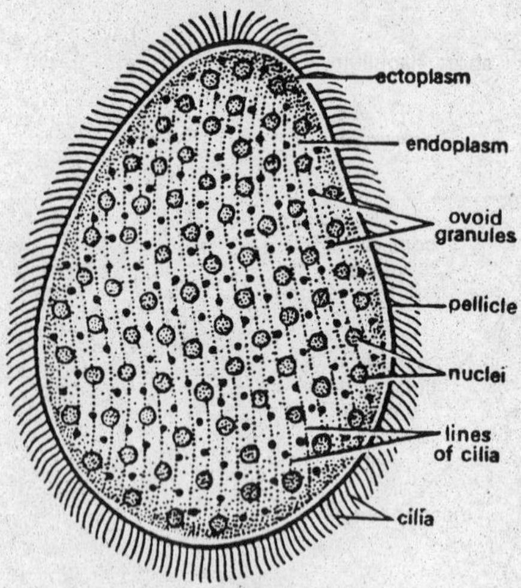

Fig. 5.5. *Opalina.*

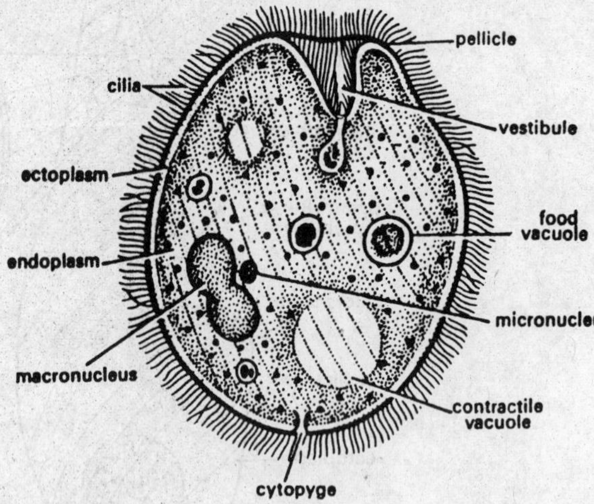

Fig. 5.6. *Balantidium.*

6. Preparation of stool smear to examine different protozoan parasites. Two types of smears are prepared to examine various protozoan parasites in stool.

1. Unstained preparation. This is useful for demonstrating actively motile trophozoite forms.

For this, take a minute portion of freshly collected stool in a cavity block and dilute it in normal saline (0.9 %). Take a drop of this diluted emulsion on a clean microscopic glass slide and cover it with a cover-slip (preferably No. 1 or 0) so that the emulsion spreads out uniformly into a thin transparent layer below the cover-slip. Examine carefully the slide under low and high powers of microscope. Note the presence of *Entamoeba histolytica*, *E. coli* or other intestinal protozoan parasites.

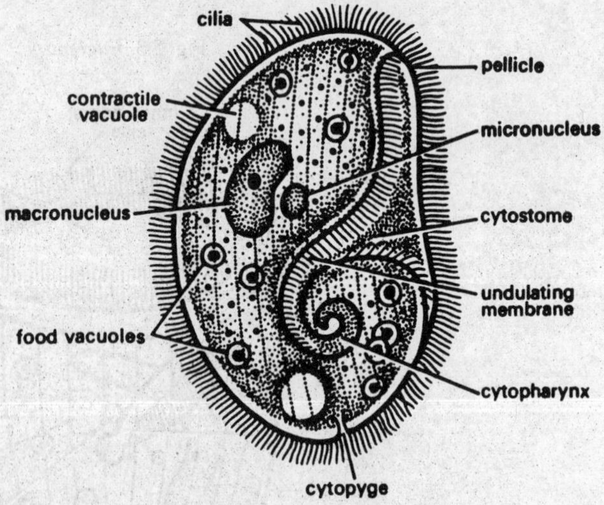

Fig 5.7. *Nyctotherus.*

2. Stained preparation. This is the best method for demonstrating cysts and dead trophozoite forms.

For this, while preparing emulsion of stool with saline add a drop of Lugol's iodine solution also in it. Then take a drop from this emulsion on a clean microscopic glass slide and cover it with a cover-slip as suggested earlier in case of unstained preparation. Excess fluid, if any, on the slide around cover-slip is removed with the help of blotting paper.

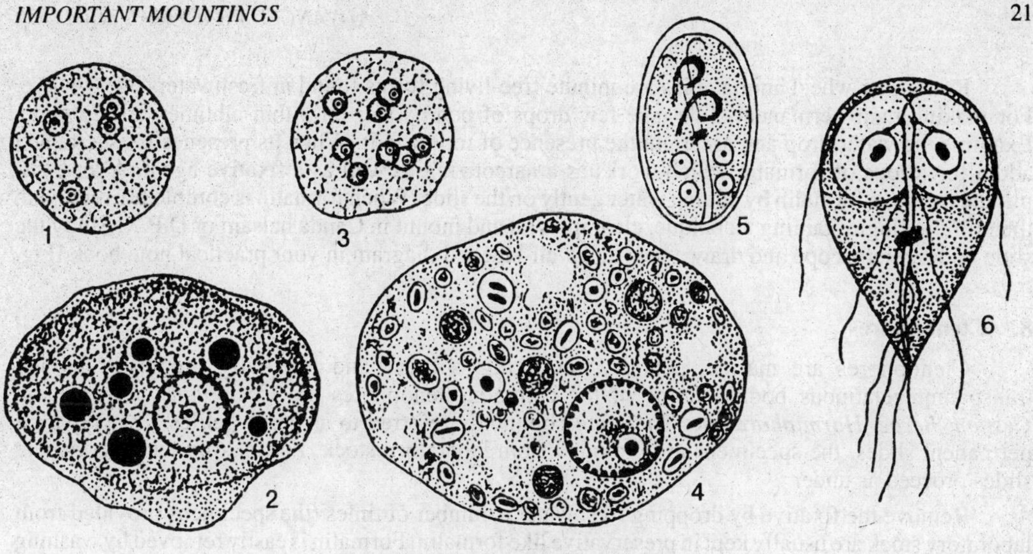

Fig. 5.8. Intestinal protozoan parasites.
1–*Entamoeba histolytica* cyst; 2–*E.histolytica* trophozoite;
3–*E.coli* cyst; 4–*E.coli* trophozoite;
5–*Giardia lamblica* cyst; 6–*G.lamblica* trophozoite.

Examine the prepared slide under low and high powers of microscope and note the presence of cysts of various protozoan parasites (See Fig. 5.8) and identify them.

7. Rotifers

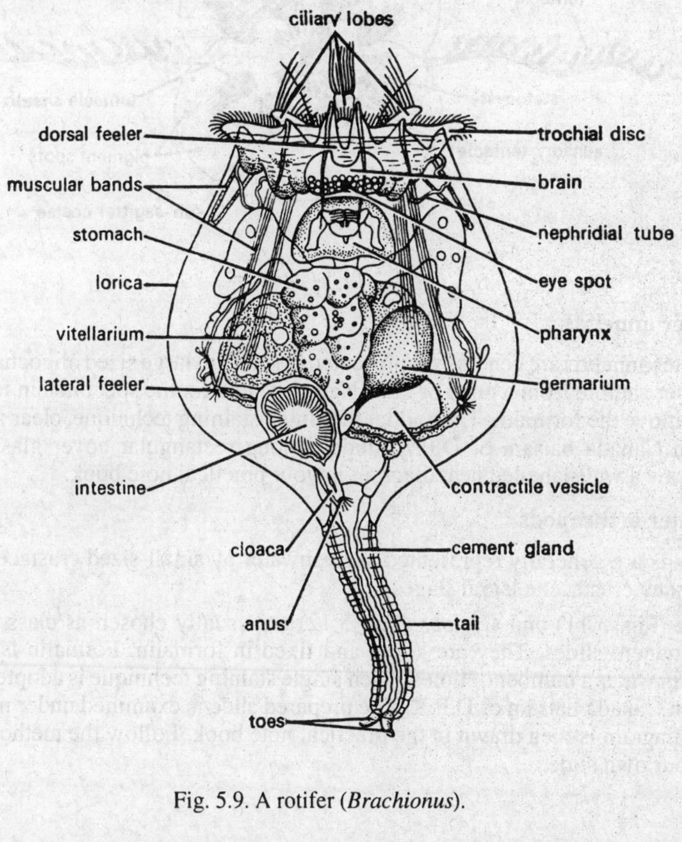

Fig. 5.9. A rotifer (*Brachionus*).

Rotifers or wheel animalcules are minute free-living forms found in freshwaters everywhere. For making their permanent slide take few drops of pond water on a thin albumen coated slide. Examine the water drop and ascertain the presence of rotifer in it. When its presence is ascertained add a few drops of formalin which works as a narcotizing as well as a fixative agent. After 8–10 minutes remove formalin by adding water gently on the slide. When formalin is completely removed, then follow single staining technique, clear in xylol and mount in Canda balsam or D.P.X. Study the slide under microscope and draw a neat and well-labelled diagram in your practical note book (Fig. 5.9).

8. Ctenophores

Ctenophores are marine, pelagic, free-living, solitary and biradially organisms having transparent gelatinous bodies. Some of the common ctenophores are *Coeloplana*, *Ctenoplana*, *Cestum*, *Beroe*, *Hormiphora*, etc. These are commonly referred to as **comb-jellies**. For preparing permanent slides, the specimens are provided from laboratory stock. For making their permanent slides proceed as under :

Remove the fixative by dropping freshwater a number of times (the specimens provided from laboratory stock are usually kept in preservative like formalin. Formalin is easily removed by washing the specimens in water). When fixative is completely removed, then follow single staining technique, clear in xylol and mount in Canada balsam or D.P.X. Study the slide under microscope and draw a neat and well-labelled diagram in your practical note book (Fig. 5.10).

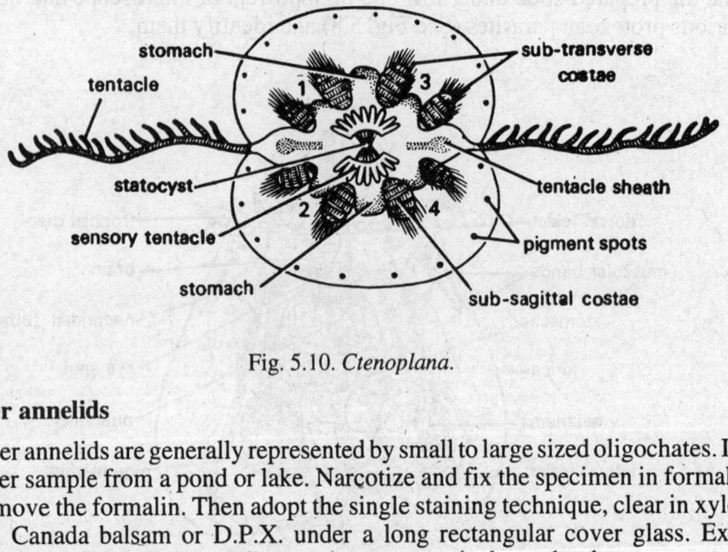

Fig. 5.10. *Ctenoplana*.

9. Freshwater annelids

Freshwater annelids are generally represented by small to large sized oligochates. Identify some annelids in water sample from a pond or lake. Narcotize and fix the specimen in formalin. Wash the specimen to remove the formalin. Then adopt the single staining technique, clear in xylol and mount on the slide in Canada balsam or D.P.X. under a long rectangular cover glass. Examine under microscope, draw a well-labelled neat diagram in your practical note book.

10. Freshwater arthropods

Arthropods are generally represented in freshwater by small-sized crustaceans like *Daphnia*, *Cyclops* and many crustacean larval stages.

Daphnia (Fig. 5.11) and *Cyclops* (Fig. 5.12) are usually chosen as class work material for preparing permanent slides. They are killed and fixed in formalin. Formalin is then removed by washing them in water a number of times. Then single staining technique is adopted, cleared in xylol and mounted in Canada balsam or D.P.X. The prepared slide is examined under microscope. A neat well-labelled diagram is then drawn in the practical note book. Follow the method described above and prepare your own slide.

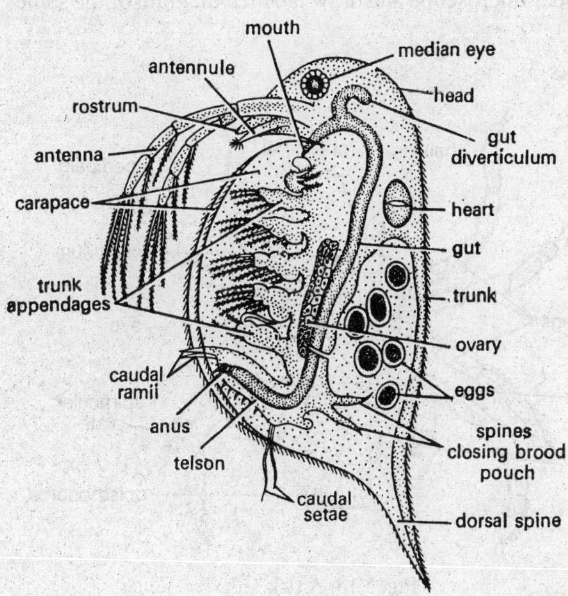

Fig. 5.1. *Daphnia.*

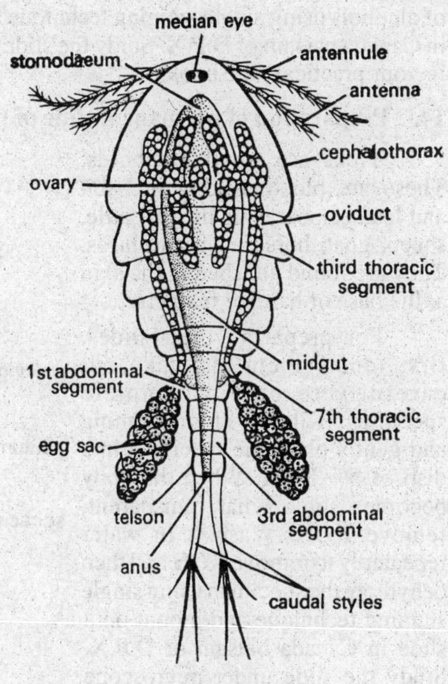

Fig. 5.12. *Cyclops.*

11. Preparation of permanent slides of *Cimex* and *Pediculus*

A. *Cimex*

Take a *Cimex* (bed bug) and kill it by chloroform and fix it in formalin. Remove formalin by washing it with water. Then keep it in a small test tube containing 10% KOH solution for over night. Then wash it thoroughly with water to remore KOH. Thereafter, dehydrate it through the grades of alcohol (staining stage may be omitted), clear in xylol and mount in Canada balsam or D.P.X. Draw a neat, well-labelled diagram in your practical note book. (Fig. 11.45).

B. *Pediculus*

Take a *Pediculus* (head louse *Pediculus humanus capitis* or a body louse *Pediculus humanus corporis*) and kill it by chloroform. Fix it in formalin. Wash in water to remove the fixative. Put it in 5% KOH solution for 10-15 minutes and then take it out from KOH. Wash it again in water so that KOH is completely removed. Thereafter, dehydrate it by passing through the grades of alcohol (staining stage may be omitted), clean in xylol and mount in Canada balsam or D.P.X. Study under microscope, draw a neat well labelled diagram in your practical note book (Fig. 11.44).

12. Preparation of permanent slides of wings, legs and mouthparts of housefly and mosquito

Carefully take out the wings or legs and keep in a cavity block containing xylol. Take out the wings or legs and mount on a clean slide in Canada balsam or D.P.X. Study the slide under microscope and draw a neat well-labelled diagram in your practical note book.

For preparing the slide of mouthparts, cut the head carefully from thorax, keep in xylol or benzene for few minutes and mount on a clean slide in Canada balsam or D.P.X. Study the slide under microscope and draw a neat well-labelled diagram in your practical note book.

13. Preparation of permanent slides of the developmental stages of mosquito and housefly

Select the larva or pupa of which slide is to be prepared. Kill them by putting in 70% alcohol or hot water. Then put the specimens in 5% KOH and leave until internal soft parts dissolve and the body becomes somewhat transparent. Then carefully draw off the KOH and add water which should be replaced several times to wash out all the KOH. Then dehydrate the specimen through usual series of alcohols using single staining technique and clear in xylol or benzene and finally mount on a slide in Canada balsam or D.P.X. Study the slide under microscope and draw labelled diagram of the same in your practical note book.

14. Preparation of permanent slide of ticks

Ticks are small arachnids. These are blood-sucking in habit and lead ectoparasitic life on cattle, sheep, goat, horse, man and birds. They are found attached to the skin at the base of hairs or feathers.

For preparing their slides, first locate them and take out carefully to obtain complete specimen. Kill by chloroforming and gently place the specimen in a dish of 5% KOH. When the body becomes some what transparent, remove KOH. Wash it in water repeatedly to remove KOH and then dehydrate the specimen using single staining technique and mount on a slide in Canada balsam or D.P.X. Study the slide under microscope and draw well-labelled diagram in your practical note book (Fig. 5.13).

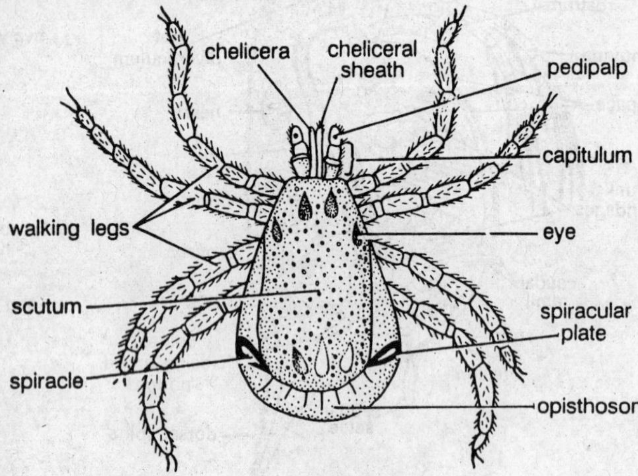

Fig. 5.13. A tick.

15. Preparation of permanent slide of mites

Mites are also arachnids but smaller than the ticks. A variety of mites parasitize vertebrates, especially birds and mammals. Some mites are very active and spend much of the time off the host, visiting it only for blood meals. The hair-clasping mites are especially common on rodents and are seen attached to the hairs. They move very slowly. Rouch, scaly or scabbed areas in the skin including inside the ears are due to mange mites. Deep scrapings of such areas are to be mixed with water and examined microscopically for mites which are easily detectable by their movements. As internal parasites, mites may occur in various parts of the body, especially in respiratory passage and in the air-sacs of birds.

So, for mange mites, proceed as under :

(i) Take pieces of mangy skin or deep scrapings in 10% KOH to soften the tissues and scrub so that mites are released in large numbers. Leave it for 4 to 8 hours and keep shaking occasionally.

(ii) Remove large pieces of tissue. Take the remaining contents in 50 ml centrifuge tubes and centrifuge at low speed to throw down the suspended material. Discard the supernatant and repeat several times with water to remove KOH.

(iii) Then centrifuge the contents once each in 30%, 50%, 70% alcohol, resuspending and letting stand a few minutes in each alcohol.

(iv) Replace the last alcohol with glycerine alcohol, shake and pour the suspension in shallow dishes. Cover the dishes with paper to avoid dust and let the alcohol evaporate until pure glycerine is left.

(v) Examine the dish under a dissecting microscope and take out carefully the stage you want to mount with the help of a micropipette. Put it on a cover glass and mount it on a slide in glycerine jelly.

(vi) Study the slide under microscope, draw a neat, well-labelled diagram of the same in your practical note book (Fig. 5.14).

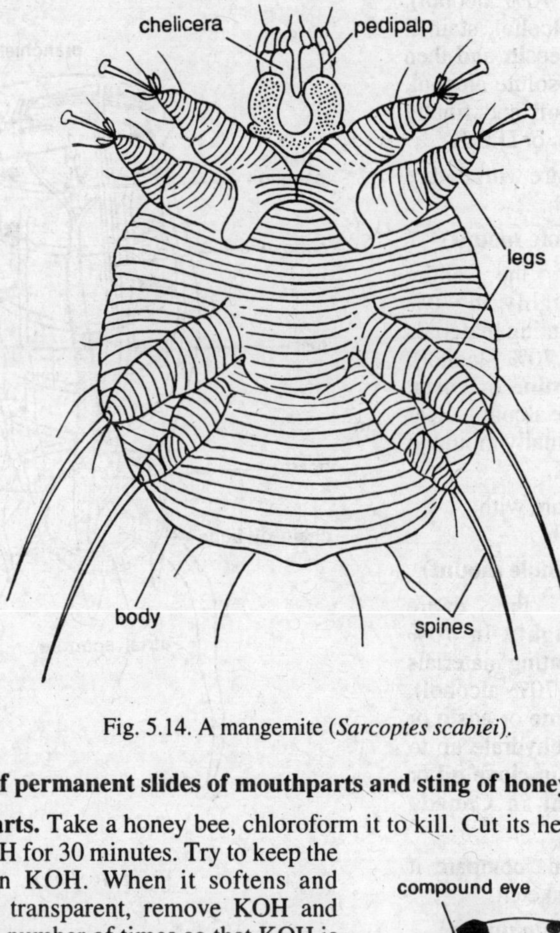

Fig. 5.14. A mangemite (*Sarcoptes scabiei*).

16. Preparation of permanent slides of mouthparts and sting of honey bee

A. Mouthparts. Take a honey bee, chloroform it to kill. Cut its head from thorax and keep the head in 10% KOH for 30 minutes. Try to keep the head submerged in KOH. When it softens and becomes somewhat transparent, remove KOH and wash it with water a number of times so that KOH is completely removed. Dehydrate through grades of alcohol, clean in xylol and mount in Canada balsam or D.P.X. Study the slide under microscope and draw a well-labelled diagram of the same in your practical note book (Fig. 5.15).

B. Sting. Take a honey bee, chloroform it to kill. Press its last abdominal segments and catch its sting carefully with the help of a fine forcep. Now pull out the sting carefully and keep in a watch glass containing KOH. After 20–30 minutes remove KOH and wash it with water to remove the traces of KOH. Dehydrate, clear and mount. Examine it under microscope and draw a well-labelled diagram of the same in your practical note book (Fig. 11.59).

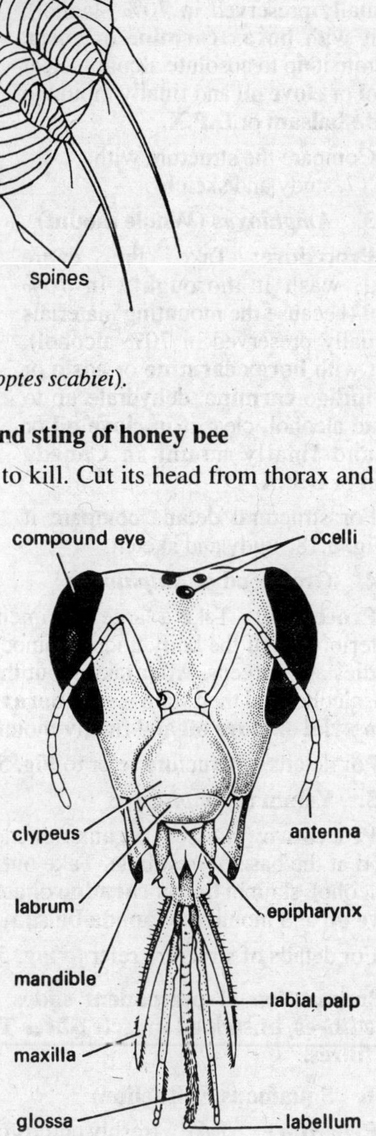

Fig. 5.15. Head and mouth parts of honey bee.

17. Preparation of permanent slides of whole mounts of *Salpa, Doliolum, Amphioxus,* oral hood of *Amphioxus*, velum of *Amphioxus*.

1. *Salpa* (Whole mount)

Procedure: Take the entire animal and wash it in 70% alcohol (because the materials for mounting are usually preserved in 70% alcohol). After washing in 70% alcohol, stain it with **borax carmine** or **eosin** and then again dehydrate up to absolute alcohol, clear in **xylol** or **clove oil** and finally mount in **Canada balsam** or **D.P.X.**

Compare the structure with Figure No. 5.16, study and sketch.

2. *Doliolum* (Whole mount)

Procedure: Take the entire animal, wash it thoroughly in 70% alcohol (because the mounting materials are usually preserved in 70% alcohol), stain it with **borax carmine** or **eosin**, dehydrate it up to absolute alcohol, clear in **xylol** or **clove oil** and finally mount in **Canada balsam** or **D.P.X.**

Compare the structure with Fig. 5.17, study and sketch.

3. *Amphioxus* (Whole mount)

Procedure: Take the entire animal, wash it thoroughly in 70% alcohol (because the mounting materials are usually preserved in 70% alcohol), stain it with **borax carmine** or **eosin** or **picro-indigo carmine**, dehydrate up to absolute alcohol, clear it in **clove oil** or **xylol** and finally mount in **Canada balsam** or **D.P.X.**

For structural details compare it with Fig. 5.18, study and sketch.

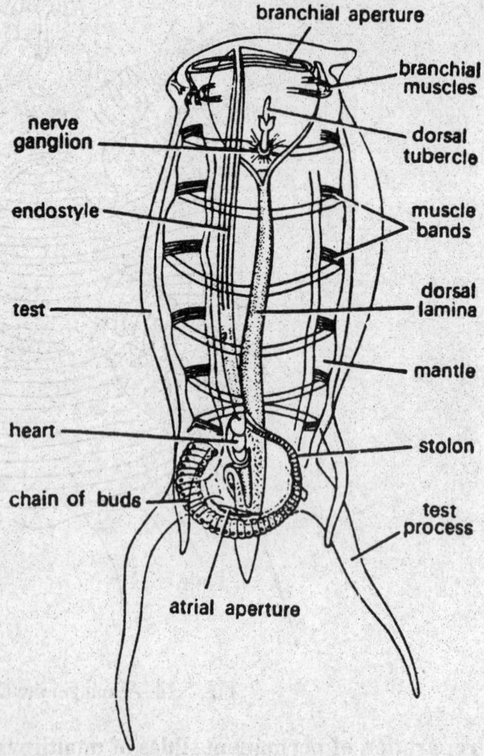

Fig. 5.16. *Salpa.*

4. Oral hood of *Amphioxus*

Procedure: Take a large specimen of *Amphioxus* and put it in water in a petridish. Dissect the anterior part of the body under a binocular microscope and take out the oral hood with the help of needles and forceps. After taking out the oral hood, wash it thoroughly in water, dehydrate it up to 70% alcohol and then stain it with **borax carmine** or **eosin** again dehydrate up to absolute alcohol, clear in **xylol** or **clove oil** and finally mount in **Canada balsam** or **D.P.X.**

For details of structure refer to Fig. 5.19, study and sketch.

5. Velum of *Amphioxus*

Procedure: Dissect the anterior end of *Amphioxus* under a binocular microscope. The velum is found at the base of vestibule. Take out the velum, wash it thoroughly in water, dehydrate up to 70% alcohol, stain in **borax carmine** or **eosin**, again dehydrate up to absolute alcohol, clear in **xylol** or **clove oil** and mount in **Canada balsam** or **D.P.X.**

For details of structure refer to Fig. 5.20, study and sketch.

18. Preparation of permanent slides of squamous epithelium, ciliated epithelium,striped muscle fibres, unstriped muscle fibres T. S. cartilage of frog, medullated and non-medullated nerve fibres.

1. Squamous Epithelium

Procedure: Take a freshly chloroformed **frog** and scrap the mucus from its dorsal and ventral surfaces by the scalpel and put the mucus in a petridish containing saline solution. The mucus contains squamous epithelium cells. Take small piece of mucus, wash in water, stain in **haematoxylin**, again

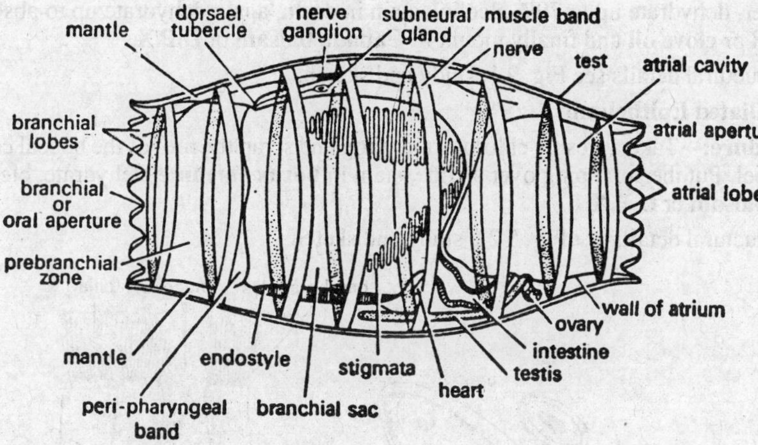

Fig. 5.17. *Doliolum.*

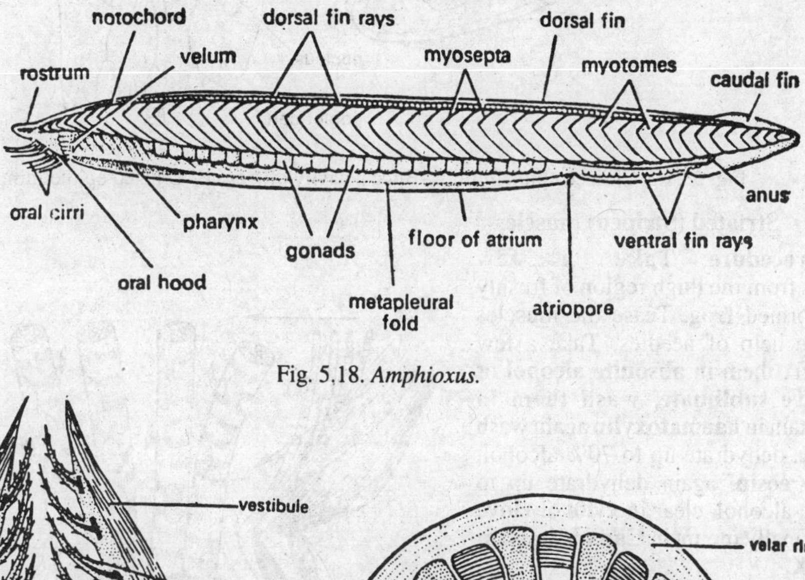

Fig. 5.18. *Amphioxus.*

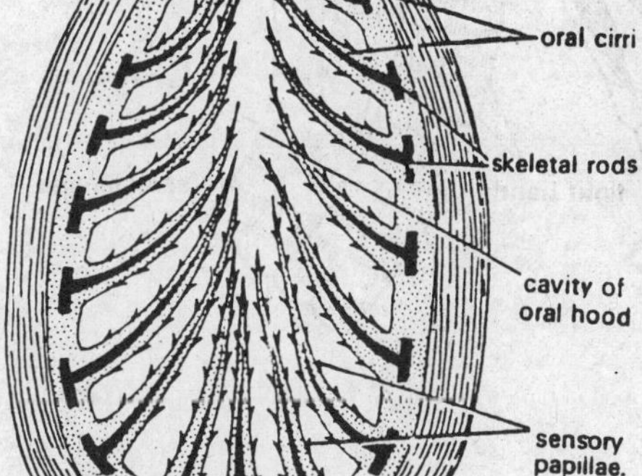

Fig. 5.19. Oral hood of *Amphioxus.*

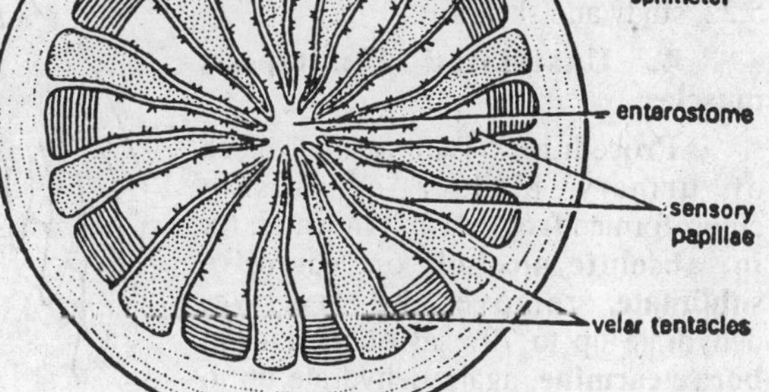

Fig. 5.20. Velum of *Amphioxus.*

wash in water, dehydrate up to 70% alcohol, stain in **eosin**, again dehydrate up to absolute alcohol, clear in **xylol** or **clove oil** and finally mount in **Canada balsam** or **D.P.X.**

For structural details see Fig. 9.1, study and sketch.

2. Ciliated Epithelium

Procedure: Take a freshly chloroformed frog and scrap the roof of the buccal cavity with the help of scalpel. Put the scrapping over a slide, stain in **borax carmine**, dehydrate, clear and mount in **Canada balsam** or **D.P.X.**

For structural details see Fig. 5.21, study and sketch.

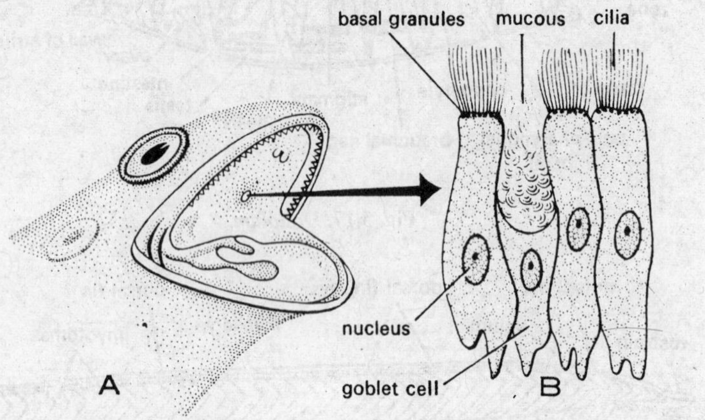

Fig. 5.21. Ciliated epithelium. A–Buccal cavity of frog; B–Ciliated epithelium.

3. Striated (Striped) muscles

Procedure: Take a few muscles from the thigh region of freshly chloroformed **frog**. Tease the muscles with the help of needles. Take a few fibres, fix them in **absolute alcohol** or **corrosive sublimate**, wash them in water, stain in **haematoxylin** again wash in water, dehydrate up to 70% alcohol, stain in **eosin**, again dehydrate up to absolute alcohol, clear in **xylol** or **clove oil** and finally mount in **Canada balsam** or **D.P.X.**

For details of structure, see Fig. 5.22, study and sketch.

4. Unstriated (Unstriped) muscles

Procedure: Take a small piece of urinary bladder of freshly chloroformed **frog**, wash it in water, fix in **absolute alcohol** or **corrosive sublimate**, remove the fixative, dehydrate up to 70% alcohol, stain in **borax carmine**, again dehydrate up to absolute alcohol, clear in **xylol** and mount in **Canada balsam** or **D.P.X.**

For structural details see Fig. 9.2, study and sketch.

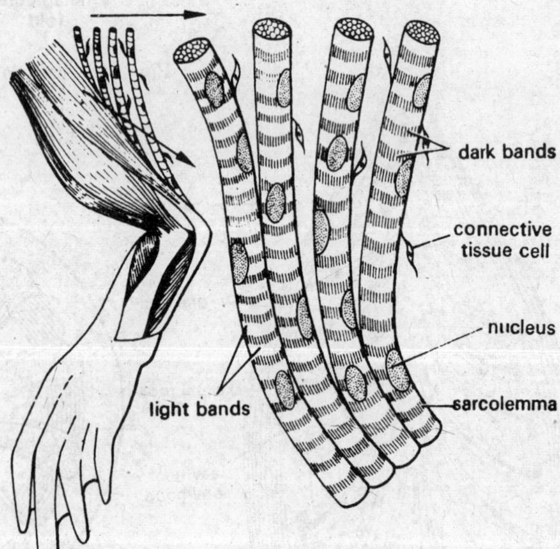

Fig. 5.22. Striated muscles.

5. T. S. of cartilage

Procedure: Take **humerus** or **femur** of frog with intact cartilage on their heads. Wash the head in water and then cut very thin transverse sections of cartilage with a sharp razor or blade. Take one or two sections, wash thoroughly in water, stain in **haematoxylin**, again wash in water, dehydrate up to 70% alcohol, stain in **eosin**, again dehydrate up to absolute alcohol, clear in **xylol** and finally mount in **Canada balsam** or **D.P.X.**

For structural details see Fig. 9.5, study and sketch.

6. Blood of Frog

Procedure: Take a drop of blood from a freshly-killed frog on a slide. Now hurriedly spread the blood over the slide by the edge of another slide in such a way that a very thin film of blood is formed. Dry the film of blood, pour two or three drops of absolute alcohol and let it be dry again. After drying, wash it in water, stain in **haematoxylin**, wash again in water, dehydrate up to 70% alcohol, stain in **eosin**, again dehydrate up to absolute alcohol, clear in xylol and finally mount in **Canada balsam** or **D.P.X.**

For structural details see Fig. 9.10, study and sketch.

7. Medullated and non-medullated nerve fibres

Procedure: For medullated nerve fibres take a small piece of **sciatic** or any other **spinal nerve** and for non-medullated nerve fibres take a small piece of **sympathetic** nerve in 0.75% NaCl solution which is a physiological saline. Tease it with a pair of fine needles to separate nerve fibres. Then select one or two fibres, wash in water and follow double staining technique, dehydrate, clean in xylol and mount on a clean slide in **Canada balsam** or **D.P.X.**

For structural details of medullated nerve fibres see Fig. 9.14 and non-medullated nerve fibres see Fig. 9.15. Study and sketch in your practical note book.

19. Preparation of permanent slide of the filoplume of pigeon

Filoplume. The filoplume is a delicate hair-like feather found sparsely distributed on the body of pigeon. These feathers are not visible unless contour feathers are removed. A filoplume is made of a short calamus and long rachis which bears few weak barbs and barbules at its free tip.

Procedure: First remove some of the contour feathers from the ventral surface of the pigeon and locate a filoplume, take it out from the skin. Wash it with water and dehydrate using borax carmine staining method as suggested earlier. Clear in xylol and mount on a clean slide in Canada balsam or D.P.X. Examine the slide under microscope and draw a neat and well-labelled diagram in your practical note book (Fig. 5.23).

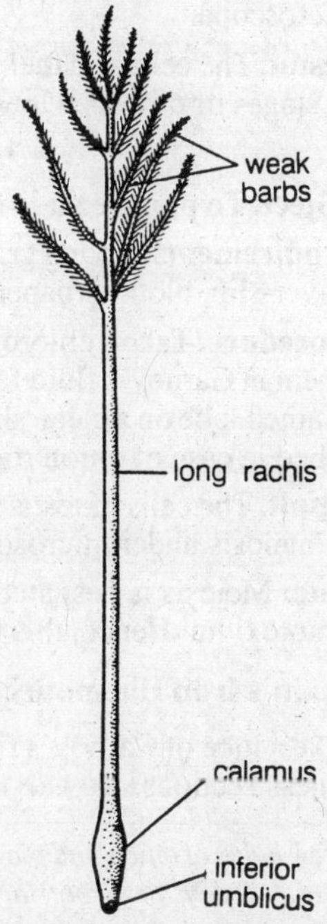

Fig. 5.23. Filoplume feather.

CHAPTER 6

CYTOLOGICAL PREPARATIONS

Cytology, study of cells, has advanced wonderfully during the last few decades. The study of cells necessarily involves sophisticated equipments and techniques. Baring few research laboratories, most of the university departments and probably all colleges in our country are still in want of such equipments and techniques. Therefore, study of cells in our country is generally confined to the reading of books and seeing micrographs available in them and research papers. However, some very simple and elementary methods are being described here to study cell division and making preparations of certain cell components. In addition, some cellular micrographs have also been given for identification and comments.

EXERCISE No. 1

Object. Squash preparation of onion root tips to observe stages of **mitosis**.

Requirements. Onion root tips* fixed in F.A.A. or Carnoy's fluid, microscopic glass slide, cover-slip, acetocarmine, spirit lamp, blotting paper and microscope.

Procedure. Take a drop of acetocarmine on a clean microscopic slide and put on it one or two root tips. Place a cover-slip over it and tap it gently by a needle. Warm the slide over the flame of a spirit lamp and then put a blotting paper over it, press it smoothly by your thumb. Examine the slide under microscope.

Result. The cells and their chromosomes are spread out and become distinct. Carefully observe different stages of mitosis. Draw their diagrams in your practical note book.

EXERCISE No. 2

Object. To prepare slide for the study of **meiosis.**

Requirements. Living grasshoppers, chloroform, normal saline, Carnoy's fluid, acetocarmine, slides, cover-slip, blotting paper and microscope.

Procedure. Take a chloroformed grasshopper and dissect it in normal saline. Take out its testes and fix them in Carnoy's fluid for 2–12 hours. Take a small lobe of testis and stain it in acetocarmine. Put the stained lobe on a clean slide and cover it with a coverslip. Adopt squash preparation technique as described in case of onion root tip. Examine the prepared slide under microscope.

Result. The cells of testis lobes are spread out and become distinct. Carefully observe different stages of meiosis under microscope (Fig. 6.2) and draw them in your practical note book.

Note: Meiosis is best studied in thin sections of testis of grasshopper stained in Heidenhain's iron haemotoxylin. Hence, this method is also described here.

Heidenhain's Iron Haemotoxylin Method

Take a lobe of Carnoy's fixed testis and follow usual procedure of microtomy to get thin about 5μ thickness sections. Dewax the sections and come down to water (xylol–absolute alcohol–90%

---

* *Take a few bulbs of onion and place them in a tray containing moist soil so that their discs are directed towards soil. After 4–5 days new roots appear and develop. Take out onion from the soil gently so that root tips are not damaged, wash them in water and cut about 3mm pieces of roots. Keep these pieces in a tube containing F.A.A. or Carnoy's fluid.*

alcohol–70% alcohol–50% alcohol–water). Keep the sections in 3% iron alum for 30 minutes to an hour. Wash the sections in distilled water and stain in 0.5% solution of Heidenhain's, iron haemotoxylin*. Wash the sections in distilled water twice. Differentiate the sections in 1.5% iron alum. Wash the sections in distilled water twice and then in tap-water for atleast 30 minutes. Counterstain the sections in a saturated solution of orange G in 90% alcohol. Dehydrate the sections, clean in xylol and mount in D.P.X. or Canada balsam.

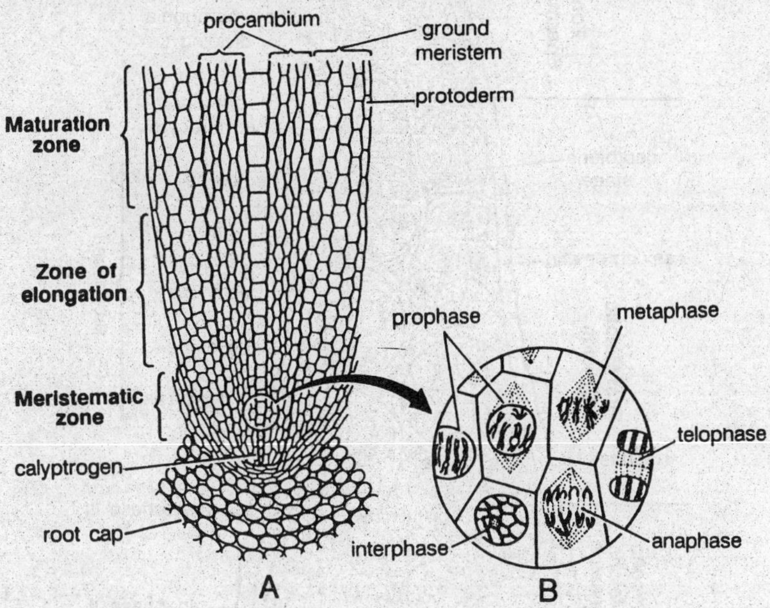

Fig. 6.1. Squash of onion root tip showing different stages of mitosis.

Result. The nuclear components are stained in deep blue colour and cytoplasm in orange colour. Examine the slide carefully and note various stages of meiosis.

EXERCISE No. 3

Object. Preparation of polytene chromosomes.

Polytene chromosome is one of the giant chromosomes found in animals and plants. In animals it is found in the salivary glands, Malpighian tubules, the epithelial cell lining of the gut and in the fatty cells of the larvae of certain Diptera. The polytene chromosomes of salivary glands in Drosophila larvae can be demonstrated easily in laboratory. It can also be demonstrated in the larva of Chironomous fly.

1. Preparation of Polytene Chromosomes from *Drosophila* Larva

Preparation of culture. Take a small specimen jar. Make a mixture of the pulp of apple, banana and lemon. Take 2 gm of moldex and add it in 40 ml boiling water. Cool it and add in pulp of fruits prepared. Place the jar in the culture of *Drosophila* flies, keep it open for 2–3 days during which some of the flies will visit the jar for feeding and lay eggs. Now cover the jar with fine muslin cloth. Within a week larvae will appear. Observe carefully for 3rd instar larvae which will be white coloured.

Dissection of 3rd instar larva for salivary gland. Take a few drops of saline on a clean slide and put the 3rd instar larva in it. Locate the junction of thorax and abdomen. Take two needles, one in each hand. Press the first needle firmly on the posterior end of thorax and other needle at the junction of thorax and abdomen. Pull the second needle so that abdomen is separated from head and thorax.

* **Heidenhain's iron haemotoxylin.** *Dissolve 10 gm of haemotoxylin in 10 ml of 90% alcohol. Keep it in a unstoppered bottle for ripening and leave it for about 6 weeks. The stain is ready to use. To make it 0.5%, take 1 part of this solution and add 20 part distilled water.*

Then press the thorax with a needle and observe that the salivary glands are seen floating in the saline water on the slide.

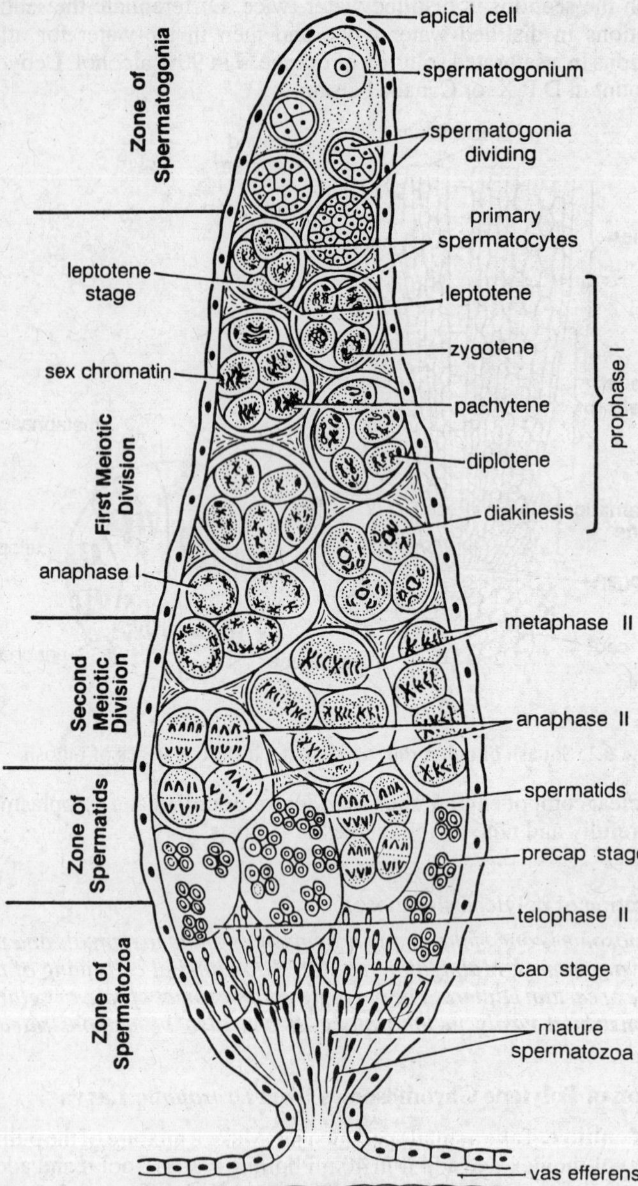

Fig. 6.2. T.S. one follicle of testis of grasshopper to show the stages of meiosis.

Preparation of slide. Take a clean slide, put a drop of acetocarmine on it. Transfer the salivary glands in acetocarmine on slide and cover it with a cover-slip. Leave it for 10 minutes and then warm it gently. Adopt squash technique as described in case of onion root tip. Observe the slide under microscope and for details observe under high power of microscope.

Comments:

1. These are large-sized, hence, called **giant chromosomes.**
2. These chromosomes present alternate pattern of **dark bands** and **light interbands.**

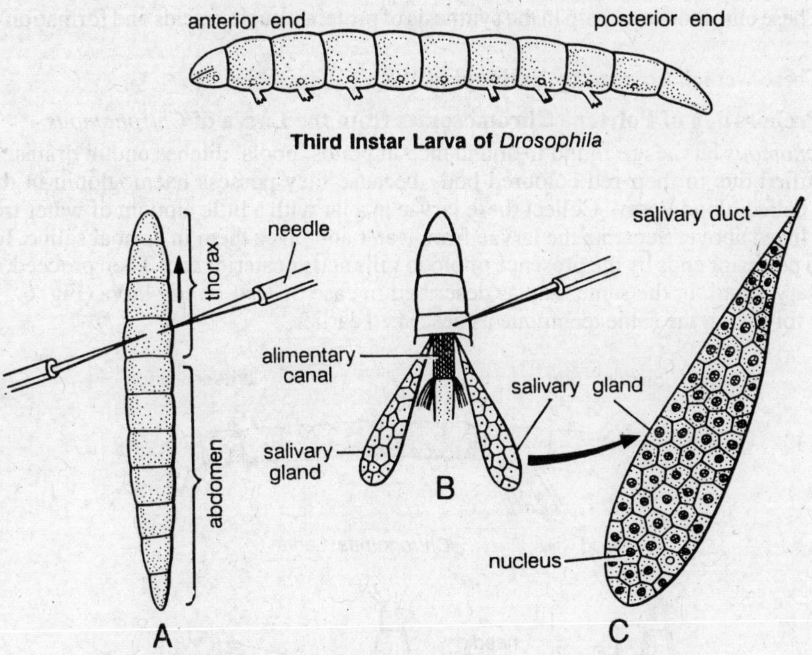

Fig. 6.3. Dissection of 3rd instar *Drosophila* larva for salivary glands.

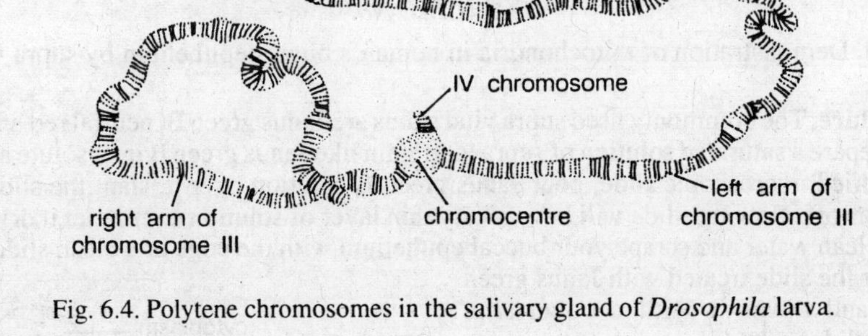

Fig. 6.4. Polytene chromosomes in the salivary gland of *Drosophila* larva.

3. The dark bands contain rich amount of DNA and RNA, and composed of much coiled chromonemal thread.

4. The light bands contain large amount of proteins and little amount of DNA and RNA.

5. A polytene chromosome is multistranded, it is formed of large number of chromonemal threads or strands.

6. A polytene chromosome exhibits **puffs** and **Balbiani rings** at certain points. The puffs are made of lateral extensions of bands of chromonemal strands into side loops.

7. The puffs and Balbiani rings are related with the metabolic activities of the chromosomes.

8. These chromosomes help in the synthesis of proteins, nucleic acids and formation of nuclear material.

9. These were discovered by **Balbiani** in 1881.

2. Preparation of Polytene Chromosomes from the Larva of *Chironomous*

Chironomous larvae are found in abundance in ponds, pools, ditches and in drains. These are easily identified due to their red coloured body because they possess haemoglobin in their blood (hence also called bloodworm). Collect these larvae in a jar with a little amount of water from any of the sources listed above. Separate the larvae from water and place them in normal saline. Identify its anterior and posterior ends by the presence of blood gills at its posterior end. Then proceed, for taking out its salivary gland, in the same way as described in case of *Drosophila* larva (Fig. 6.5). Prepare the slide by following the same technique as described earlier.

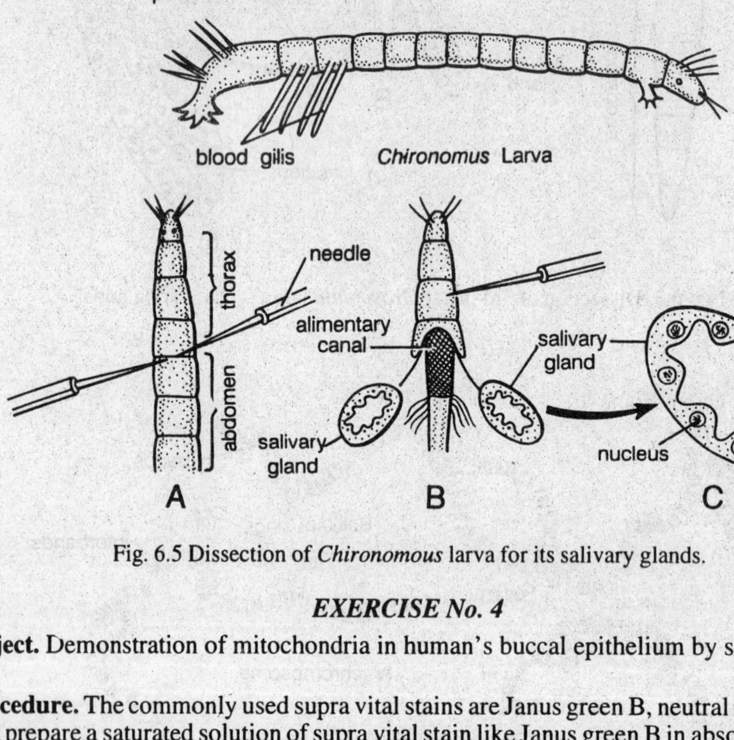

Fig. 6.5 Dissection of *Chironomous* larva for its salivary glands.

EXERCISE No. 4

Object. Demonstration of mitochondria in human's buccal epithelium by supra vital staining method.

Procedure. The commonly used supra vital stains are Janus green B, neutral red and methylene blue. First prepare a saturated solution of supra vital stain like Janus green B in absolute alcohol. Take a clean and dried microscopic slide, pour Janus green B solution over it, slant the slide so that the solution drains off. Now the slide will have a very thin layer of solution over it, let it dry. Wash your mouth with clean water and scrape your buccal epithelium with the edge of a clean slide. Spread the scraping over the slide treated with Janus green B solution and cover it with a cover-slip. Examine the slide under microscope.

Result. Observe and note dot-shaped dark green coloured mitochondria in light green or colourless cytoplasm in the epithelial cells.

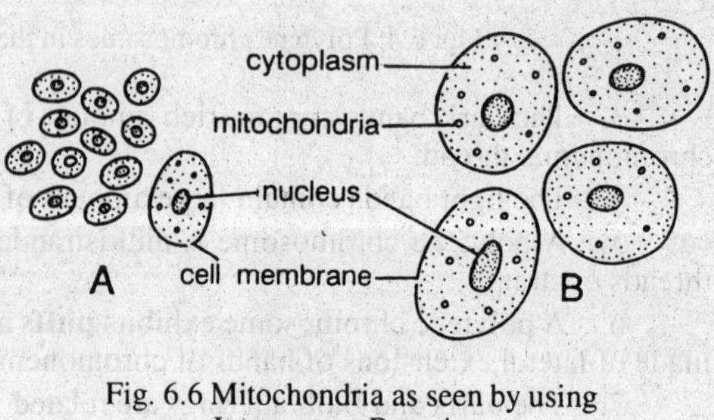

Fig. 6.6 Mitochondria as seen by using Janus green B strain.

EXERCISE No. 5

Object. To comment upon the slides or microphotographs showing ultrastructures of some cell types and cell organelles like endoplasmic reticulum, mitochondria, Golgi body, nucleus, etc.

1. Mycoplasma cell

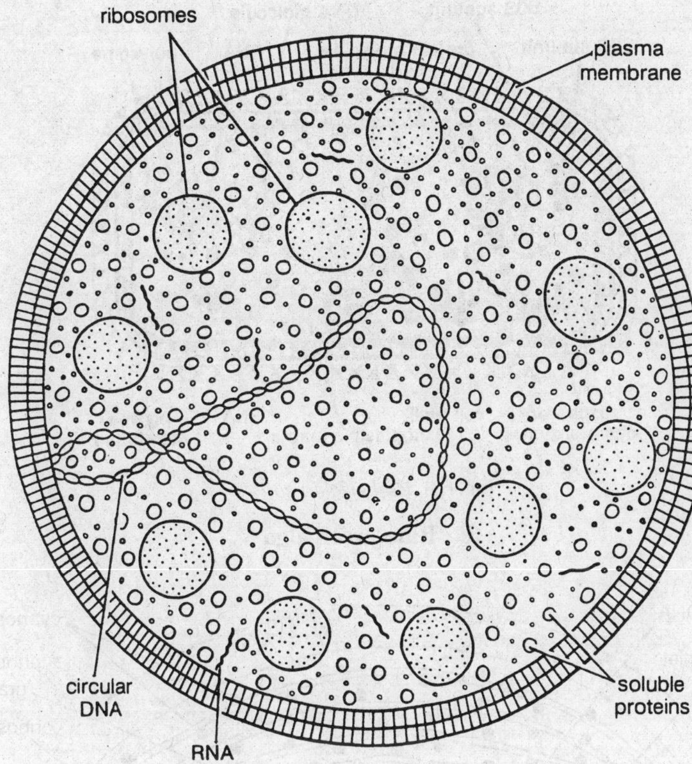

Fig. 6.7. Mycoplasma cell.

Comments:

1. It is the smallest living **prokaryotic cell** found in sewage and soil.
2. Cell-shape is rounded having a thick cell wall formed of phospholipid protein layer.
3. Ribosomes in form of large spherical bodies.
4. Soluble proteins in form of small rounded bodies.
5. DNA double helical loop-like or twisted.

2. *Escherichia coli*

Comments :

1. *Escherichia coli* (Fig. 6.8) is well-known **prokaryotic cell,** studied extensively, and found in the intestine of man.
2. It is rectangular or rod-shaped, measuring about 2 μm in length and 0.5 μm in thickness.
3. It possesses thick and rigid cell wall being made of polysaccharides, protein and lipid molecules.
4. Its cell wall is followed by plasma membrane which is lipoprotein in nature.
5. It possesses free enzymes in the cytoplasmic matrix as well as polyribosomes or polysomes.
6. Its DNA is circular and coiled, RNA is also noticed.
7. It is a nonpathogenic symbiotic bacterium of the intestinal flora of man.
8. It has been used extensively in the studies of molecular biology.
9. It constitutes the main bacterial fauna of man's intestine and non-pathogenic.

10. It is a favourite experimental organism for cell biologists, geneticists, biochemists and molecular biologists.

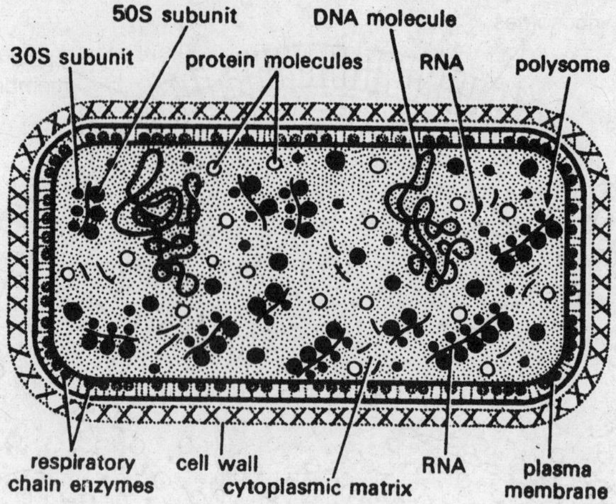

Fig. 6.8. *Escherichia coli.*

3. Blue-green alga

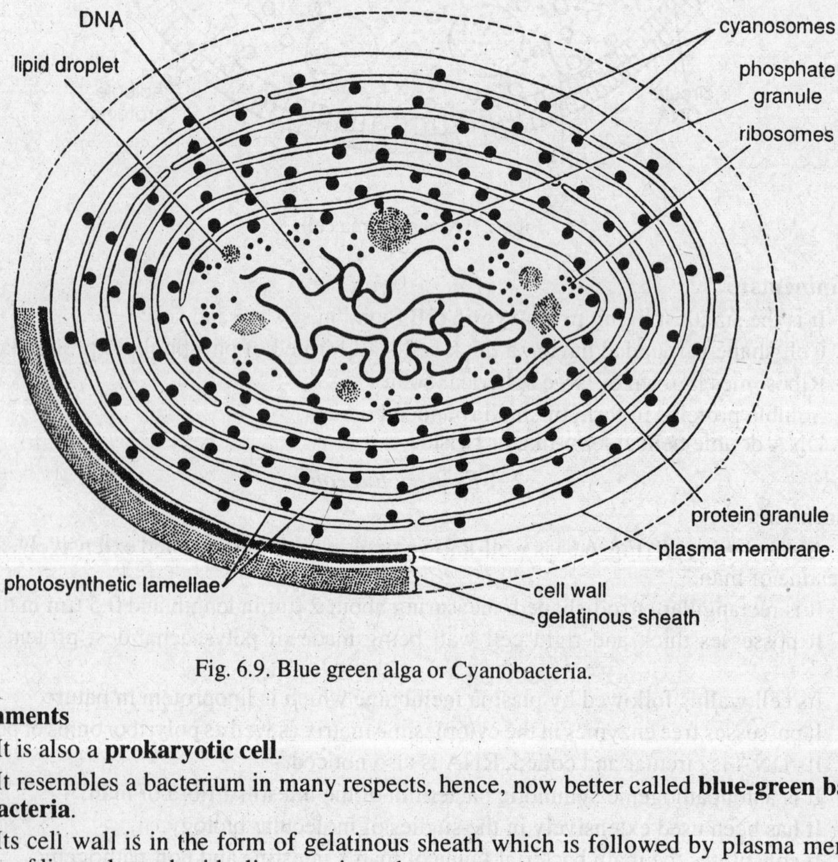

Fig. 6.9. Blue green alga or Cyanobacteria.

Comments

1. It is also a **prokaryotic cell**.

2. It resembles a bacterium in many respects, hence, now better called **blue-green bacteria** or **cyanobacteria**.

3. Its cell wall is in the form of gelatinous sheath which is followed by plasma membrane being made of lipopolyssacharide, lipoprotein and mucoprotein.

4. It contains photosynthetic pigments like chlorophyll and carotenoids.
5. Its DNA is circular and double helical.
6. Its storage products are cyanophycean starch and protein.
7. Some of them are found in symbiotic association with other organisms, while some of them possess the ability to fix atmospheric nitrogen in soil.

4. Plant cell

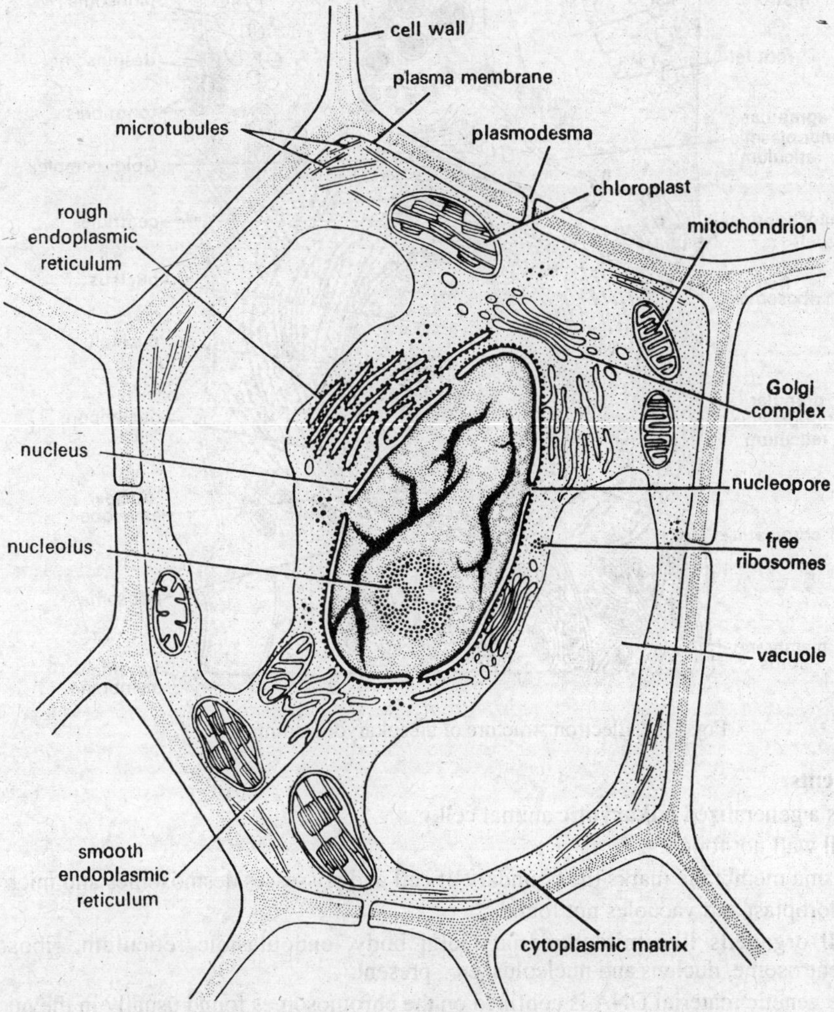

Fig. 6.10. Electron structure of a plant cell.

Comments:

1. It is a generalized **eukaryotic** plant cell.
2. Cell-shape not definite.
3. Cell wall present which is thick and formed of cellulose.
4. Chloroplast present, vacuoles large.
5. Cell organelles like mitochondria, Golgi body, endoplasmic reticulum, ribosomes, nucleus and nucleolus present.
6. Chromosomes within the nuclear substance.
7. Centrosome not found.

5. Animal cell

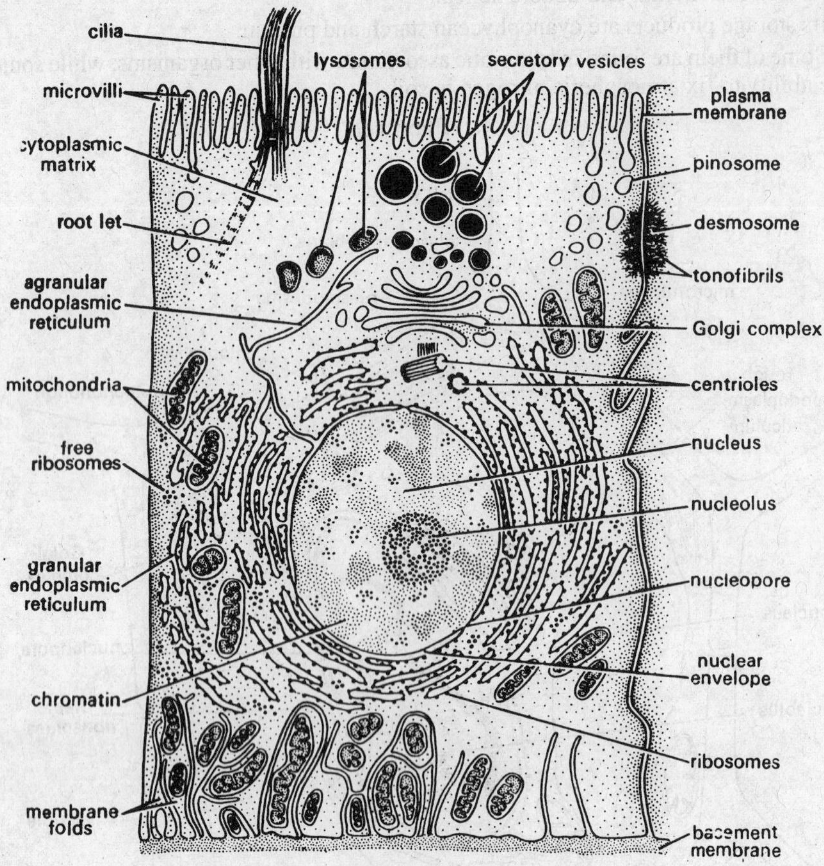

Fig. 6.11. Electron structure of an eukaryotic animal cell.

Comments:

1. It is a generalized **eukaryotic** animal cell.
2. Cell wall not found.
3. Plasma membrane marks the boundary of cell and possesses desmosomes and microvilli.
4. Chloroplast and vacuoles not found.
5. Cell organells like mitochondria, Golgi body, endoplasmic reticulum, ribosomes, lysosomes, centrosome, nucleus and nucleolus, etc., present.
6. The genetic material DNA is confined on the chromosomes found usually in the nucleus.
7. Pinocytic and secretory vesicles are seen.
8. Cell-shape definite.

6. Endoplasmic reticulum (E.R.)

Comments:

1. Endoplasmic reticulum is found in all eukaryotic cells except R.B.Cs. and mature egg cells.
2. E.R. in secretory cells like pancreas, liver and endocrine glands are highly developed.
3. It is a network of tubules, cisterane and vesicles traversing in the cytoplasm; cisternal membrances are interconnected by interconnecting tubules.
4. E.R. is either smooth or rough; the tubular surfaces of smooth endoplasmic reticulum are devoid of ribosomes, while those with ribosomes are rough or granular endoplasmic reticulum.

5. It was first reported by **Porter** in 1948.

6. E.R. provides mechanical support to the cell, provides increased surface area for cell metabolism and attachment of ribosomes which are the sites of protein synthesis.

7. E.R. also gives rise to the other membranous structures of the cell like mitochondria and nuclear membrane, and helps in conducting intracellular informations.

7. Mitochondria

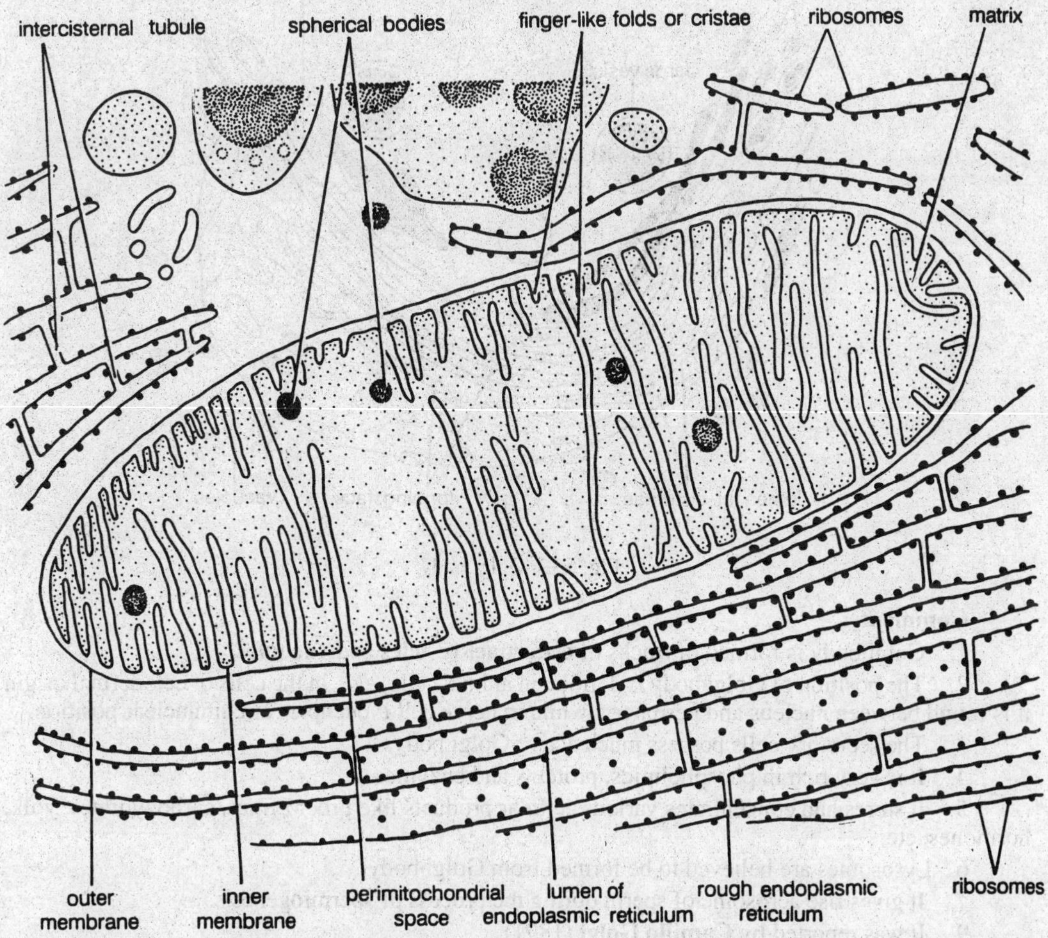

Fig. 6.12. Endoplasmic reticulum and mitochondria.

Comments:

1. Mitochondria are found in all **eukaryotic cells.**

2. It is sausage-shaped having outer and inner membranes. The inner membrane projects into matrix in the form of folds called **cristae.** The space between outer and inner mitochondrial membranes is called **perimitochondrial space** (Fig. 6.12).

3. The inner membranc surrounds the **inner chamber** which is filled with matrix.

4. The cristae bear several **F₁ particles.**

5. The mitochondria are said to be originated from E.R.

6. The outer and inner mitochondrial membranes, outer and inner chambers bear several respiratory enzymes.

7. The mitochondria is called powerhouse of the cell because they release energy by the oxidation of food stuffs and store it in the form of ATP.

8. The name mitochondria was given by **Benda** (1897–98) to the bioblast of **Altmann** (1890).

8. Golgi Body

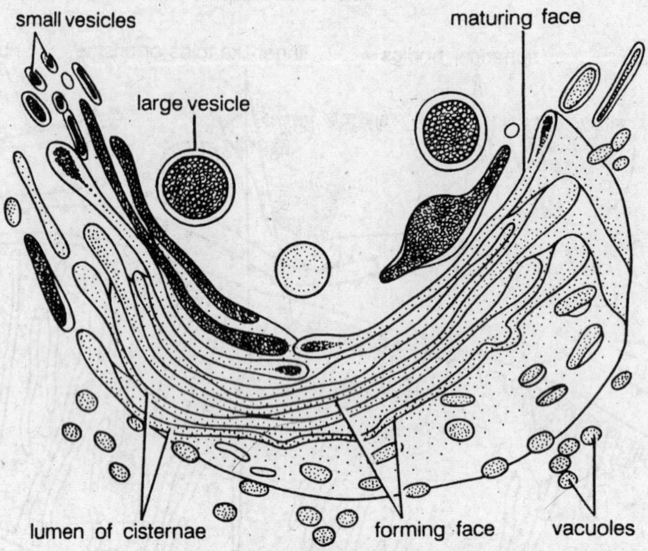

Fig. 6.13. Golgi body.

Comments:

1. Golgi body is formed of stacks of **cisternae, vesicles** and **vacuoles.**

2. The position of Golgi body is definite in animal cells, *e.g.*, in the cells of ectodermal origin it is found between nucleus and periphery, while in nerve cell it occupies circumnuclear position.

3. The secretory cells possess much active Golgi body.

4. It is very rich in phospholipids, proteins and enzymes.

5. It stores and concentrates various cellular products like proenzymes, carbohydrates, yolk, hormones, etc.

6 Lysosomes are believed to be formed from Golgi body.

7. It gives rise acrosome of sperm during the process of spermiogenesis.

9. It was reported by **Camillo Golgi** (1891).

9. Nucleus

Comments:

1. Nucleus (Fig. 6.14) is found in all eukaryotic cells except mammalian R.B.Cs.

2. It is usually spherical and formed of two membranes which are separated by a **perinuclear space.**

3. Both the membranes are fused together at a number of places so as to form **nuclear pores** through which nuclear and cytoplasmic contents remain in communication to each other.

4. It contains nucleoplasm, chromatin net and nucleolus.

5. The nucleolus is a spherical body containing RNA and nucleoprotein.

6. It controls all the vital activities of the cell and carries hereditary material from generation to generation.

7. It was first described by **Robert Brown** (1831).

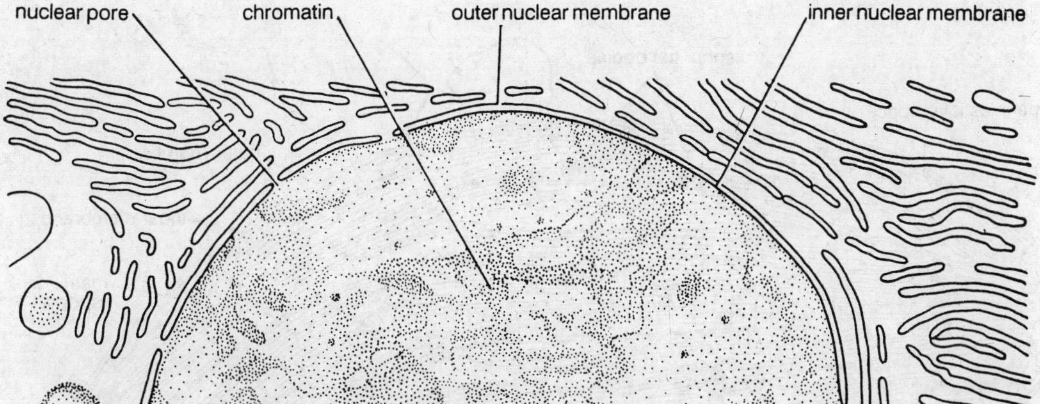

Fig. 6.14. Nucleus.

10. Chloroplast

Comments:

1. Chloroplasts are found in the cells of green plants.
2. It is a kind of chlorophyll-bearing **plastid.**
3. These are spheroidal or ovoidal in the leaves of higher plants, but may be club-shaped, stellate, collar-like or like a spiral band in algae.
4. On an average the chloroplasts in higher plants measure 4–8 microns in diameter.
5. In the cells of higher plants about 20–40 chloroplasts are found in each cell, but in algae usually one chloroplast is found in each cell.
6. Each chloroplast is covered by a double unit membrane. Each unit membrane is 40–60 Å in thickness.
7. The space between the two unit membranes is 200–300 Å.
8. This space is called **periplastidial space**.
9. The inner matrix of chloroplast is known as **stroma.**
10. Several chlorophyll bearing double membrane **lamellae** are found scattered throughout the length of chloroplast.

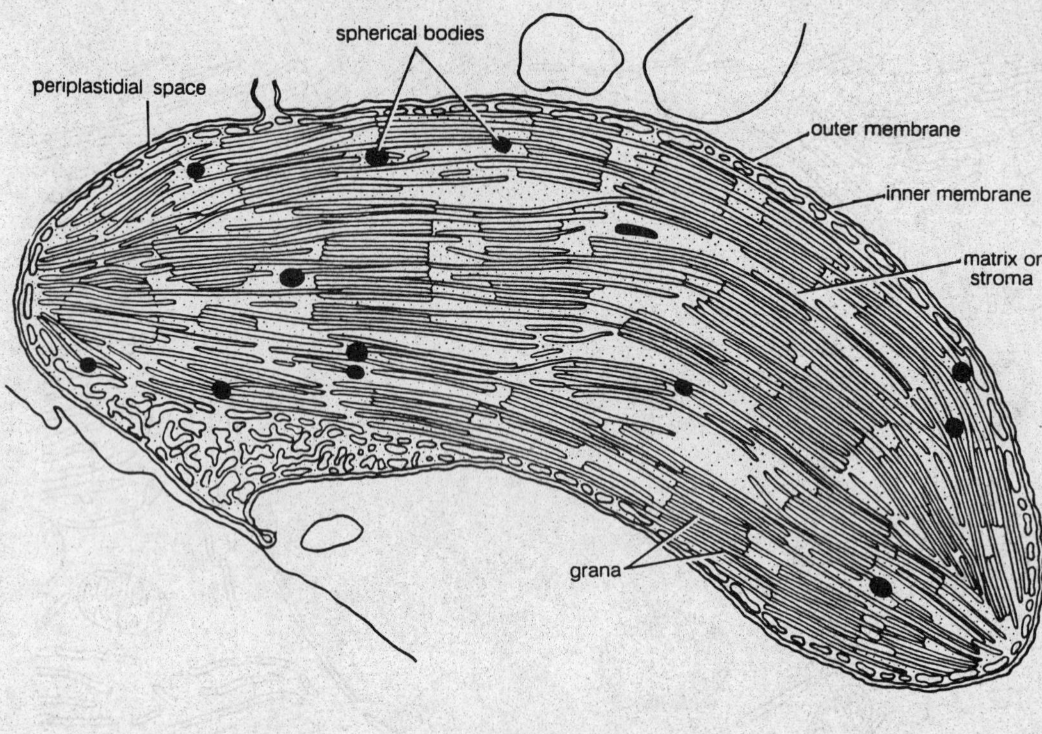

Fig. 6.15. Chloroplast.

11. Lamellae are of two types : (i) disc-like **grana lamellae** which are piled up at several places; (ii) **stroma lamellae** connecting the adjacent grana.

12. Each granum consists of about fifty superimposed membranous compartments, the **thylakoids.**

13. Chloroplasts carry out photosynthesis by trapping solar energy. These are said to contain enzymes which are required during Krebs cycle and during synthesis of fatty acids.

CHAPTER 7

MICROBIOLOGY

The study of **microorganisms,** organisms of microscopic dimensions, is called **microbiology.** In fact, microbiology includes the studies of viruses, bacteria, fungi, algae and protozoa, *i.e.,* virology, bacteriology, mycology, phycology and protozoology. Since microorganisms (=microbes) are of immense importance to man because they are causative agents of various diseases, source of various foods and medicines, and dispose wastes, hence, special techniques and methodology have been developed to study them. Some of the simpler experiments to isolate and cultivate them, and simple staining method are described here.

EXPERIMENT No. 1

Object. Isolation and cultivation of soil microorganisms.

Requirements. Soil sample, petridishes, potato dextrose agar medium*, incubator, metal loop, double distilled water, sieve and pipette.

Soil dilution plate method. Proceed step-wise as described below :

(i) Take 5 gm of air dried sieved soil, add 500 ml of sterilized distilled water and shake the container in a mechanical shaker for 30 minutes (dilution 1:100).

(ii) With the help of a pipette take out 10 ml of soil water suspension, add 90 ml of sterilized water in it and shake thoroughly (dilution 1:1000).

(iii) Take 20 ml from (ii) and add an equal volume of sterilized distilled water (dilution 1:2000).

(iv) From each of the above dilutions (i, ii and iii) spread 1 ml of clear supernatant of soil suspension in separate well sterilized petridishes.

(v) Add 20 ml of cooled potato dextrose agar medium in each petridish and then rotate the petridishes so that the sample gets dispersed uniformly in the medium (hereafter referred to as petriplates or plates only).

(vi) Incubate a few petriplates at 20 ± 2 °C and rest of them at 37 ± 2 °C for 5–6 days.

(vii) Observe these plates regularly. Microbial colonies will be observed. The plates at 20 ± 2 °C will have predominately fungal colonies, while those at 37 ± 2 °C will have mainly bacterial colonies.

Precautions :

1. Keep the plates inverted while incubating when medium gets solidified.
2. Keep covered the plates after soil suspension is taken in step (iv) to prevent contamination of any air borne microorganism.

---

* **Preparation of medium.** *Take cork peeled and sliced potatoes in 500 ml of distilled water. After sometime filter the potato extract and store. Dissolve 20 gm agar-agar in 500 ml of distilled water by heating. Add potato extract in agar-agar solution just prepared and make it to 1000 ml by adding distilled water. Then add 20 gm dextrose sugar in the extract so prepared. Sterilize the medium, thus, prepared and now the medium is ready for use.*

EXPERIMENT No. 2

Object. To isolate and cultivate bacteria from leaf sample.

Requirements. GYCA medium, sterilized scissors, sterilized distilled water, 0.1% mercuric chloride solution, petriplates, infected leaves, incubator, metal loop.

Preparation of GYCA medium. Take 1000 ml of double distilled sterilized water, add 5 gm of glucose, 5 gm of yeast extract + 4 gm calcium carbonate and 15 gm agar. Sterilize the solution so formed and then GYCA medium becomes ready for use.

Procedure. Prepare petriplates with GYCA medium in the manner described earlier. Then collect bacterial blight infected leaves with tiny black spots and old brown lesions. Sterilize the surface of these leaves by dipping them in 0.1% mercuric chloride solution and then wash them thoroughly with sterilized distilled water 3–4 times. Cut the leaves into small pieces with the help of sterilized scissors. Place these leaf pieces over GYCA plates. Keep these plates in an incubator at 37 °C for 4–5 days. Observe, you will note watery, pale yellow coloured growths at the cut ends of the leaf pieces after the incubation. Take fresh petriplates having GYCA medium and transfer a small part of the growth so formed on the petriplates with the help of metal loops. Keep these petriplates in an incubator at 37 °C for 4–5 days again and observe. Cultivation of bacteria occurs.

EXPERIMENT No. 3

Object. To stain bacterial smear by using Gram staining technique.

Requirements. Crystal violet solution, Lugol's iodine, 1% Basic fuchsin, 95% alcohol, cover-slips, microscopic glass slides, sterilized distilled water, metallic loop and Bunsen burner.

Preparation of Reagents :

Crystal violet solution. Take 0.5 gm crystal violet in 20 ml of absolute alcohol and add 80 ml of 1% ammonium oxalate.

Lugol's iodine. Take 1000 ml of distilled water add 1 gm iodine and 2 gm potassium iodide.

1% Basic fuchsin. Dissolve 100 mg of basic fuchsin in 100 ml of boiling distilled water.

Preparation of bacterial smear. Take microscopic glass slides and cover-slips. Wash them carefully and rinse them with chromic acid solution. Wash them in tap water several times and then in distilled water. Make the cover-slips and slides sterilized and then prepare smear of any one of them as suggested below :

Sputum. Take a drop of sterilized distilled water on the slide. Touch the sputum by the sterilized metallic loop and transfer it to the drop of water. Put the cover-slip over the water drop gently so that the water drop spreads uniformly under the cover-slip. Dry the smear so formed by gently warming it over Bunsen's burner and stain (process given below).

Milk. Take a drop of milk on the microscopic glass slide cleaned as above. Spread the milk drop over the slide by a sterilized needle to form a thin film. Dry the smear so formed over a Bunsen's burner by gently warming it and stain.

Staining of Smear. Stain the smear in crystal violet solution for 1 minute. Rinse it in water. Then put Lugol's iodine solution over the smear. Rinse it again in water and pour 95% alcohol over it till smear shows clear colour. Then rinse it again in water. Stain the smear again in basic fuchsin solution for 30 minutes, wash in water. Let it dry and observe under microscope.

Result. The Gram positive bacteria is stained deep violet or blue and Gram negative bacteria is stained pink or red.

CHAPTER 8

HISTOCHEMICAL TESTS

Histochemistry deals with the composition of the cells or tissues in terms of the chemical elements and their compounds in them. Histochemistry has long been used as a research tool in all the branches of the life sciences, and is now being increasingly used in diagnostic histopathology and pathophysiology in medical and veterinary sciences. However, histochemical techniques or test methods have enabled us in identifying and localizing specific chemical constituents in cells and tissues. These testing methods depend on chemical reactions between the substance to be identified and localized in tissue sections, and one or more reagents in which the tissue sections are subjected.

Different histochemical test methods are applied for the identification and localization of different chemical constituents in the tissues. Some of the tests which are simple but correct are described below :

1. HISTOCHEMICAL DETERMINATION OF PROTEINS

The proteins constitute the bulk of all biological tissues. They may be **simple proteins**, *viz.*, albumins, globulins, protamines and histones, or **conjugated proteins**, *viz.*, nucleoproteins, mucoproteins, glycoproteins, lipoproteins and phosphoproteins. The proteins can be demonstrated in paraffin sections by using any one of the following methods :

A. Millon Reaction (Modified from Baker)

Sections. Fix the tissues in 10% formalin. Embed the tissues in paraffin wax and cut sections of 4–5 μ thickness.

Preparation of the reagent. Add 10 gm mercuric sulphate to 100 ml of 10% H_2SO_4 and heat till mercuric sulphate dissolves completely. Then add distilled water to make it 200 ml. Take 5 ml of above solution and add 0.5ml of 0.25% sodium nitrate solution in it. This is Millon's reagent.

Procedure of staining. Dewax the sections and bring down to water through various grades of alcohol. Then place the sections in a small beaker containing Millon's reagent and boil gently. Stop heating after a while and allow the solution to come down to room temperature.

After this, wash the sections three times in distilled water (give 2 minutes time in each wash).Then either mount in glycerine jelly or dehydrate through various grades of alcohol, clear in xylol and mount in D.P.X. or Canada balsam.

Result. Proteins containing tyrosine are stained **red, pink** or **yellowish red.**

B. Mercuric Bromophenol Blue Method (After Bonhag, 1952)

Sections. Fix the tissues in Carnoy or 10% formalin. Embed the tissues in paraffin and cut sections as suggested earlier.

Preparation of the reagent. Take 1% mercuric chloride solution in distilled water and add in it 0.05% bromophenol blue in 2% aqueous acetic acid.

Procedure of staining. Dewax the sections and bring down to water through various grades of alcohol. Stain the sections in the reagent prepared above for two hours at room temperature. Then rinse the sections for 5 minutes in 0.5% acetic acid. Thereafter, transfer the sections directly into a beaker containing tertiary butyl alcohol (give two changes of 10 minutes each). Clear the sections in xylol and mount in D.P.X. or Canada balsam.

Result. Deep blue colours in the sections suggest the presence of proteins.

2. HISTOCHEMICAL DETERMINATION OF LIPIDS

The lipids are the major component reserve food in animal body and are the chief source of energy. They also have structural importance because they are associated with membrane system. They are found in the form of **simple lipids**, *viz.*, fats, oils and waxes, **complex lipids**, *viz.*, phospholipids, glycolipids, chromolipids and aminolipids, and **derived lipids**, *viz.*, cholesterol and ergosterol. The Sudan dyes (Sudan III, Sudan IV, Sudan Black and Oil Red O) have long been used to demonstrate lipids. However, alcoholic Sudan Black-B and Oil Red O are the most recent.

Sudan Black-B Method (After McManus, 1946)

Sections. Fix the tissues for 1–5 weeks in a solution of 1 gm of cobalt nitrate in 80 ml of distilled water, 10 ml of 10% calcium chloride and 10 ml of 40% commercial formalin. Formalin preserves lipid portion in tissues. For getting better results, post-chroming treatment is also suggested. For this the sections are treated for 24–28 hours in 3% potassium dichromate solution. Then dehydrate the tissues in acetone for one and half hours with 3 changes and then embed the tissues in paraffin wax and cut sections of 4–5 μ thickness.

Procedure of staining. Pass your sections after dewaxing through 70% alcohol. Stain the sections for 30 minutes in a saturated Sudan Black-B prepared in 70% alcohol at room temperature. Remove excess dyes from the sections by rinsing quickly in 70% alcohol. Then wash the sections in running water. Counterstain in Mayer's carmalum for 16 hours or in 1% aqueous neutral red for 1 minutes. Wash in water and mount the sections either in glycerine jelly or in D.P.X. or Canada balsam.

Result. The black stained portions in the sections shows the presence of lipids.

3. HISTOCHEMICAL DETERMINATION OF CARBOHYDRATES

Carbohydrates are stored as glycogen in animal tissues particularly in muscles and liver. The carbohydrates are the main energy producers and they are also involved in the structural organisation of several tissues of animals and plants. They are found in form of **monosaccharides,** *viz.,* glucose, fructose, galactose and pentoses, **disaccharides,** *viz.*, maltose, sucrose, lactose, and **polysaccharides,** *viz.,* starch, insulin, dextrin, glycogen, lignin and chitin, etc. However, the method described below is the best for identifying and localizing carbohydrates.

A. Periodic Acid Schiff (PAS) Method (After Hotchikiss, 1948)

Sections. For demonstrating hexoses fix the tissue (liver or muscle) in alcoholic Bouin's fluid and after passing through microtomy procedures sections of 4–5 μ thickness are cut.

Preparation of required reagents :

(i) Periodic acid. Dissolve 0.4 gm of periodic acid (HIO_4, $2H_2O$) in 35 ml of absolute alcohol and add 5 ml of 0.2 M sodium acetate (M= molar solution; 27.2 gm of hydrated salt in 1000 ml of distilled water) and 10 ml of distilled water. Label the solution, thus, prepared and keep it in dark after wrapping black paper around the container at 17–22 °C. Discard the above solution if brown colour appears and proceed to prepare the solution afresh.

(ii) Reducing bath. Take 30 ml of absolute alcohol, add 20 ml distilled water in it and then dissolve 1 gm potassium iodide and 1 gm sodium thiosulphate in it. Add 0.5 ml of 2% conc. HCl. A deposit of sulphur may appear but ignore it. Keep the solution at a temperature between 17–22 °C and use it within 2 weeks. Do not use older solution.

(iii) Schiff's reagent. Dissolve 1 gm of basic fuchsin in 400 ml of boiling water, cool it to 50 °C and filter. Then add 1 ml of thionyl chloride ($SOCl_2$) in the filtrate, wrap the container with black paper and keep it in dark for 12-hours. Then clean by shaking for 1 minute with 2 gm of activated charcoal. Filter it and keep in dark at 0–4 °C. Use this solution in dark at room temperature.

(iv) Celestine blue solution. Dissolve 2.5 gm of iron alum in 50 ml distilled water by keeping it stand overnight at room temperature. Add 0.25 gm celestine blue B and boil for 3 minutes. When solution cools, filter it and add 7 ml glycerine in it.

(v) Orange G solution. Dissolve 2 gm Orange G in 100 ml of 5% aqueous phosphotungstic acid. Allow it to stand for 24 hours and use only supernatant.

Procedure of staining. After dewaxing the sections bring them down to water through various grades of alcohol. Rinse the sections in 70% alcohol and immerse them in periodic solution for 5 minutes. Rinse them again in 70% alcohol and immerse in reducing bath for a minute. Rinse the sections once again in 70% alcohol and immerse in Schiff's reagent for 20 minutes. Then wash the sections in running water for 10 minutes. (For washing, keep the slides in an enamel tray along its edges and place one end of a rubber tube in the tray which is connected to water tap on the other end. Thus, water enters the tray, washes the slides and overflows from the tray. Do not wash the slides directly with running water). Stain the sections with celestine blue solution for 2–3 minutes (this stage may be omitted). Differentiate the sections for a very short time in acid alcohol. Wash again with running water for 30 minutes and then stain with orange G for about 10 seconds (this stage may also be omitted). Wash the sections in water for about 30 seconds to 1 minute till the sections become pale yellow in colour. Now dehydrate the sections in alcohol, clean in xylol and mount in D.P.X.

Result. Hexose containing mucopolysacchasides stain deep **purplish red.** (Nuclei stains blue black and acidophil proteins yellow. These results are obtained only when stages mentioned for omission are not omitted).

B. Periodic Acid Schiff (PAS) Method (After Mc Manus)

Sections. For demonstrating glycogen fix the tissues (liver or muscles) in Gendre fluid (Gendre fluid is prepared by mixing saturated picric acid 85 parts + 40% formaldehyde 10 parts and glacial acetic acid 5 parts). Cut the sections as usual passing through microtomy procedures of 4–5 μ thickness.

(The method of preparing reagents has already been described above.)

Procedure of staining. After dewaxing the sections bring them down to water through grades of alcohol. Keep the sections for 10 minutes in 1% aqueous perodic acid. Wash the sections for 5 minutes in running water, then immerse them in Schiff's reagent for 10 minutes and wash in running water for 5 minutes again. Stain nuclei with celestine blue (a stage which can be omitted). If necessary, differentiate the sections in 1% acid alcohol. Wash them thoroughly in running water for about 30 minutes and then dehydrate in alcohol, clear in Xylol and mount in D.P.X. or Canada balsam.

Result. Hexose containing mucopolysaccharides stain in various shades of **purplish red.** Glycogen stains deeply.

4. HISTOCHEMICAL DETERMINATION OF NUCLEIC ACIDS

Nucleic acids are complex organic molecules found in the cells. They play a key role in all the biosynthetic and hereditary functions of the living organisms. The nucleic acids are of two types : **Deoxyribonucleic acid** (DNA) and **Ribonucleic acid** (RNA). They are determined by the following methods :

A. Methyl Green Pyronin Y Method for DNA and RNA (After, Kurnick, 1955)

Sections. Fix the tissue in Carnoy solution for about 8 minutes and follow the general procedures of microtomy to obtain sections of 4–5 μ in thickness in paraffin wax.

Preparation of reagents :

(i) Pyronin Y solution. Prepare 2% pyronin Y solution in water. Extract with excess of chloroform by shaking the solution in a separating funnel till the solution becomes colourless.

(ii) Methyl green solution. Prepare 2% methyl green solution in water. Extract with chloroform till the violet shade of the solution lasts.

(At the time of use, mix 12.5 ml of pyronin Y solution and 7.5 ml of methyl green solution in 30 ml of distilled water. Thus, prepared reagent should be used within a week and thereafter it should be discarded).

Procedure of staining. Bring the sections to water stage after dewaxing and passing through grades of alcohol. Stain the sections for 6 minutes in freshly prepared methyl green pyronin Y solution. Blot dry with a filter paper and immerse the sections in *n*-butyl alcohol (give two changes, each after 5 minutes). Then immerse the sections in xylol and keep in it for 5 minutes. Finally immerse the sections in cedar wood oil for 5 minutes and mount in D.P.X.

Result. Chromatin containing DNA stains **clear green.** RNA of nucleolus and cytoplasm stains **bright red.**

Note: Follow methods suggested below to determine only DNA or RNA. For DNA, Feulgen Reaction (Feulgen and Rossenbeck, 1924) and for RNA, methyl green pyronin and ribonuclease method (Brachet, 1942).

5. HISTOCHEMICAL DETERMINATION OF ACID AND ALKALINE PHOSPHATASES

The enzymes phosphatases are hydrolysing esters of phosphoric acid. They play an important role in metabolic processes and are widely distributed in the tissues. They are of 3 classes; **pyrophosphatases, phosphomonoesterases** and **phosphodiesterases.** The phosphatases working in acidic medium are called **acid phosphatases** and those in alkaline medium are called **alkaline phosphatases**. They may be demonstrated in sections of animal tissues as described below :

A. HISTOCHEMICAL DEMONSTRATION OF ACID PHOSPHATASE

Lead Nitrate Method (After Gomori, 1950)

Sections. Fix the tissue in cold acetone and adopt general microtomy procedure to cut sections in paraffin wax of 4–5 μ thickness.

Procedure of staining. Bring the sections down to the water stage after dewaxing and passing through different grades of alcohol. Incubate the sections at 37 °C, for 15 minutes to 4 hours depending upon the reaction, in freshly prepared 0.01 M sodium β-glycerophosphate in 0.05 M acetone buffer (pH 5) containing 0.004 M lead nitrate. Wash the sections briefly in water and then immerse them in dilute yellow ammonium sulphide solution for 1–2 minutes. Then wash the sections in water and mount directly in D.P.X. without undergoing the process of dehydration.

Result. Presence of **black** precipitate of lead sulphide shows the presence of acid phosphatase in the sections.

B. HISTOCHEMICAL DEMONSTRATION OF ALKALINE PHOSPHATASE

Alkaline Phosphatase Method (After Fredricson, 1956)

Sections. Fix the tissues in 90% alcohol for 24 hours at 22 °C. Transfer in 96% alcohol for an hour and transfer in absolute alcohol giving two changes for 15 minutes each. Then transfer the tissues in benzene (give 2 changes of 30 minutes each). Embed in paraffin wax at 56 °C and cut sections of 4–5 μ of thickness.

Procedure of staining. Dewax the sections in xylol and pass the sections through pure acetone and then 40% acetone to water. Incubate the sections for 10–60 minutes in a closed couplin jar at 37 °C in a mixture of solution prepared by mixing 2% sodium β-glycerophosphate 25 ml, 2% sodium vernol 25 ml, 2% calcium nitrate 5 ml, 0.8% magnesium chloride 5 ml and acetone 40 ml. After incubating the sections, rinse them in 40% acetone. Now treat the sections with 2% cobalt nitrate in 40% acetone for 5 minutes. Again rinse with 40% acetone and treat with dilute yellow ammonium sulphide in 40% acetone for 3 minutes. Rinse the sections in pure acetone and dehydrate the sections in the grades of alcohol, clear in xylol and mount in D.P.X.

Result. Black deposits of cobalt sulphide in the sections indicate the presence of alkaline phosphatase.

CHAPTER 9

HISTOLOGICAL SLIDES

Histology refers to the study of tissues and in fact deals with the cells and non-cellular components of tissues, and formation of organs. This study is useful in understanding the relationship between structure and function of an organ. The present chapter deals with various animal tissues and the tissues organisation of various organs.

1. Simple squamous epithelium

Comments:

1. It consists of a single layer of flattened plate-like cells.
2. The cells are closely fit together by their edges and form a sort of mosaic.
3. The cells frequently have a roughly hexagonal, irregular and wavy outline.
4. Each cell contains a nucleus.

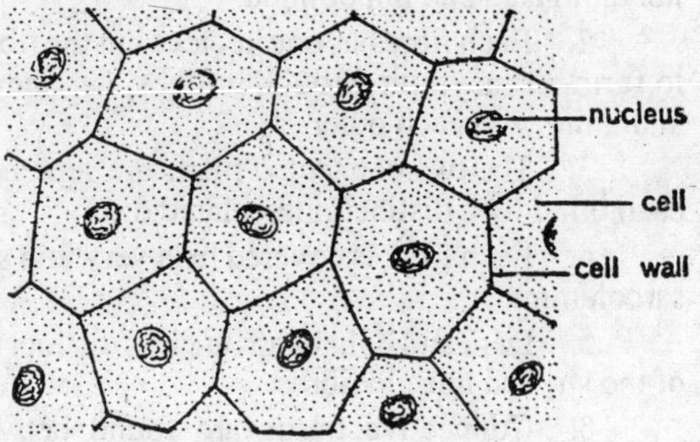

Fig.9.1. Simple squamous epithelium.

2. Unstriated muscle fibres

Comments:

1. The unstriated muscles consist of elongated cells bound together by delicate connective tissues.
2. Each muscle fibre is an elongated cell, pointed at both the ends and thickened out in the middle.
3. In the thickened middle part of each fibre lies a nucleus.
4. A number of threads, the **myofibrillae** or **sarcostyles** extend lengthwise through the fibres.
5. The fibres are bound together by connective tissue forming small bundles.
6. Unstriated muscle fibres are found in the wall of the digestive tract, in the ducts of glands connected with digestive tract and in the urinary bladder of frog. Involuntary in nature.

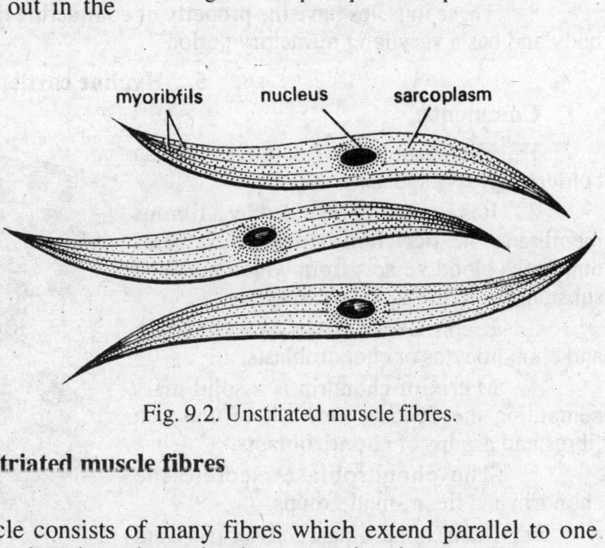

Fig. 9.2. Unstriated muscle fibres.

3. Striated muscle fibres

Comments:

1. The striated or voluntary muscle consists of many fibres which extend parallel to one another but do not branch or anastomose and are bound together by connective tissue.

2. Each muscle fibre is elongated and cylindrical and contains cross strations, some of which are light, while others are dark and both are arranged alternately.

3. These striations are known as **light** and **dark bands.**

4. Each fibre is surrounded by a thin sheath called the **sarcolemma.**

5. Numerous nuclei are present in each fibre making it a **syncytium.** The nuclei are oval and elongated in shape.

6. Each fibre is formed of a number of **myofibrillae** or **sarcostyles** which run longitudinally parallel to each other.

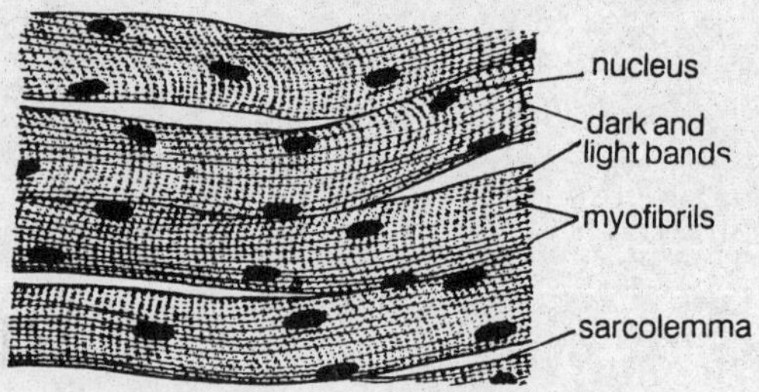

Fig. 9.3. Striated muscle fibres.

7. The myofibrils are composed of bundles of myosin filaments.

4. Cardiac muscle fibres

Comments:

1. The cardiac muscle fibres are cylindrical but not elongated and not pointed.

2. Each fibre has one nucleus but the fibres unite to form a syncytium through bridges, *i.e.*, they branch and unite with each other.

3. The two connected fibres are separated from each other by a distinct intercalated disc.

4. Each fibre is covered by a very thin sheath of sarcolemma.

5. Dark and light bands are less distinct than that of the striated muscle fibres.

6. Connective tissue are found between the fibres.

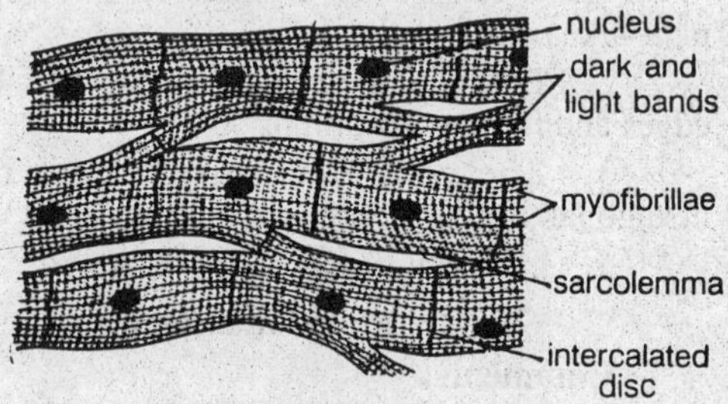

Fig.9.4. Cardiac muscle fibres.

7. Cardiac muscles are found in the walls of the heart of all vertebrates.

8. These muscles have the property of contracting rhythmically, even when isolated from the body and has a very long refractory period.

5. Hyaline cartilage

Comments:

1. Microscopically it is a clear bluish coloured, glossy substance.

2. It is externally bounded by a fibrous membrane the **perichondrium** which bears numerous blood vessels from which nutritive substances diffuse to the cartilage.

3. It consists of **matrix of chondrin** and chondrocytes or chondroblasts.

4. Matrix of chondrin is a solid mass containing the delicate network of collagen fibres and groups of **chondroblasts.**

5. The chondroblasts secrete the chondrin and lie in small groups.

6. Each chondroblast cell lies in a fluid filled space, the lacuna.

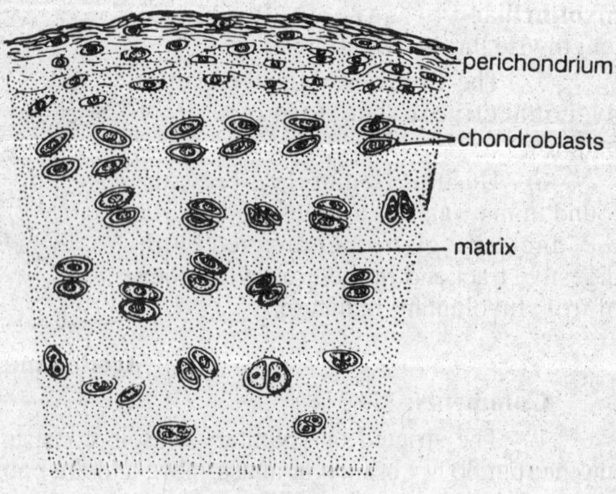

Fig.9.5. Hyaline cartilage.

7. Hyaline cartilage is found as the costal cartilages joining the ventral ends of the ribs to the sternum, on the ends of bones, in the cartrilages of larynx and trachea.

6. Elastic cartilage

Comments:

1. Elastic cartilage is fundamentally similar to hyaline cartilage.
2. Numerous **yellow fibres** are present which run through the matrix in all directions.
3. **Chondrocytes** are somewhat rounded in shape and lie within the lacunae. These cells are scattered in the matrix in groups of two or singly.
4. It is yellowish in colour and more flexible and elastic.
5. It is found in mammals in the pinna of ear, the ends of nose, in the epiglottis and certain other regions.

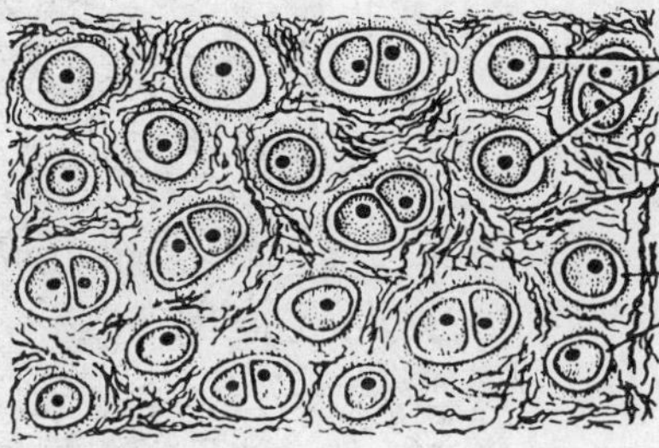

Fig. 9.6. Elastic cartilage.

7. Fibro-cartilage

Comments:

1. The matrix of **fibro-cartilage** contains thick compact bundles of white fibres.
2. The **white fibres** run parallel with one another by narrow spaces in which lie chondrocytes.
3. Chondrocytes are more or less oval and slightly elongated cells which are enclosed within the lacunae.
4. It may be regarded as transitional type between hyaline cartilage and areolar tissue.
5. It is typically found in the intervertebral cartilaginous discs in mammals.

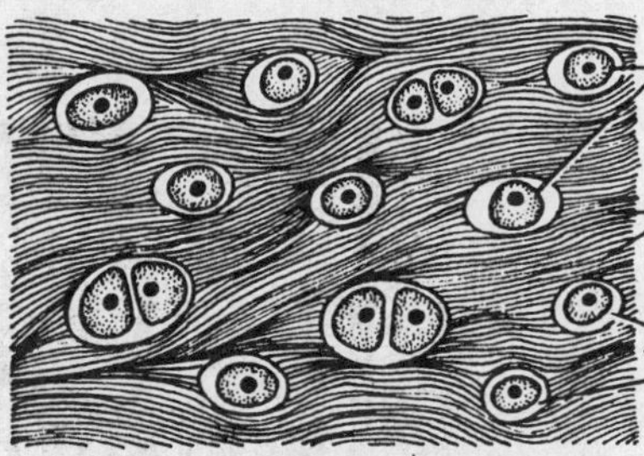

Fig. 9.7. Fibro-cartilage.

8. T.S. of bone

Comments:

Transverse section of bone (Fig. 9.8) exhibits the following important characters :

1. Haversian system

(i) Each consisting of a central **Haversian canal** surrounded by rings of osteocytes lying each in a lacuna.

(ii) The lacunae are connected together by five **canaliculi.**

(iii) Among the rings of lacunae lie very thin concentric layers of bone lamellae which compose the matrix of tissue.

2. Some bone lamellae, bone lacunae and canaliculi are present among the Haversian system but are not arranged around Haversian canals. These are called interstitial lamellae or non-Haversian system.

3. Haversian canals are about 22–110 microns in diameter.

9. L.S. of bone

Comments:

Longitudinal section of bone (Fig. 9.9) reveals the following details :

1. **Haversian canals** are connected with each other by **connecting canals** called **Volkman's canals.**

2. Osteocytes, bone lacunae, bone canaliculi and bone lamellae are arranged lengthwise.

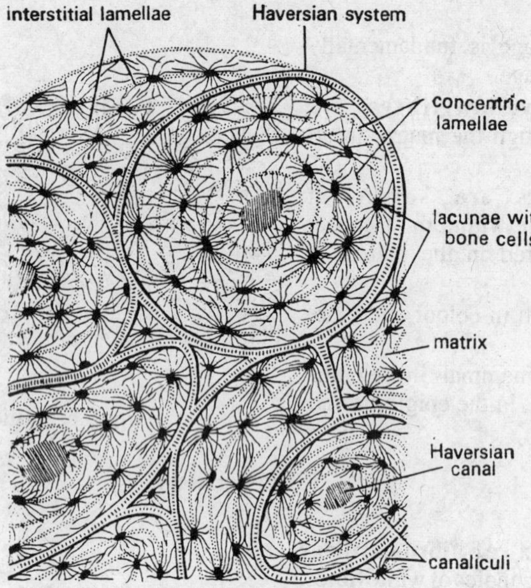

Fig. 9.8. T.S. of bone.

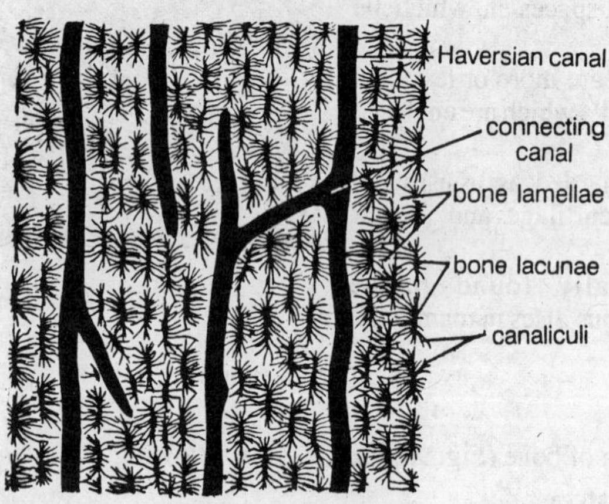

Fig. 9.9. L.S. of bone.

10. Blood of frog

Comments:

The blood film of frog on the slide (Fig. 9.10) exhibits the following important structures :

1. It shows numerous **erythrocytes** (R.B.Cs.) and l**eucocytes** (W.B.Cs.) of different shapes and sizes suspended in the plasma of blood.

2. The **erythrocytes** (R.B.Cs.) are oval, biconvex and have a nucleus.

3. The l**eucocytes** (W.B.Cs.) are colourless and nucleated.

Four types of leucocytes are seen, *viz.*,

(i) **polymorphonuclear leucocytes,**

(ii) **macrocytes,**

(iii) **lymphocytes,** and
(iv) **blood platelets**.

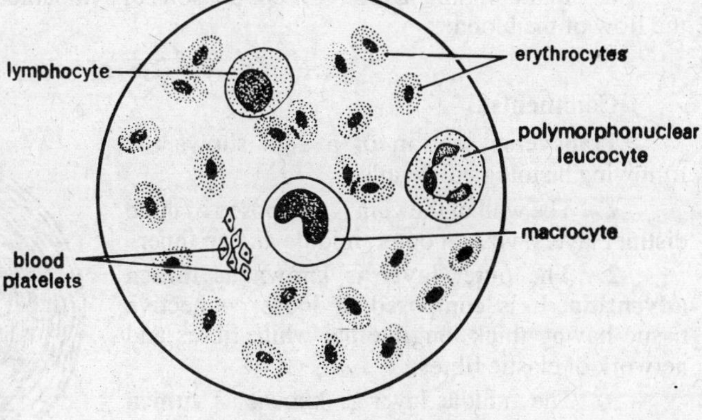

Fig. 9.10. Blood of frog.

11. Blood of man (mammal)

Comments:

The film of mammalian blood on the slide (Fig. 9.11) exhibits the following structures :

1. It shows numerous blood corpuscles of different shapes and sizes suspended in the plasma.

2. The **erythrocytes** (R.B.Cs) are round, non-nucleated and in side view they appear biconvex. The size of erythrocytes is 7.74 to 8 μ in diameter.

3. The **leucocytes** (W.B.Cs) are colourless, nucleated and exercise an amoeboid movement. Three types of leucocytes are seen in the slide.

(i) **Polymorphonuclear leucocyte:** The nucleus of this type of leucocytes is divided into a number of segments (3–5) connected with one another by fine thread. The cytoplasm of this type of cells is granular and stain well with most dyes.

(ii) **Macrocytes** are largest leucocytes and possess a horse-shoe-shaped nucleus. The cytoplasm of these cells is without granules.

(iii) **Lymphocytes** are small with large nucleus and little cytoplasm.

4. The number of erythrocytes in normal adult man and woman is about four and half million or five million in 1 cubic millimeter of blood.

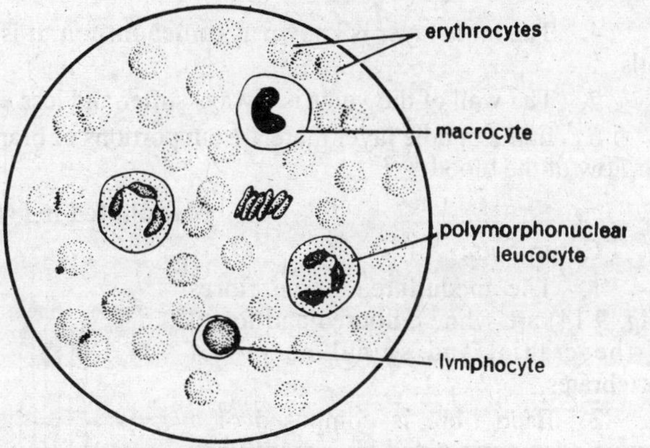

Fig. 9.11. Blood of man.

12. T.S. of artery

Comments:

Transverse section of an **artery** shows the following histological details :

1. The outer wall of an artery is composed of three distinct layers or coats, an outer, middle and an inner.

2. The outer layer is known as **tunica adventitia.** It is composed of connective tissue mixed up with elastic fibres which form a short of network.

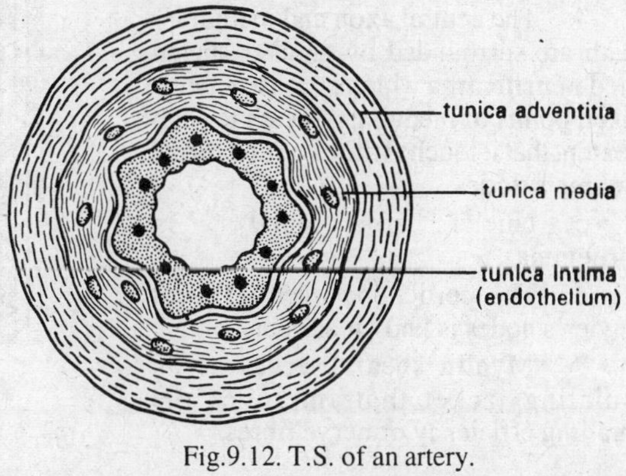

Fig.9.12. T.S. of an artery.

3. The middle layer is known as **tunica media.** It is composed of smooth muscles and elastic connective tissue fibres. It is many layered in thickness.

4. The inner layer is known as **tunica intima.** It is composed of a single layer of endothelial cells.

5. The elasticity of the arteries is mainly due to the presence of elastic fibres in its walls.

6. In the middle layer a few cut portions of sympathetic nerves are seen. This nerve controls the flow of the blood.

13. T.S. of vein

Comments:

Transverse section of a vein shows the following histological details :

1. The wall of the vein is composed of three distinct layers, *viz.*, an outer, middle and an inner.

2. The outer layer is known as **tunica adventitia.** It is composed of loose connective tissue having thick longitudinal white fibres and network of elastic fibres.

3. The middle layer is known as **tunica media.** It is poorly developed and, hence, the veins are thin-walled. It consists of circular smooth muscle fibres which are separated by longitudinal fibres.

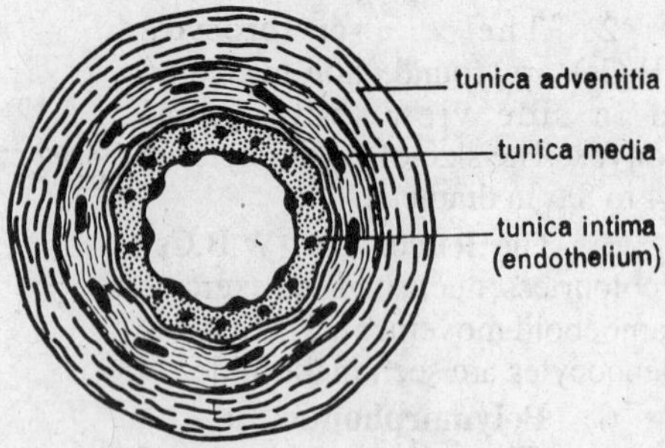

Fig. 9.13. T.S. of a vein.

4. The inner layer is known as **tunica intima.** It is composed of a single layer of endothelial cells.

5. The wall of the veins is always softer and less elastic in comparison of arteries.

6. In the middle layer there are cut portions of branches of sympathetic nerve which controls the flow of the blood.

14. Medullated nerve fibres

Comments:

1. The medullated nerve fibres (Fig. 9.14) are white coloured and found in the cranial and spinal nerves of vertebrates.

2. Each fibre is composed of a central **axon** surrounded by a **medullary sheath** or **myelin sheath.**

3. The medullary sheath is broken at definite intervals.

4. The central axon and medullary sheath are surrounded by another sheath called **neurilemma** which is constricted at broken points of medullary sheath to such an extent that it touches the axon and forms **Ranvier's nodes.**

5. Nuclei are found in neurilemma.

6. The portion between adjacent Ranvier's nodes is known as neurilemma.

7. Myelin sheath serves as an insulating jacket that improves the signalling efficiency of nerve fibres.

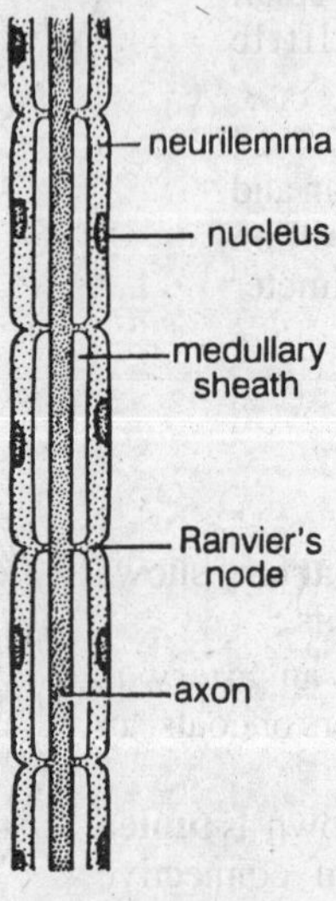

Fig. 9.14. Medullated nerve fibres.

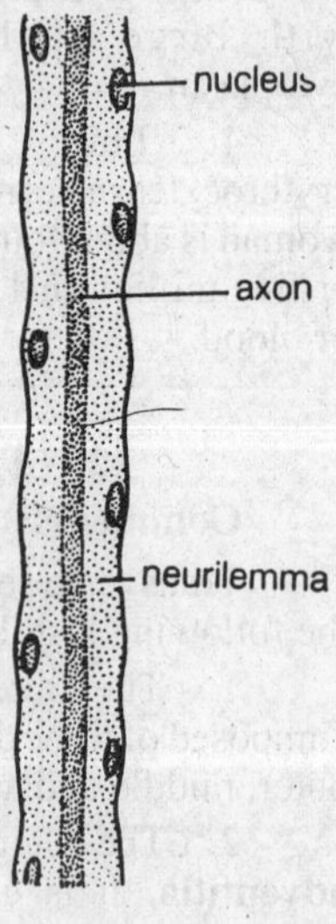

Fig. 9.15. Non-medullated nerve fibres.

15. Non-medullated nerve fibres

Comments:

1. The non-medullated nerve fibres (Fig. 9.15) are grey in colour and found in sympathetic nerves.
2. Each fibre is composed of a central axon surrounded by the **neurilemma.**
3. The **neurilemma** is wavy sheath and contains nuclei.
4. The medullary sheath, Ranvier's nodes and internodes are entirely absent.

16. V.S. of skin of a mammal

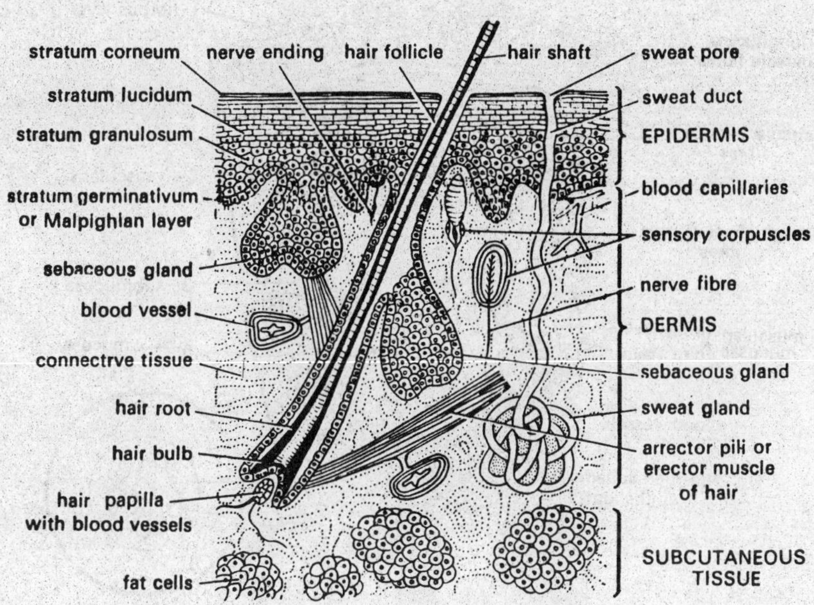

Fig. 9.16.. V.S. of skin of a mammal.

Comments:

Vertical section of **skin** of a **mammal** shows the following histological structures :

1. The skin is composed of two layers, *i.e.*, an outer layer **epidermis** and an inner layer **dermis.**
2. The **epidermis** comprises four layers namely outer **stratum corneum,** next to it **stratum lucidium** then **stratum granulosum** and innermost layer is **stratum germinativum.**
3. **Stratum corneum** consists of horny cells and periodically moulted.
4. **Stratum granulosum** is made up of granular cells.
5. The **dermis** consists of dense areolar connective tissue, muscle fibres, blood vessels, nerves and glands.
6. The mammalian skin is characterized by the presence of **hairs** and **glands.**
7. The hair root is lodged in the hair follicle and hair follicle swells up at the base forming the hair bulb.
8. Blood vessels, nerves and connective tissue from the dermis project into the hair bulb forming the **hair papilla.**
9. Some unstriated muscle fibres connect the hair with the epidermis. These muscle fibres move the hair involuntarily and are known as arrector muscles of the hair.
10. The glands are of two types namely **sebaceous glands** and **sweat glands.**
11. The **sebaceous glands** are small glands of the simple branched alveolar type. Usually each gland is connected to a hair and opens by a short duct close to it.

12. The **sweat glands** are coiled, tubular and much longer. Each gland opens on the surface through a long coil duct.

13. The main function of sweat glands is temperature regulation of body.

Functions of skin: (i) Protection of body; (ii) Conservation of heat; (iii) Excretory; (iv) Secretory; and (v) Sensory.

17. T.S. of oesophagus of a mammal

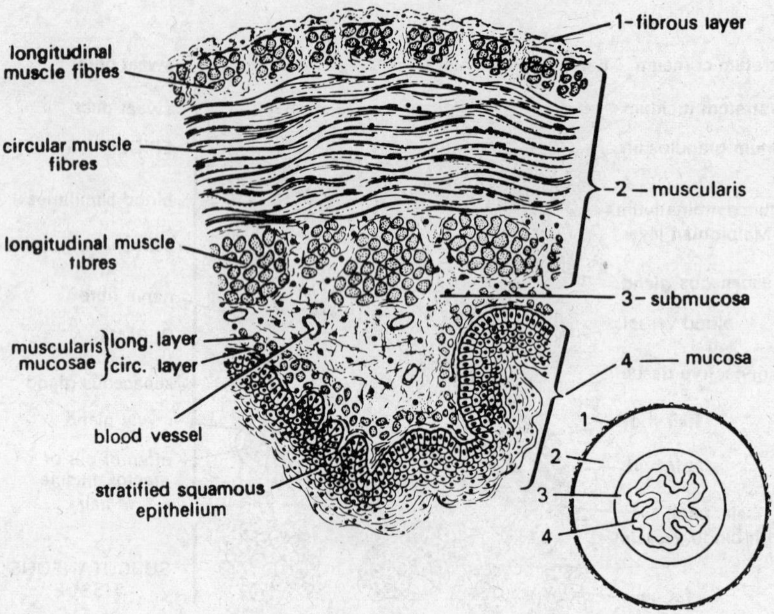

Fig. 9.17. T.S. of oesophagus of a mammal.

Comments:

Transverse section of **oesophagus** of a **mammal** shows the following histological details :

1. Outermost layer is composed of fibrous connective tissue and is known as **fibrous layer.**
2. The **muscularis** consists of three layers of muscle fibres, the outer and inner layers are composed of **longitudinal muscle fibres** and the middle layer is made up of **circular muscle fibres.**
3. The **submucosa** is composed of connective tissue containing blood vessels, nerves and lymph vessels.
4. The **muscularis mucosae** is also present and mainly composed of longitudinal muscle fibres and extends close to the stratified epithelium.
5. The **mucosa** is the innermost layer and forms 4–6 major folds.
6. It is formed of a **stratified squamous epithelium**.

18. T.S. of stomach of a mammal

Comments:

Transverse section of **stomach** of a **mammal** (Fig. 9.18) shows the following histological details :

1. The **serosa** is the outermost layer which is composed of simple squamous epithelium.
2. The **muscularis** consists of two layers, the outer layer is composed of **longitudinal muscle fibres,** while the inner layer is composed of **circular muscle fibres.**
3. The muscle fibres in both the layers are arranged in bundles which are bound together by connective tissue.

4. The **submucosa** is made up of connective tissue containing blood vessels, nerves and lymph vessels.

5. **Muscularis mucosae** is a thin layer and its fibres are arranged in two layers.

6. The **mucosa** is the thickest layer. It consists of **simple columnar epithelium** and **tubular** or **gastric glands.**

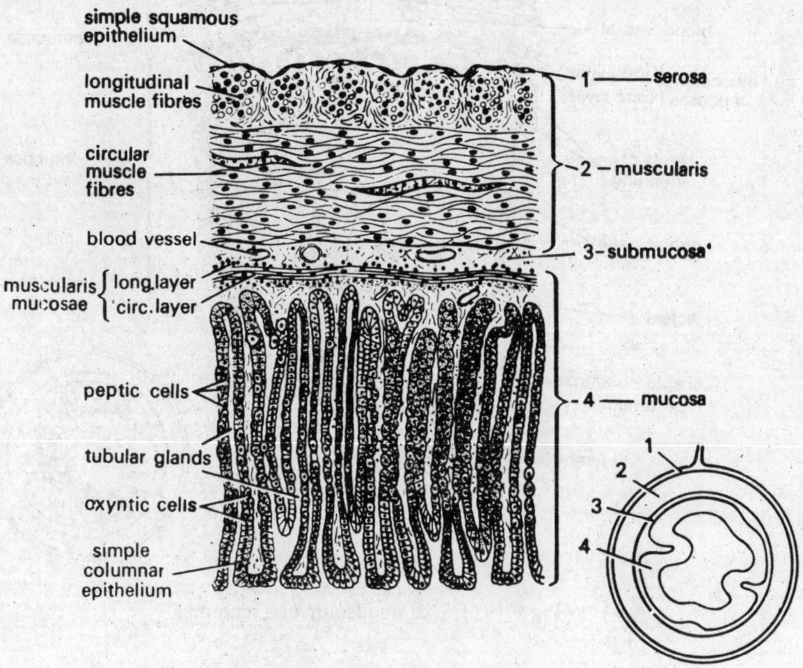

Fig. 9.18. T.S. of stomach of a mammal.

7. The gastric glands are very long and are arranged parallel to each other.
8. They are simple or simple branched tubular type.
9. The gastric glands composed of two types of cells namely **peptic cells** and **oxyntic cells.**
10. The **peptic cells** are found at the base of gastric glands.
11. These are polygonal and granular and secrete digestive enzymes.
12. The **oxyntic cells** (parietal cells) are found towards the luminal part of gastric glands.
13. These are circular or oval, non-granulated and secrete HCl.

19. T.S. of duodenum of mammal

Comments:

Transverse section of **duodenum** of a **mammal** (Fig. 9.20) shows the following histological structures :

1. Outermost layer is **serosa** which usually consists of a simple squamous epithelium.

2. The **muscularis** consists of **longitudinal muscle fibres** to the outside and **circular muscle fibres** to the inside.

3. The **submucosa** consists of connective tissue holding blood vessels, nerves and lymphatic vessels.

4. The **muscularis mucosae** is thin and double layered consisting of outer layer of longitudinal fibres and inner layer of circular fibres.

5. The **mucosa** is thrown into numerous large and small finger-like folds called **villi** which are all covered by **simple columnar epithelium** with scattered **goblet cells.**

6. Each villus contains **blood vessels, lymphocytes** and a **lacteal.**

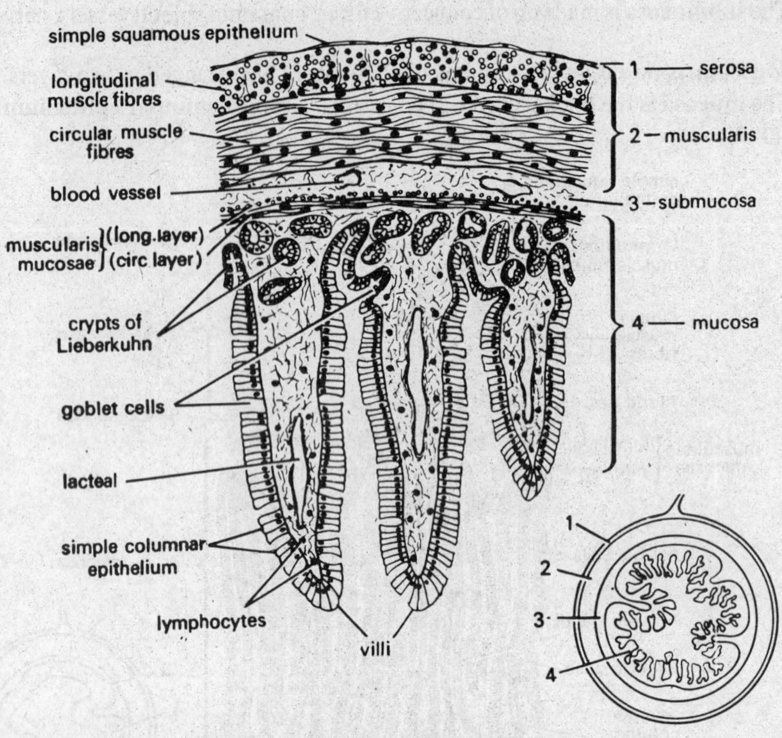

Fig.9.19. T.S. of duodenum of a mammal.

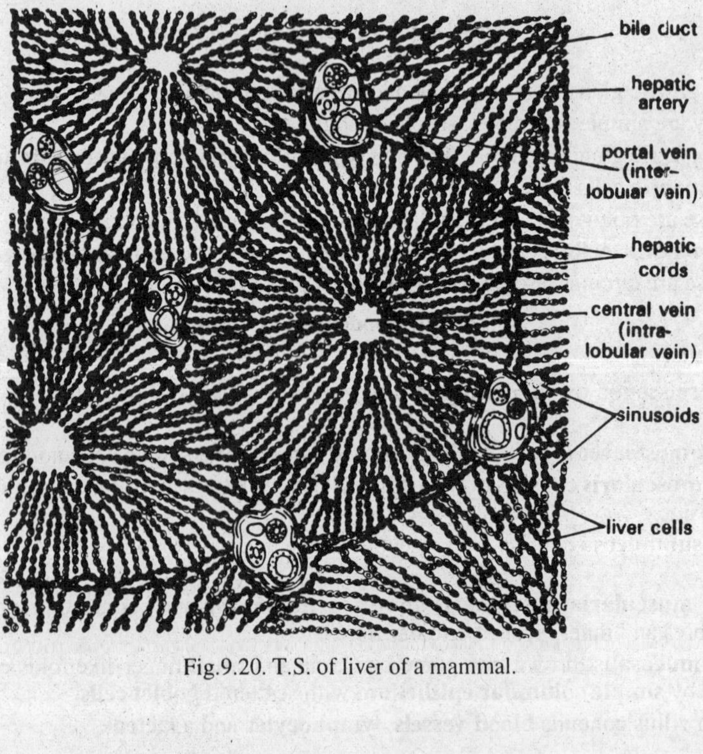

Fig.9.20. T.S. of liver of a mammal.

7. At the base between the villi are present **crypts of Lieberkuhn.** These are lined by columnar epithelium which contains **goblet cells.**

8. The crypts of Lieberkuhn lead into **Bruner's glands.**

9. The secretions of the crypts of Lieberkuhn and Bruner's glands form the intestinal juices.

20. T.S. of liver of a mammal

Comments:

Transverse section of liver of a mammal (Fig. 9.20) shows the following histological structures:

1. The liver is composed of polygonal **lobules** containing a **central vein** (intra-lobular vein) in the centre and **portal canals** at the corners.

2. Each portal canal consists of connective tissue strand and contains a branch of **portal vein** (inter-lobular vein), **hepatic artery, bile duct** and **lymph** vessel.

3. The liver cells are polyhedral or rectangular and arranged in single celled long chains extending radially from the central vein to the periphery of the lobule.

4. Each liver cell has granular cytoplasm and a prominent nucleus.

5. The **sinusoids** are formed from branches of the hepatic portal veins and empty into central veins.

6. Liver has several functions which are as follows :

(i) It produces bile which plays an important role in the digestion of food.

(ii) It stores the soluble products of digestion and metabolize them for assimilation.

(iii) Oxidation of sugar takes place in it.

(iv) Toxic substances are detoxicated in the liver.

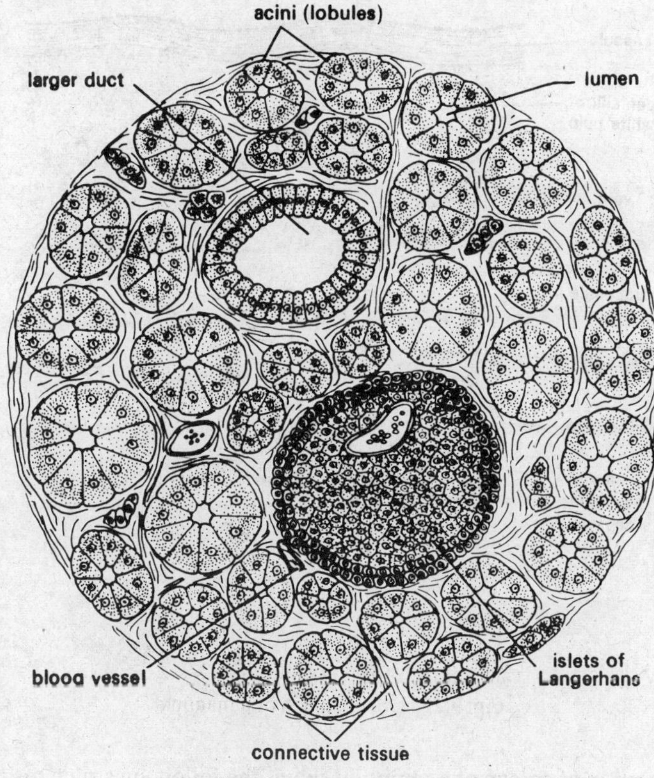

Fig. 9.21. T.S. of pancreas of mammal.

21. T.S. of pancreas of a mammal

Comments:

Transverse section of **pancreas** of a **mammal** (Fig. 9.21) shows the following histological details :

1. The pancreas consists of two portions namely, **exocrine portion** and **endocrine portion.**
2. The **exocrine portion** consists of a series of **lobules** or **acini.**
3. The lobules or acini are bound together by loose connective tissue containing blood vessels, nerves and lymph vessels.
4. Each **lobule** or **acinus** is made up of few pyramidal pancreatic cells having granular cytoplasm and prominent nuclei.
5. The lobules or acini open into small ductules which join large ducts and eventually the main pancreatic ducts.
6. The exocrine portion produces pancreatic juice which contains **trypsin, amylase** and **lipase** enzymes.
7. The **endocrine portion** is composed of **islets of Langerhans** found between the acini.
8. The islets of Langerhans are compact masses of cells and secrete two hormones namely **insulin** and **glucagon.**
9. The insulin is said to be produced by the **beta cells** of islets of Langerhans. It reduces the sugar contents of the blood.
10. The glucagon is said to be produced by **alpha cells**. It increases the sugar contents of the blood and, thus, causes **diabetes.**
11. Pancreas acts both as an exocrine as well as an endocrine gland.

22. T.S. of spleen of a mammal

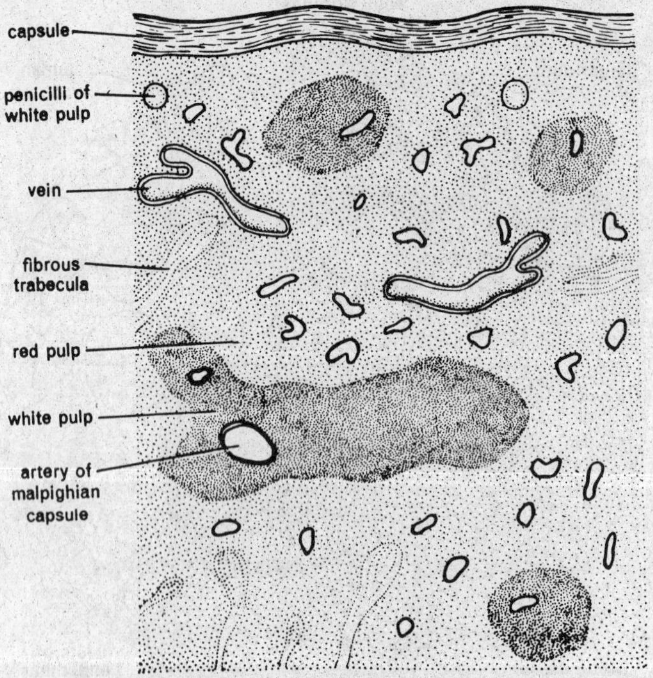

Fig. 9.22. T.S. of spleen of a mammal.

Comments:

Transverse section of **spleen** of a **mammal** shows the following structures :

1. The spleen is surrounded by a thin **capsule** made up of fibrous connective tissue and involuntary muscles.

2. From the capsule arises branching **trabeculae** which penetrate into the tissue of the spleen.

3. The trabeculae contain fibrous connective tissue.

4. The spleen is composed of lymphatic tissue which is distinguished into **white pulp** and **red pulp.**

5. The **white pulp** consists of recticular fibres and form sheath around the arteries.

6. The **red pulp** contains **erythrocytes, capillaries** and **venous sinuses.**

7. **Macrophages** or **phagocytes** are produced in the spleen which eat up foreign bodies and dead cells.

8. It also produces **lymphocytes.**

9. The spleen is very important organ of the body and serves the following functions :
   (i) Produces erythrocytes and lymphocytes.
   (ii) Eats foreign bodies, bacteria and dead cells.
   (iii) Produces antibodies which prevent the entrance of bacteria.
   (iv) It acts as a filtering organ of the blood.

23. T.S. of lung of a mammal.

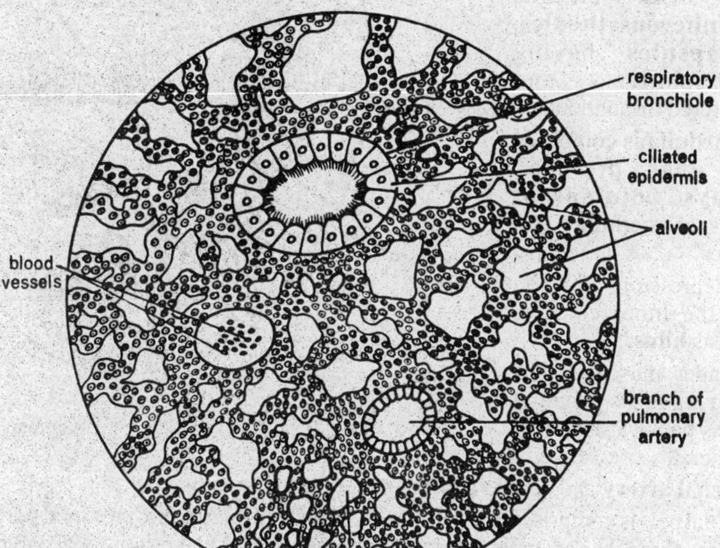

Fig.9.23. T.S. of lung of a mammal.

Comments:

Transverse section of **lung** of a **mammal** shows the following structures :

1. Histologically it consists of numerous **alveoli.**

2. The alveoli communicate with one another by apertures in their walls.

3. Around each alveolus is a network of capillary blood vessels in connection with **pulmonary artery** or **vein** of the lung.

4. Numerous alveoli form clusters which open in a alveolar duct.

5. Each bronchus as it enters the lungs, divides and subdivides into finer branches, the bronchioles.

6. The bronchioles are subdivided into **respiratory bronchioles.**

7. The respiratory bronchiole gives rise to several **alveolar ducts** which open into alveoli or air-sac.

8. The alveoli which are richly supplied with blood vessels form the seat of respiration.

9. The air is taken into the alveoli by the respiratory bronchioles through alveolar ducts which get it from bronchioles which in their turn get it from the bronchus; air → trachea → bronchus → bronchioles → respiratory bronchioles → alveolar ducts → alveoli → gaseous exchange takes place and CO_2 is then taken out.

24. L.S. of kidney of a mammal

Comments:

Longitudinal section of **kidney** of a **mammal** shows the following histological details :

1. The kidney is surrounded by a **capsule** of dense connective tissue.

2. The glandular part of the kidney composed of outer **cortex** and inner **medulla.**

3. The **cortex** contains numerous **uriniferous tubules, malpighian capsules** having **Bowman's capsules** and **glomerulus** scattered throughout.

4. The **medulla** is composed of several renal **pyramids, medullary rays, columns of Bertini, tubules of medulla** and connective tissue.

5. The depression found in the middle of the inner concave region is known as **hilus.**

6. A slender muscular tube known as **ureter** takes its origin at the hilus and runs backwards to join the urinary bladder.

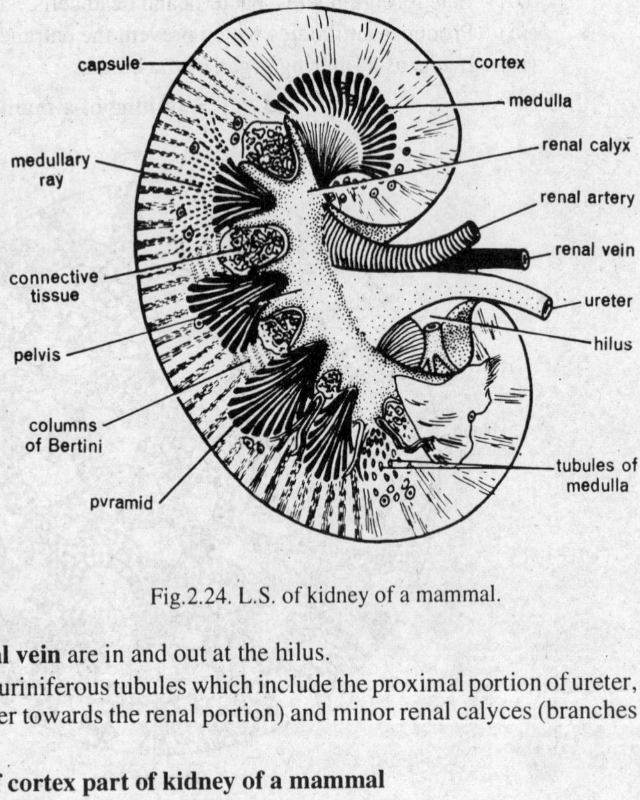

Fig.2.24. L.S. of kidney of a mammal.

7. The **renal artery** and **renal vein** are in and out at the hilus.

8. The renal pelvis comprises uriniferous tubules which include the proximal portion of ureter, major renal calyces (branches of ureter towards the renal portion) and minor renal calyces (branches of the major calyces).

25. T.S. of cortex part of kidney of a mammal

Comments:

Transverse section of cortex part of kidney (Fig. 9.25 and 9.26) shows the following histological details :

1. Cut ends of large number of **uriniferous tubules** of various shapes and sizes are present. The uriniferous tubules are distinguished into two types namely–**proximal uriniferous tubules** and **distal uriniferous tubules.**

2. The **proximal uriniferous tubules** are lined by ciliated glandular epithelium.

3. The **distal uriniferous tubules** are lined by glandular epithelium but its border is not provided with cilia.

4. In between the uriniferous tubules are found numerous rounded structures, the malpighian capsules.

5. Each **malpighian capsule** consists of a **glomerulus** and **Bowman's capsule.**

6. The **glomerulus** is formed by the network of arteries and veins.

7. Medullary rays and collecting ducts are also present.

8. The inner lining of Bowman's capsule absorbs waste products from glomerulus and converts it into urine which passes out through the uriniferous tubules.

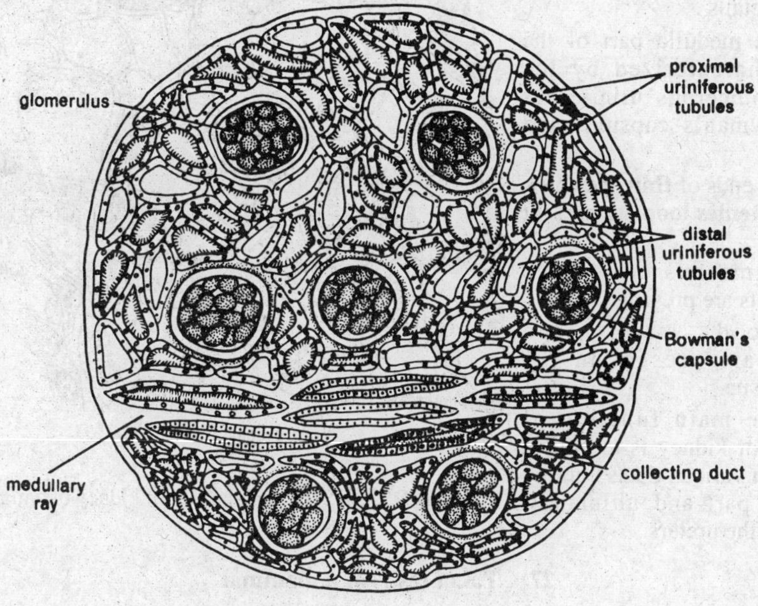

Fig. 9.25. T.S. of cortex part of kidney of a mammal.

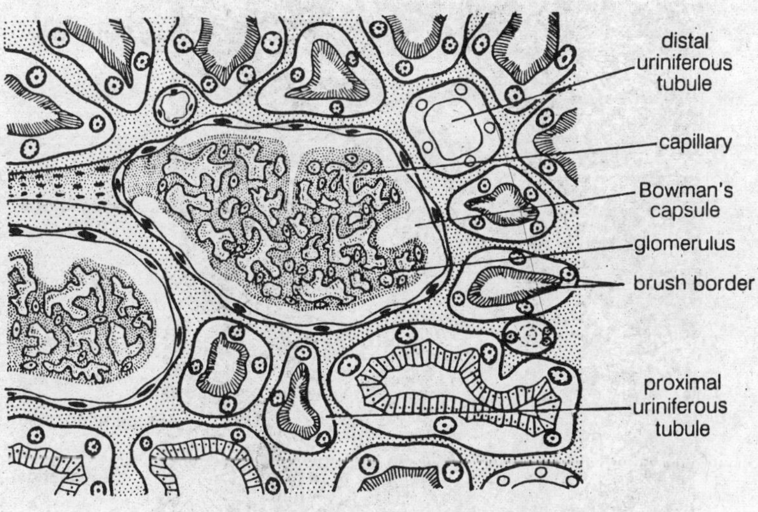

Fig. 9.26. T.S. of cortex part of kidney of a mammal (highly magnified).

26. T.S. of medulla part of kidney of a mammal

Comments:

Transverse section of **medulla part of kidney** of a **mammal** shows the following histological details :

1. The medulla part of the kidney is characterized by the absence of numerous uriniferous tubules, Bowman's capsules and glomerulus.
2. Cut ends of **thin** and **thick** segments of **Henles loop** are present in the section.
3. Numerous large collecting ducts are present.
4. Blood vessels and capillaries are present in the connective tissue.
5. The main function of medulla part of kidney is to collect the urine from various ducts coming from cortex part and ultimately release it into the ureters.

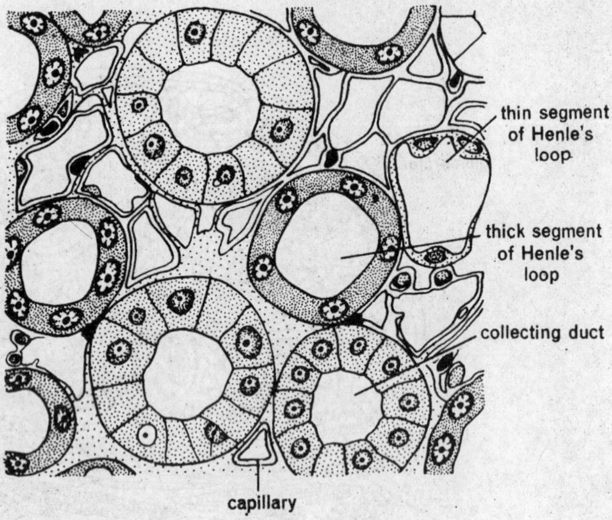

Fig.9.27. T.S. of medulla part of kidney of a mammal.

27. T.S.of testis of a mammal

Comments:

Transverse section of **testis** of a **mammal** shows the following details:

1. The testis is somewhat rounded or oval in shape and surrounded by peritoneum followed by a layer of fibrous connective tissue, the **tunica albuginea.**

Fig. 9.28. T.S. of testis of a mammal.

2. Histologically each testis is composed of a mass of coiled **seminiferous tubules.**

3. The seminiferous tubules are separated from one another by **intertubular tissue.**

4. The intertubular tissue is formed of connective tissue which holds the tubules together and contains blood vessels and **interstitial cells.**

5. The interstitial cells secrete a hormone **testosterone** responsible for male secondary sexual characters.

6. Each seminiferous tubule appears rounded or oval in section surrounded by basement membrane and lined by germinal epithelium.

7. In between the germinal cells certain larger cells called **Sertoli cells** are usually seen. These cells have the role of supplying nourishment to the developing sperms.

8. The germinal epithelium gives rise to sperms which are seen in various stages of development in a seminiferous tubule and are as follows :

(i) The **spermatogonia** lie along the periphery of the tubule and appear closely packed together.

(ii) The **spermatocytes** lie just below the spermatogonia and comprise primary and secondary spermatocytes.

(iii) The **spermatids** aggregate in clusters below the spermatocytes.

(iv) The **spermatozoa** lie in the cavity of the tubule, grouped in clusters and appear connected with Sertoli cells.

9. A **spermatozoon** or sperm has an elongated head and long delicate tail. Its nucleus lies in the head.

10. The groups of spermatocytes, spermatids and spermatozoa are separated from each other by **Sertoli cells.**

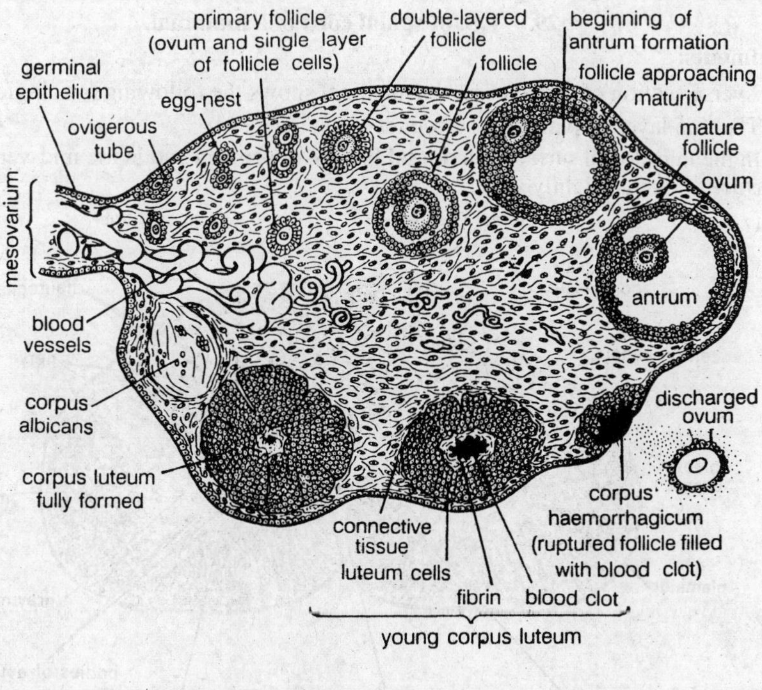

Fig. 9.29. T.S. of ovary of a mammal.

28. T.S. of ovary of a mammal

Comments:

Transverse section of **ovary** of **mammal** (Fig. 9.29) shows the following important structures :

1. The ovary is lined by germinal epithelium which is bounded by the connective tissue, the **tunica albuginea.**

2. It consists of a mass of connective tissue and spindle-shaped cells, the two together forming the **stroma.**

3. Lying in the stroma are egg cells in various stages of development, each surrounded by a nourishing epithelial layer, the follicle and blood vessels.

4. In the section **egg nest, primary follicle** (ovum and single layer of follicle cells), **double layered follicle, follicle approaching maturity** showing **antrum formation, mature follicle, ruptured follicle, young** and **fully formed corpus luteum** and **corpus albicans** are seen.

5. The primary follicles arise from ingrowth of the germinal epithelium into the stroma.

6. One group of cells enlarges to form a developing **ovum,** while the other forming a single layer of cells the **follicle** around it.

7. The follicle and ovum slowly move deeper into the **stroma** and become larger.

8. Later ovum and cells around it become separated by fluid-filled space from the rest of the follicle cells except at one point forming the **antrum.**

9. Further the enlargement of the ovum and follicle results in the production of a **mature follicle** or **Graafian follicle.**

10. A mature follicle (Graafian follicle) is made up of three layers, an outer **theca externa,** middle layer **theca interna** and the inner layer **membrana granulosa.**

11. The mature Graafian follicle then migrates to the surface of the ovary and ruptures releasing the ovum into the Fallopian tube.

12. After discharging of ovum, the follicle cells undergo proliferation and change in structure to form **corpus luteum** which secretes hormone, **progesterone.**

29. T.S. of spinal cord of a mammal.

Comments:

Transverse section of spinal cord of a mammal shows the following histological structures :

1. The thin layer of **piameter** surrounds the spinal cord.

2. In the mid-dorsal surface is a **dorsal fissure** or **septum** and in the mid-ventral surface is a **ventral fissure** which is slightly wider.

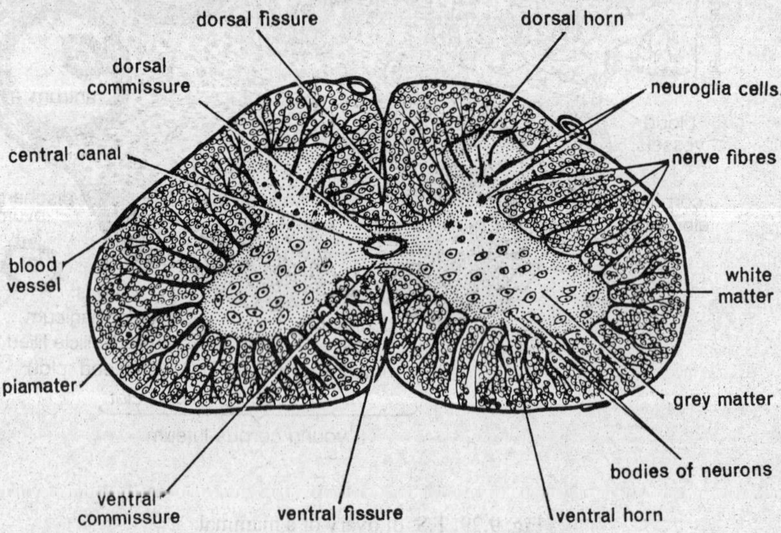

Fig.9.30. T.S. of spinal cord of a mammal.

3. In the centre there is a small cavity known as central canal. It is lined by simple epithelial cells.

4. The substance of the cord is differentiated into two zones, *i.e.*, the central zone surrounding the central canal and is called **grey matter** and a peripheral zone called the **white matter.**

5. The grey matter is H-shaped projecting dorsally into two **dorsal horns** and ventrally into two **ventral horns.**

6. The grey matter shows the presence of **bodies of neurons** with tree-like branching of their **dendrons** and **neuroglial cells.**

7. The white matter is composed of obliquely running medullated nerve fibres supported by prolongations of the neuroglia.

8. The bands of fibres which extend transversely, one dorsal and other ventral to the central canal, are known as **dorsal** and **ventral commissures** respectively.

CHAPTER 10

EMBRYOLOGICAL SLIDES

All sexually reproducing multicellular animals fundamentally develop from the single-celled zygote by the process of embryonic development or embryogenesis. Embryogenesis in all such animals occurs through fundamentally similar events one after another to give rise new individual. In present chapter the embryogenesis of **frog** and **chick** have been described from a view point of practical work which will enable the students to understand the embryonic processes involved to which they have already studied in their theory classes.

DEVELOPMENT OF FROG

1. Unfertilized egg of frog

Comments:

1. The egg or ovum is a spherical cell about 1.6 mm in diameter.
2. It is covered by albumen layer and vitelline membrane.
3. One half of its surface is deep blackish brown in colour due to the presence of a superficial layer of pigment.
4. The other half is lighter in colour and it is in this half that the bulk of the yolk is aggregated, *i.e.*, **telolecithal.**
5. The large egg nucleus lies in the pigmented half which is, therefore, the animal hemisphere of the egg.
6. The dark area is known as vegetative pole.

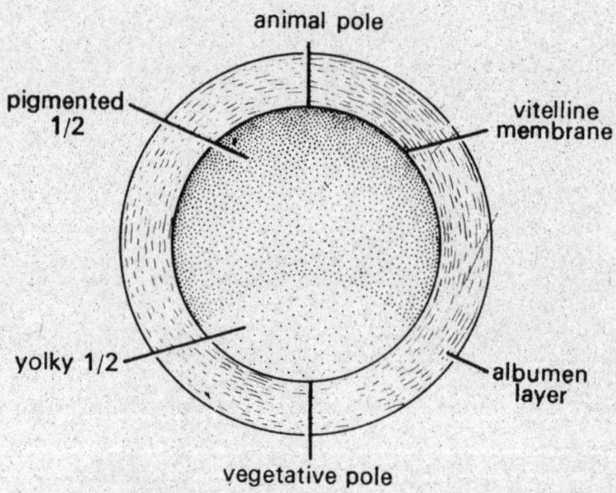

Fig 10.1. Unfertilized egg of frog.

2. Sperm of frog

Comments:

1. The sperm is microscopic, thread-like, motile body having typical **head, middle piece** and **tail.**
2. The head is elongated having a large haploid nucleus capped with conical acrosome.
3. The head is followed by a short middle piece having two centrioles and mitochondria.
4. The middle piece is followed by a long tail having a **main piece** and an **end piece.**
5. The tail bears an axial filament.
6. The axial filament is enclosed in a sheath forming main piece and its naked terminal part forms the end piece.
7. The sperm is funtionally mature and is capable of fertilizing the ovum.
8. The sperms are produced in the seminiferous tubules of the testes as a result of spermatogenesis.

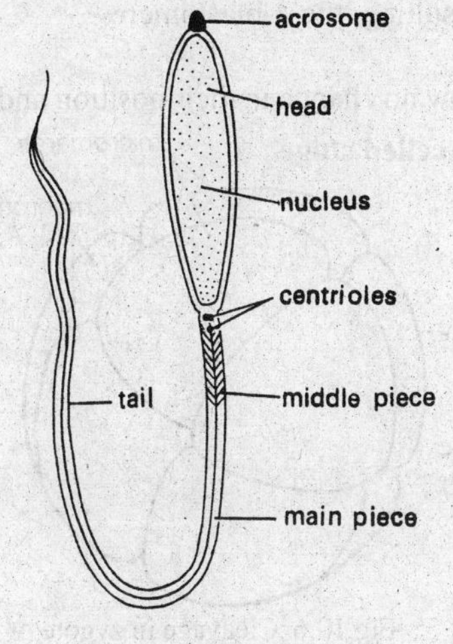

Fig. 10.2. Sperm of frog.

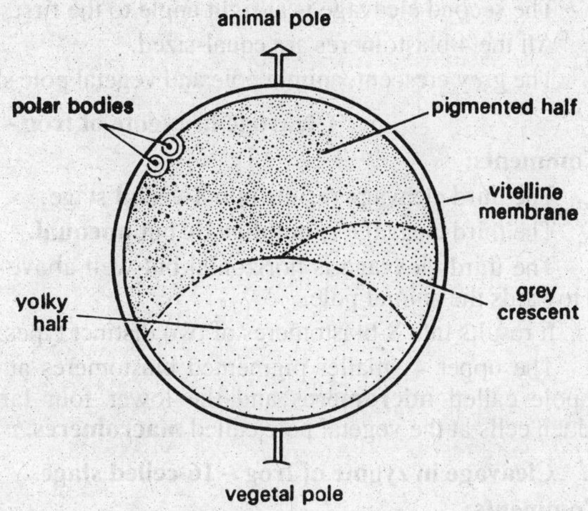

Fig.10.3. Fertilized egg (zygote) of frog.

3. Fertilized egg of frog

Comments:

1. It is a **diploid** cell.
2. In a fertilized egg 'rotation of orientation' occurs.
3. Due to rotation of orientation, the ooplasm differentiates into upper pigmented half or **animal henisphere**, light black-coloured crescent-shaped **grey crescent** and whitish yolky lower part or **vegetal hemisphere.**
4. It is an **activated cell,** ready to undergo the process of cleavage.

4. Cleavage of zygote of frog – 2-celled stage

Comments:

1. The first cleavage plane is **meridional**.
2. The pattern of cleavage is **holoblastic** and **unequal** type.
3. The first cleaving furrow appears near the animal pole and progressively extends towards the vegetal pole.
4. The cleaving furrow cuts the egg through its medium animal-vegetal polar axis vertically into two equal-sized **blastomeres.**
5. The mode of division of the zygote is **mitotic.**

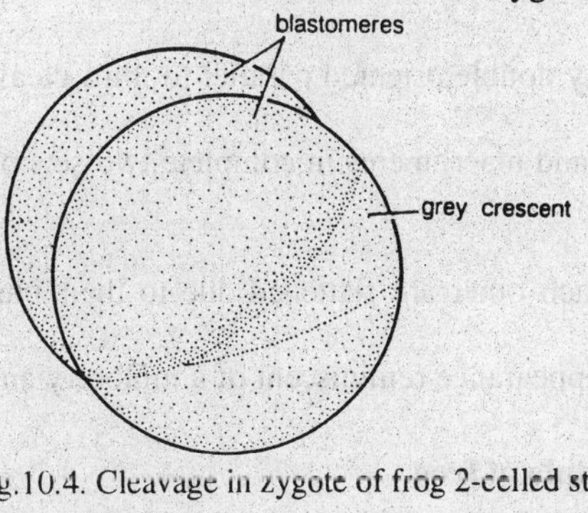

Fig.10.4. Cleavage in zygote of frog 2-celled stage.

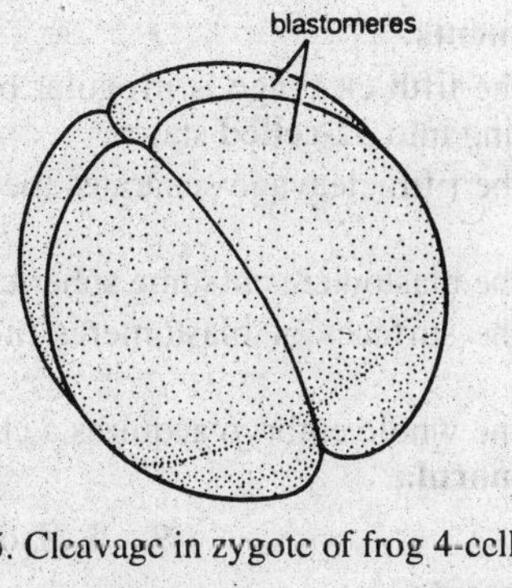

Fig.10.5. Cleavage in zygote of frog 4-celled stage.

5. Cleavage of zygote of frog – 4-celled stage

Comments:

1. The second set of cleavage is also **meridional** which starts before the completion of the first.

2. The second cleavage is at right angle to the first, resulting into 4-blastomeres.
3. All the 4 blastomeres are equal-sized.
4. The grey crescent, animal pole and vegetal pole show no change in their position and size.

6. Cleavage in zygote of frog – 8-celled stage

Comments:

1. The third cleavage results into **8-celled** stage.
2. The third cleavage is **holoblastic** but **unequal.**
3. The third cleavage is horizontal but well above the equator towards the animal pole.
4. It results into 8-blastomeres of two distinct types.
5. The upper 4 smaller pigmented blastomeres at the animal pole called **micromeres** and the lower four larger yolk-ladden cells at the vegetal pole called **macromeres.**

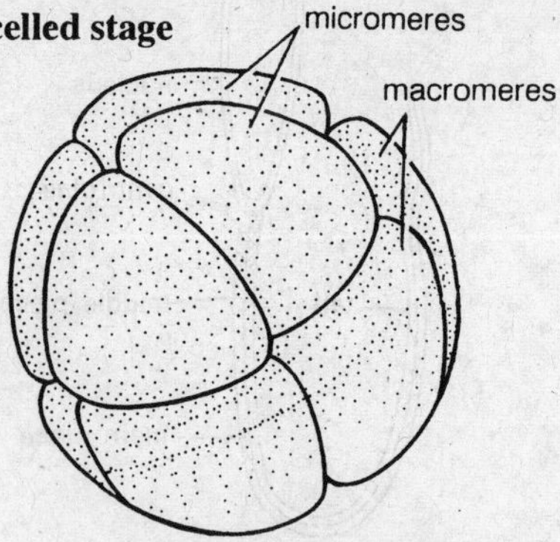

Fig.10.6. Cleavage in zygote of frog 8-celled stage.

7. Cleavage in zygote of frog – 16-celled stage.

Comments:

1. The fourth cleavage plane is also oriented from animal to vegetal pole.
2. The fourth set of cleavage is also vertical, therefore, produces a 16-celled stage.
3. The micromeres divide more rapidly.
4. The yolk-ladden macromeres divide slowly.
5. The yolk shows its influence on cleavage rate.

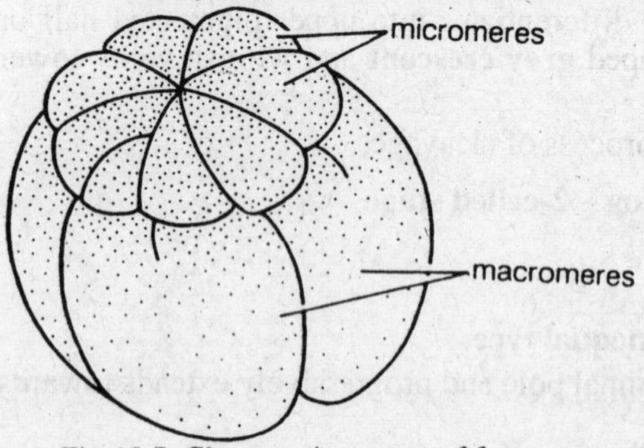

Fig.10.7. Cleavage in zygote of frog 16-celled stage.

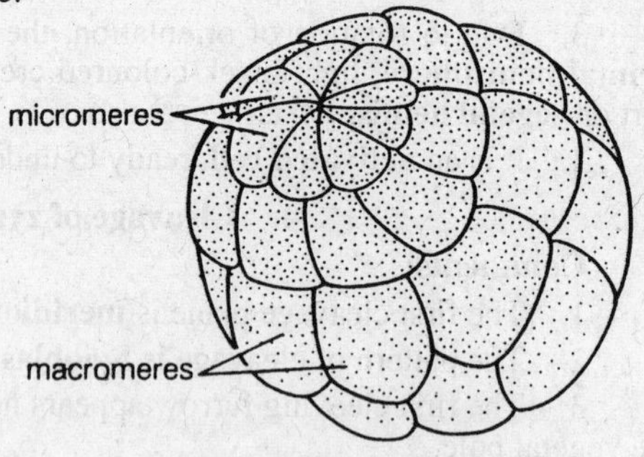

Fig.10.8. Cleavage in zygote of frog 32-celled stage

8. Cleavage in zygote of frog – 32-celled stage

Comments:

1. The fifth cleavage is irregular but it is usually double oriented parallel to third cleavage plane resulting into **32-celled** stage.
2. The fifth cleavage separates the micromeres and macromeres in complete two sets of 16 each.
3. The blastomeres assume spherical shape.
4. The surfaces of blastomeres in contact to each other are flattened due to their mutual pressure.
5. The whole embryo acquires a characteristic appearance reminiscent of a mulberry and so it is called **morula.**

9. L.S. of early balastula of frog

Comments:

1. The fertilized egg undergoes two vertical, third horizontal and then less regular divisions and gives rise to blastula.
2. The section of blastula shows that it contains a cavity known as **blastocoel.**

3. The blastocoel is surrounded by the blastomeres.

4. The blastomeres are distinguished into two types, *i.e.* **micromeres** and **macromeres.**

5. The **micromeres** are smaller, found in the anterior animal pole and contain dark pigments.

6. The **macromeres** are larger, found in the posterior vegetative pole and are laden in the yolk.

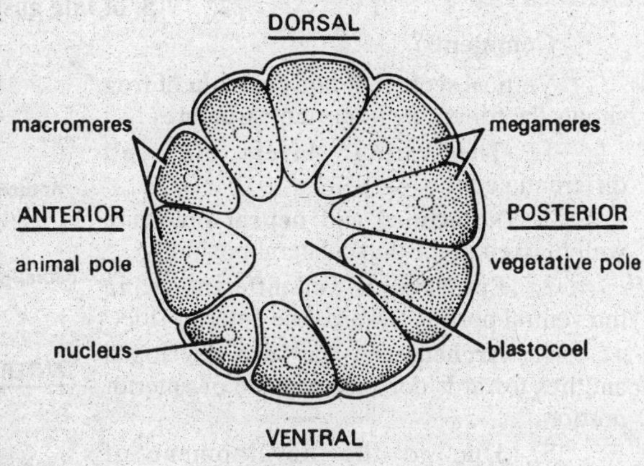

Fig.10.9. L.S. of early blastula of frog.

10. V.S. late blastula of frog

Comments:

1. It is a hollow ball of different sized cells representing **blastula** stage in the development of frog.

2. The hollow space is called **blastocoel.**

3. The blastocoel is eccentric placed towards the animal pole.

4. The blastocoel remains filled with a watery albuminous fluid secreted and infiltered from the surrounding blastomeres.

5. The micromeres of animal pole have arranged themselves into two layers called **blastoderm** forming a sort of roof of the blastocoel.

6. The sides and floor of the blastocoel are occupied by multilayered large yolky blastomeres.

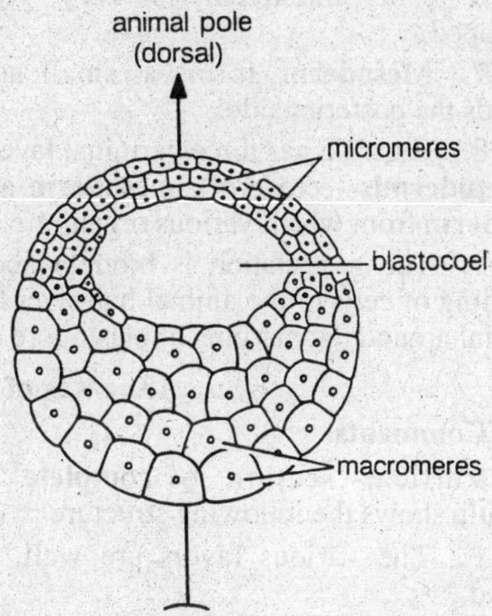

Fig.10.10. V.S. late blastula of frog.

11. V.S. of early gastrula of frog

Comments:

The vertical section of **early gastrula** shows the following important structures :

1. The various layers are more or less well differentiated.

2. **Notochord** and **neural plate** are well marked in the dorsal and dorso-lateral position respectively.

3. Towards the anterior animal pole the **epidermis** is also well differentiated in the ventro-lateral position.

4. The **blastocoel** is large and surrounded by a layer of cells.

5. On the dorso-lateral side towards the posterior vegetative pole a small depression appears which gives rise to **blastopore** guarded by dorsal and ventral lips.

6. The **endoderm** is very well developed.

7. A small area comprising the **mesoderm** is also seen on the ventral side towards the posterior pole.

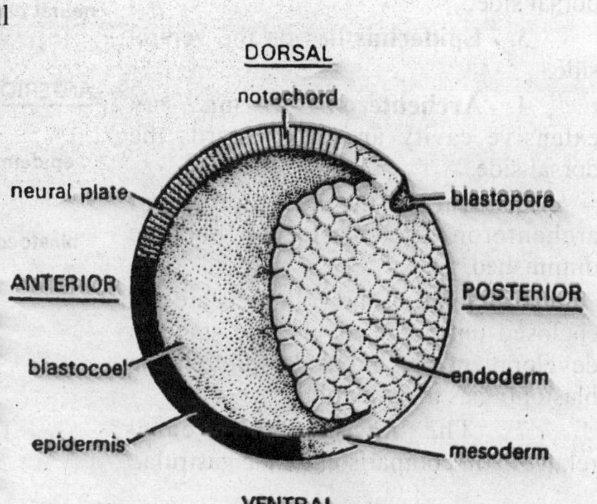

Fig. 10.11. V.S. of early gastrula of frog.

12. V.S. of late gastrula of frog

Comments:

Vertical section of late **gastrula** of frog shows the following important structures :

1. The various layers are well differentiated.
2. **Notochord** and **neural plate** are well marked in the dorso-lateral position.
3. **Epidermis** is well differentiated in the ventral portion towards the anterior side.
4. **Archenteron** has also developed and lies towards dorsolateral side of anterior portion.
5. Due to the development of archenteron, the blastocoel is reduced.
6. The **endoderm** is very well developed.
7. **Mesoderm** forms a small area towards the posterior side.
8. Gastrula has three germinal layers, *viz.*, **epidermis** (ectoderm), **mesoderm** and **endoderm** from which various organs are derived.

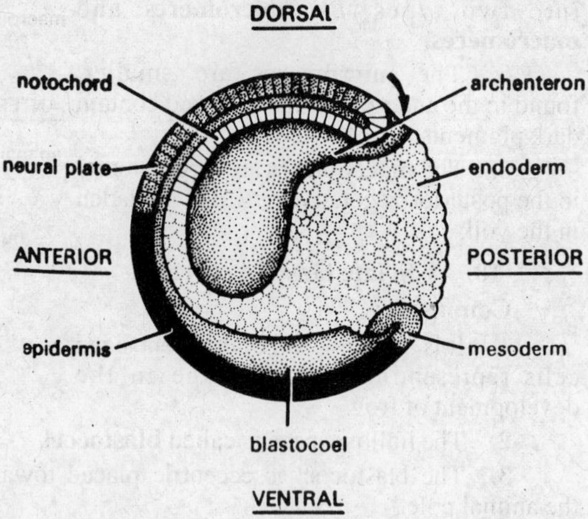

Fig.10.12. V.S. of late gastrula of frog.

9. The gastrulation is brought about by three main processes **overgrowth (epiboly)** or spreading of cells of the animal half over those of the vegetative half, **intucking (invagination)** of material around the margins of blastopore and **contraction** of margins (lips) of blastopore.

13. V.S. of complete gastrula of frog

Comments:

Vertical section of **complete gastrula** shows the following structure :

1. The various layers are well marked.
2. The **notochord** and **neural plate** are well differentiated and lies at the dorsal side.
3. **Epidermis** lies on the ventral side.
4. **Archenteron** forms an extensive cavity and lies towards the dorsal side.
5. Due to the development of archenteron the **blastocoel** become diminished.
6. Some of the yolk cells are not enclosed until a relatively late stage in development and protrude through the blastopore as the yolk plug.
7. The mesoderm increases relatively in comparison to late gastrula.

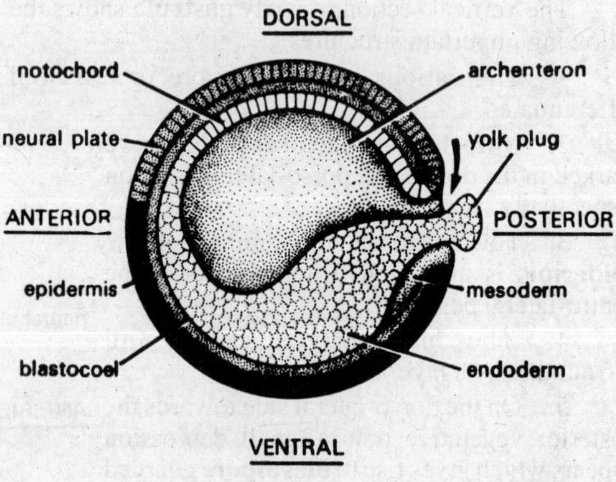

Fig. 10.13. V.S.of complete gastrula of frog.

14. L.V.S. of an early embryo of frog (5mm)

Comments:

1. The oval gastrula elongates quickly.

2. The embryo has acquired distinct head and an ovoid body.
3. The **notochord** extends from anterior to posterior end of the embryo
4. A definite gut is formed below the notochord.

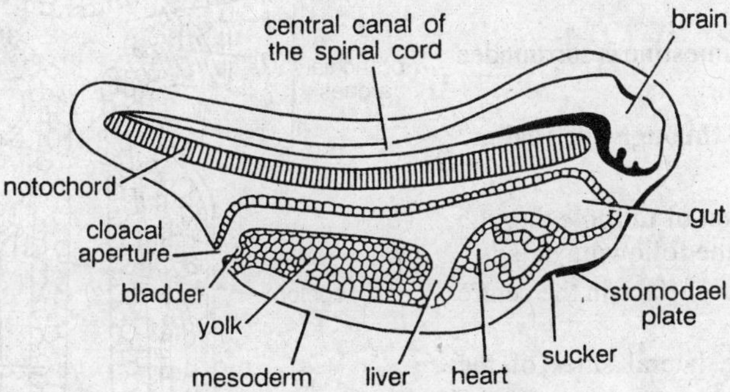

Fig.10.14. L.V.S.of an early embryo of frog (5 mm).

5. The floor of gut bears a thick layer of large yolk-filled cells.
6. The liver and heart have developed.
7. The neural tube, dorsal to notochord, flexed anteriorly to form the brain.

15. Tadpole larva of frog

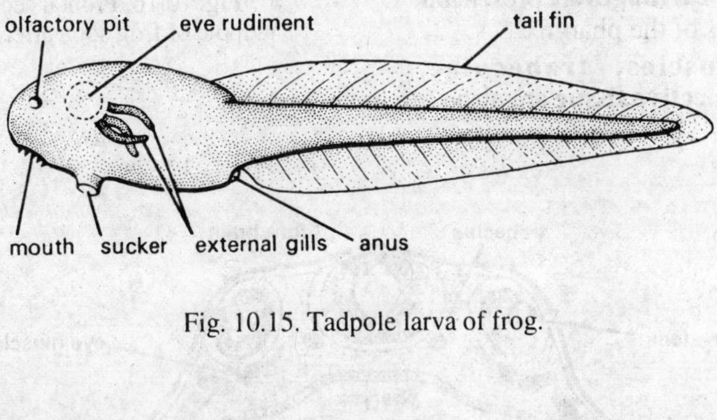

Fig. 10.15. Tadpole larva of frog.

Comments:

1. Egg develops into tadpole larva.
2. Tadpole larva measures about 5–7 mm in length.
3. It is small blackish fish-like animal.
4. The larval body is differentiated into body proper and tail.
5. Body proper contains **olfactory pits, eye rudiments, mouth, sucker, external gills, gut** and **anus.**
6. **Tail** is long and bears a **tail fin** on its dorsal and ventral surfaces.
7. There are three pairs of external feathery gills which act as functional respiratory organs.
8. Mouth bears **horny jaws** or **horny teeth** and intestine is coiled.
9. Tadpole larva metamorphoses into adult.

16. Frontal section of tadpole of frog

Comments:

1. The head region is distinct having a part of fore-brain.

2. The **olfactory placodes** and **optic vesicles** are clearly visible.

3. The pharynx is spacious bearing hyoid arch and branchial arches.

4. The pronephic tubules and their ducts are clearly seen.

5. The tubular intestine is surrounded by yolky cells.

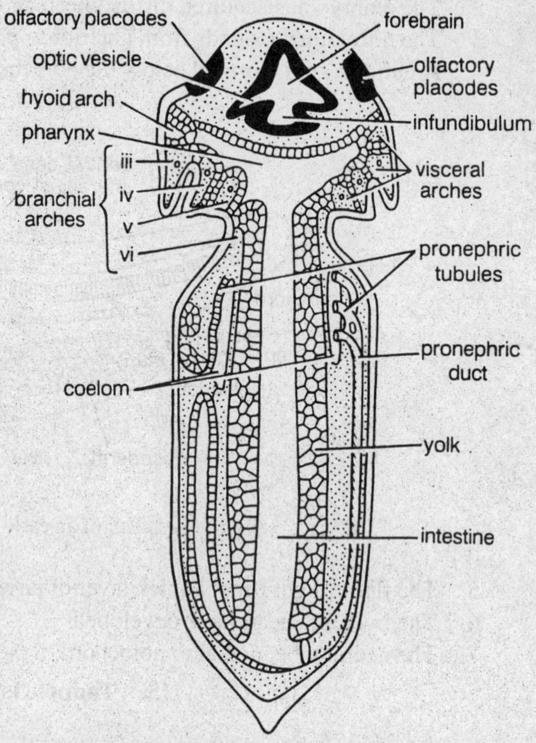

Fig. 10.16. Frontal section of tadpole of frog soon after hatching.

17. T.S. of tadpole through eye region

Comments:

Transverse section of **tadpole** through the **eye region** shows the following details :

1. On the dorsal surface in the centre lies the **fore-brain.**

2. On both the lateral sides of the fore-brain are present the eyes.

3. **Eyes** are conspicuous and large. Each eye is composed of lens, retina and eye muscles.

4. **Pharynx** lies just in the centre of the section.

5. **Hyoid arch** lies just below the pharynx.

6. **Palatoquadrate bars** are present on both the lateral sides of the pharynx.

7. **Meckel's cartilages** are present on both the lateral sides of the pharynx.

8. **Jaw muscles, trabeculae, epidermis** and **connective tissue** are also seen in the section.

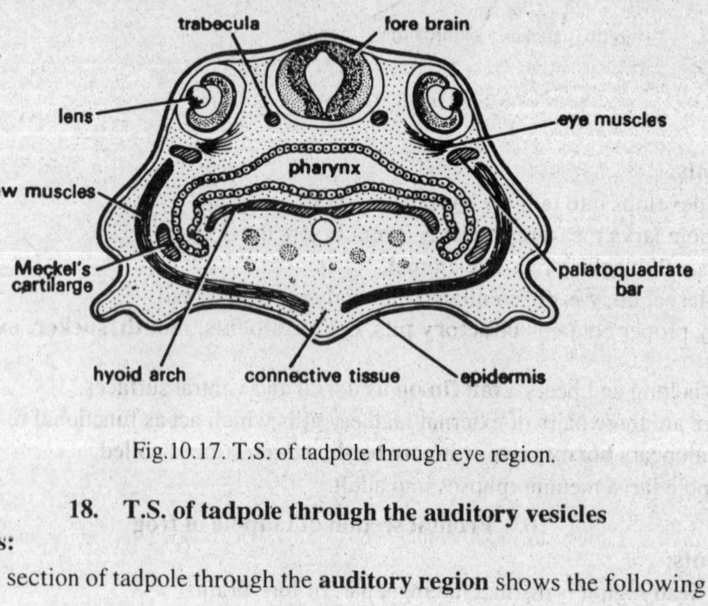

Fig.10.17. T.S. of tadpole through eye region.

18. T.S. of tadpole through the auditory vesicles

Comments:

Transverse section of tadpole through the **auditory region** shows the following structures :

1. On the dorsal surface hind-brain showing **medulla oblongata** enclosing the fourth ventricle is seen.

2. Below the hind-brain lies the **notochord** which is surrounded by the **parachordals** on both the lateral sides.

3. **Auditory vesicles** lie on the lateral sides below the hind-brain.

4. **Branchial chamber** having the **laryngeal chamber** lies below the auditory vesicles and notochord.

5. **Dorsal aorta,** first, second, third and fourth branchial arches are present on the lateral sides of the section.

6. **Pericardial cavity** enclosing the **auricle** and **ventricle** lies on the ventral side below the laryngeal chamber.

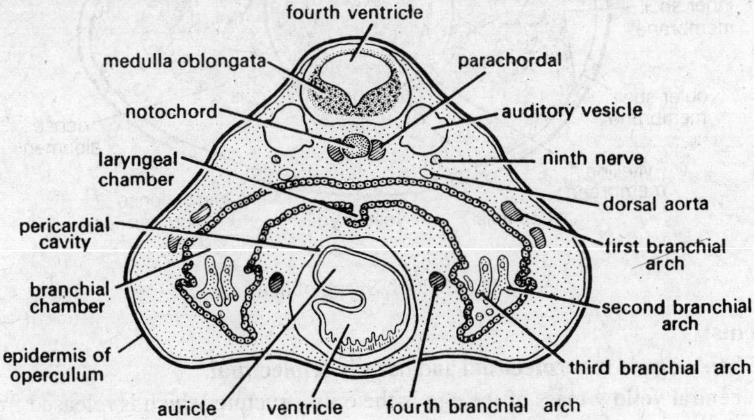

Fig.10.18. T.S. of tadpole through the auditory vesicles.

19. T.S. of tadpole through the head and gills

Comments:

Transverse section of tadpole through the head and gills shows the following details:

1. On the dorsal surface in the centre lies the medulla oblongata.

2. On both the lateral sides of the medulla oblongata are present the membranous labyrinth.

3. **Notochord** is seen just below the medulla oblongata.

4. **Pharynx** lies in the centre of the section.

5. **Larynx** is also seen in continuation with the pharynx.

6. Heart enclosed in the pericardium is seen just below the larynx.

7. On the lateral sides of the pericardium the **branchial arches** and **internal gills** are visible.

8. **Lateral dorsal aortac** are lying above the pharynx.

9. **Lymphatic spaces** are lying on the lateral sides.

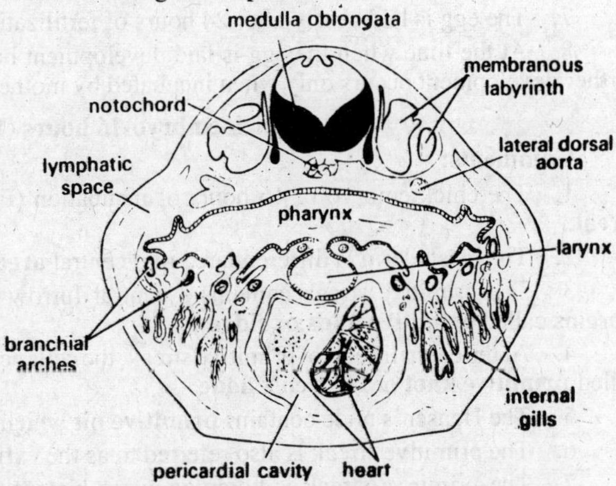

Fig.10.19. T.S. of tadpole through the head and gills.

DEVELOPMENT OF CHICK

20. Hen's egg

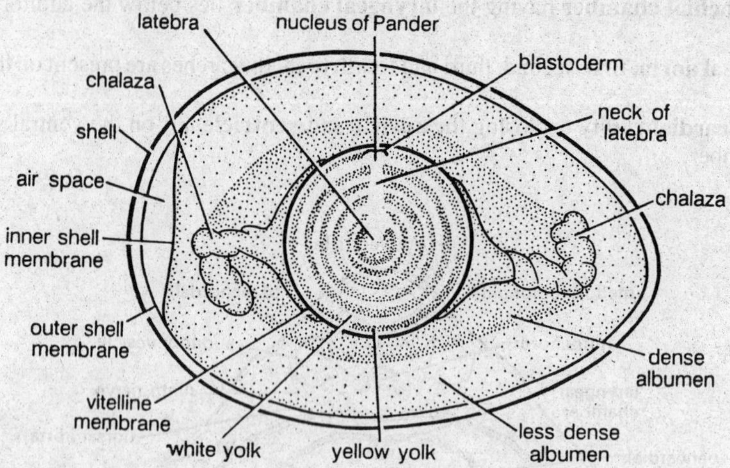

Fig.10.20. L.S. egg of hen.

Comments:

1. The hen's egg is **macrolecithal** and heavily **telolecithal.**
2. The central yellow mass of the egg is the only structure which is released from the ovary enclosed in a vitelline membrane.
3. The yellow mass of the egg has basically white and yellow variety of yolks arranged in alternating concentric rings.
4. The innermost layer of the yellow mass is the yolk of white variety called **latebra** which is produced into neck and **nucleus of Pander.**
5. The germinal vesicle containing female pronucleus is placed anterior to the nucleus of Pander called **blastodisc.**
6. After fertilization in the magnum of oviduct, the yellow mass descends through the oviduct and acquires the layers of dense and thin albumen, chalazae, shell membranes and the egg shell.
7. The egg is laid nearly after 24 hours of fertilization.
8. At the time when the egg is laid development has already taken place up to blastula and further development occurs only when incubated by mother.

21. Chick embryo-16 hours (Whole mount)

Comments:

1. The chick embryo of 16 hours of incubation (Fig. 10.21) shows a prominent **primitive streak.**
2. The blastoderm is differentiated into central **area pellucida** and marginal **area opaca.**
3. The primitive streak contains a central furrow called **primitive groove** and thickened margins called **primitive folds** or **ridges.**
4. At the caphalic end of primitive streak, the cells accumulate in the form of a local thickening called **primitive knot** or **Hensen's node.**
5. The Hensen's node contains **primitive pit** which is continuous with the primitive groove.
6. The primitive streak is also referred to as the **extra embryonic axial structure.**
7. The primitive streak is homologous to blastopore of frog's embryo as it is the site of mesodermal invagination

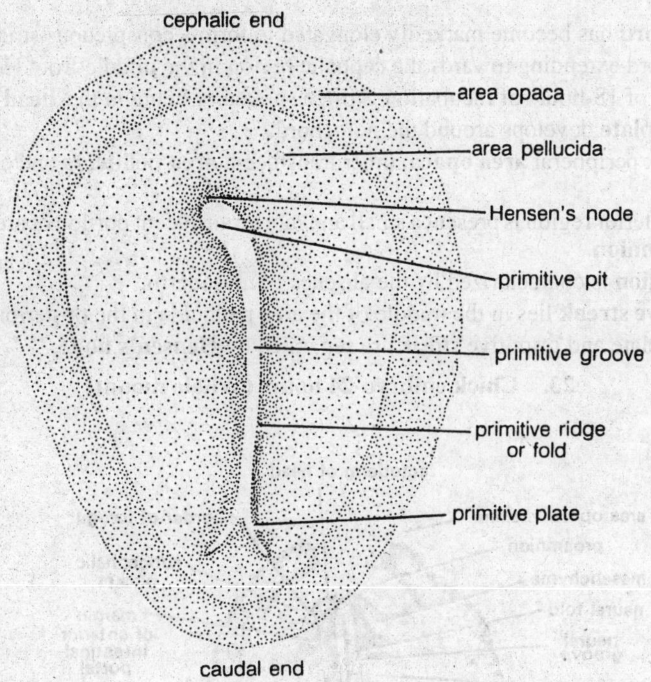

Fig.10.21. Chick embryo-16 hours (Whole mount).

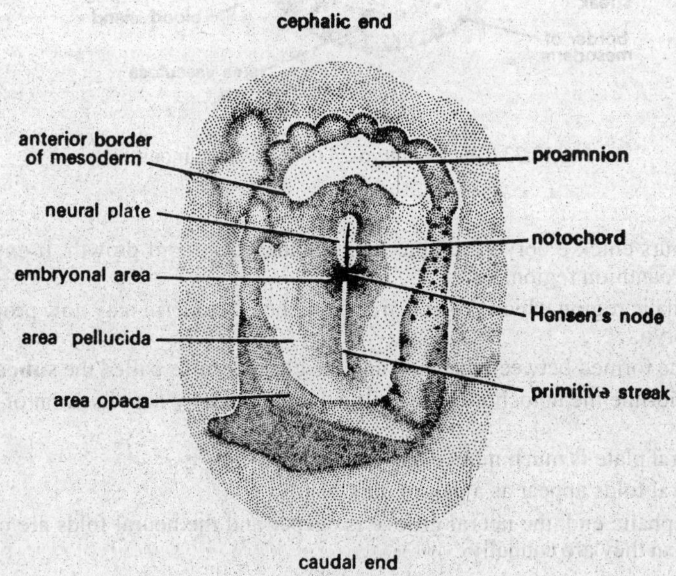

Fig.10.22. Chick embryo-18 hours (Whole mount).

22. Chick embryo-18 hours (Whole mount)

Comments:

1. **Notochord** has become markedly elongated to form a conspicuous structure.
2. Notochord extending towards the cephalic region in the middle from Hensen's node.
3. Embryo of 18-hours of incubation is often spoken of being in the"**head process stage.**"
4. **Neural plate** develops around the notochord.
5. The dark peripheral **area opaca,** inner translucent **area pellucida** and central **embryonal area** are seen.
6. In the anterior region is present a small and more translucent portion of area pellucida which is known as **proamnion.**
7. **Proamnion** is characterized by the absence of mesoderm.
8. **Primitive streak** lies in the middle of the area pellucida in the posterior half.
9. Neural plate and primitive streak are separated by **Hensen's node.**

23. Chick embryo-24 hours (Whole mount)

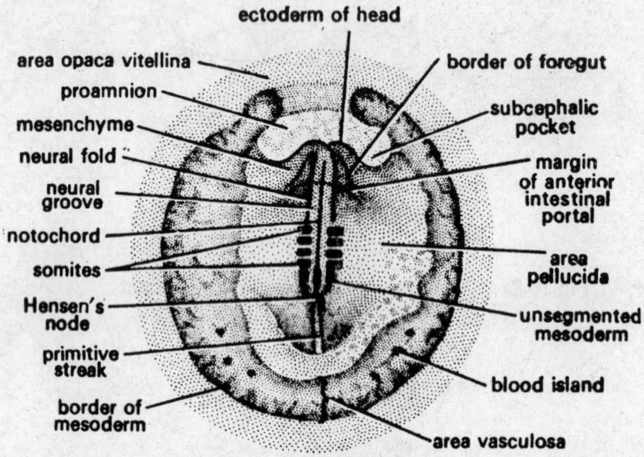

Fig.10.23. Chick embryo-24 hours (Whole mount).

Comments:

1. In 24-hours chick embryo cephalic region undergoes rapid growth. It extends anteriorly overhanging the proamnion region.
2. The cephalic region which projects free from the blastoderm may now properly be termed as the head of embryo.
3. The space formed between the head and the blastoderm is called the **subcephalic pocket.**
4. In the mid-line the notochord is seen. It is larger caudally near its point of origin than it is cephalically.
5. The neural plate is much more clearly marked.
6. The neural folds appear as a pair of dark bands.
7. At its cephalic end, the neural groove is deeper and the neural folds are correspondingly more prominent than they are caudally.
8. Four pairs of **somites** are seen in the mid-line.
9. **Primitive streak** gradually decreases in size.
10. Foregut is also formed. The part of the gut caudal to the foregut is termed the midgut and opening from the midgut into the foregut is called the **anterior intestinal portal.**

11. Besides the above structures, **area opaca vitellina, area pellucida, proamnion, Hensen's node, area vasculosa, blood islands** and **unsegmented mesoderm** are also seen.

24. Chick embryo-28 hours (Whole mount)

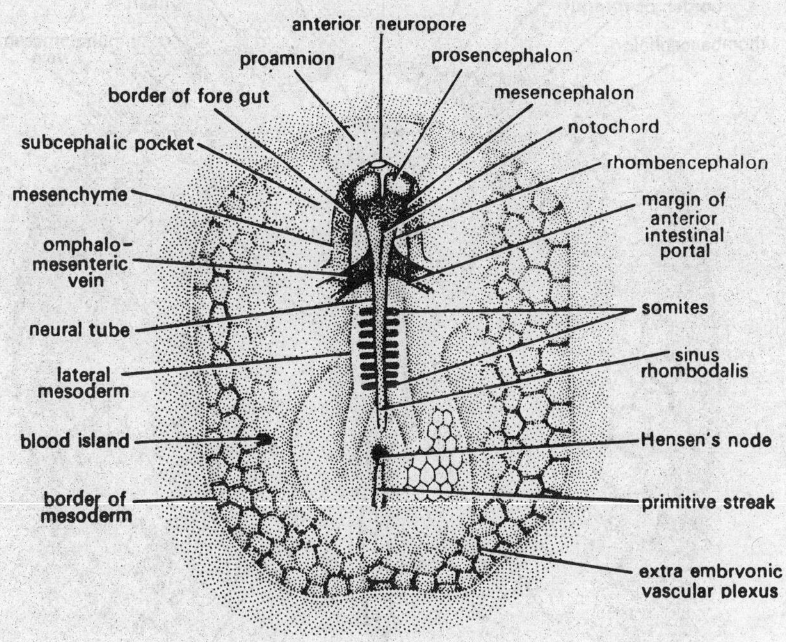

Fig.10.24. Chick embryo-28 hours (Whole mount).

Comments:

1. The embryo of 28-hours of incubation shows marked advances in the development of cephalic region.
2. The head has elongated rapidly and projects free from blastoderm for a considerable distance.
3. The depth of **subcephalic pocket** and the length of foregut increases.
4. The **neural folds** in the cephalic region meet in the mid-dorsal line and their edges become fused forming the **neural tube.**
5. The cephalic part of neural tube is markedly enlarged as compared with the more caudal parts.
6. The three primary brain vesicles namely **prosencephalon** (fore-brain), **mesencephalon** (mid-brain) and the **rhombencephalon** (hind-brain) can be distinguished.
7. The undilated caudal part of the neural tube gives rise to the spinal cord.
8. The **anterior neuropore** is an opening communicating with the outside.
9. Eight pairs of **somites** are also developed in this stage.
10. **Hensen's node** moves towards the caudal end and the **primitive streak** becomes shorter.
11. Besides the above mentioned structures, **proamnion, notochord, margin of anterior intestinal portal, sinus rhomboidalis, omphalomesenteric vein, extra-embryonic vascular plexus** and **blood islands** are also seen.

25. Chick embryo-33 hours (Whole mount)

Comments:

1. The embryo of 33-hours of incubation shows some of the fundamental steps in the formation of central nervous system and circulatory system.

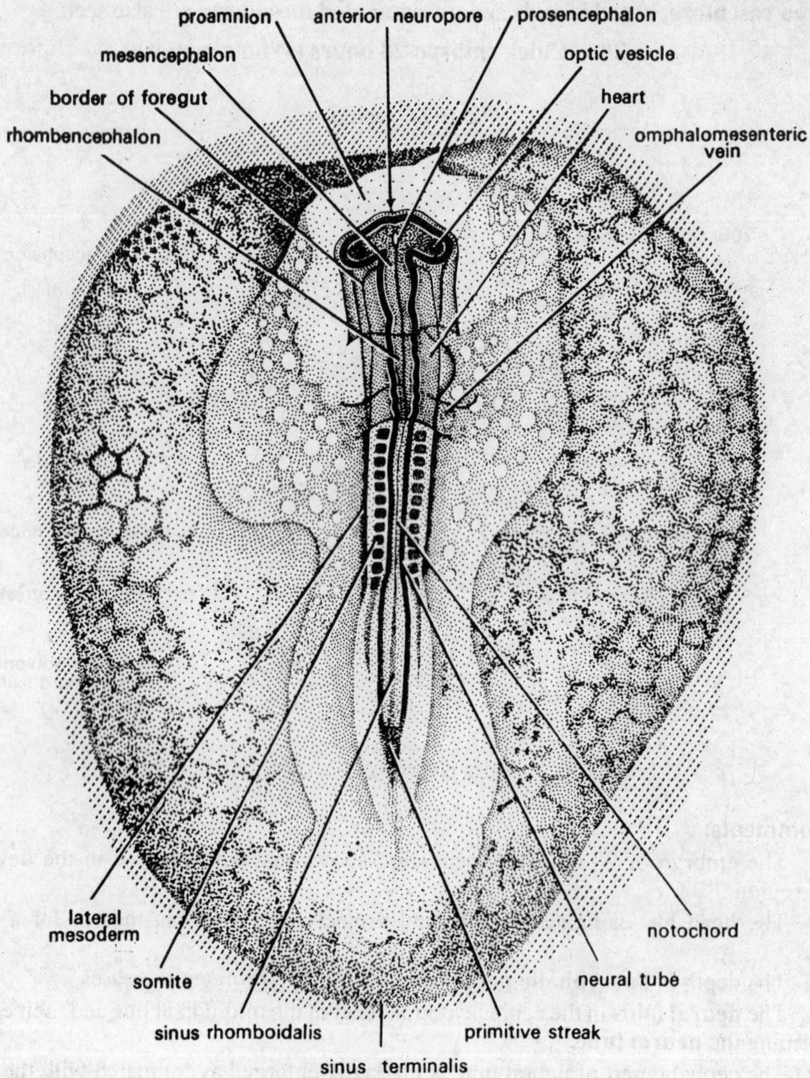

Fig.10.25. Chick embryo-33 hours (Whole mount).

2. Brain is differentiated into **prosencephalon** (fore-brain), **mesencephalon** (mid-brain) and **rhombencephalon** (hind-brain).
3. The **optic vesicles** are established as paired lateral outgrowths of the prosencephalon.
4. The optic vesicles soon extend to occupy the full width of the head.
5. Infundibulum is formed in the floor of the prosencephalon.
6. Mid-region of the **heart** is considerably dilated and bent to the right.
7. **Twelve pairs** of **somites** are formed.
8. Anterior **omphalomesenteric veins** have developed.
9. **Primitive streak** becomes shorter because of the lengthening of the neural tube.
0. **Proamnion, neural tube, notochord, sinus rhomboidalis** and **sinus terminalis** are also present.

26. Chick embryo-36 hours (Whole mount)

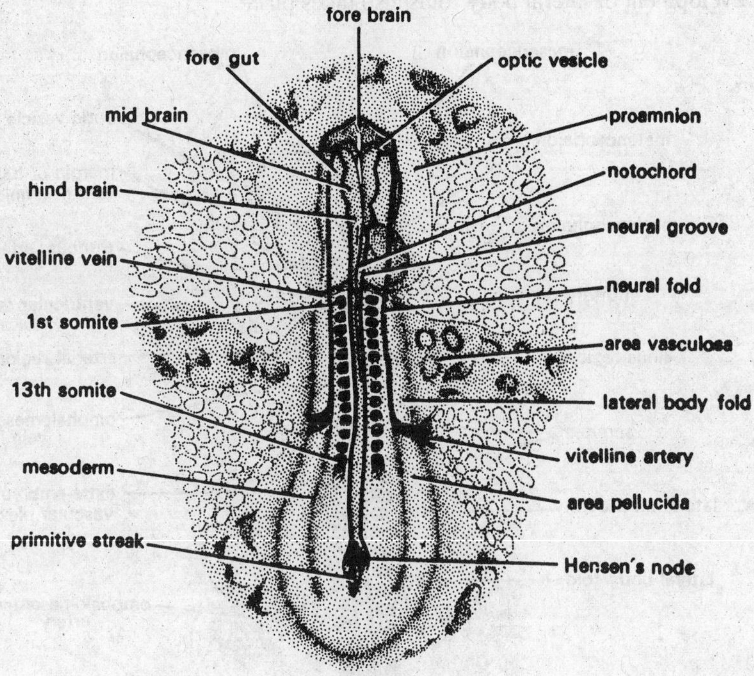

Fig.10.26. Chick embryo-36 hours (Whole mount).

Comments:

The embryo of 36-hours of incubation shows the following important peculiarities :

1. The brain is differentiated into **fore-brain**, **mid-brain** and **hind-brain.**
2. The **optic vesicles** are established as paired lateral outgrowths of the fore-brain.
3. **Thirteen pairs** of **somites** are formed.
4. **Primitive streak** becomes shorter because of the lengthening of the neural tube.
5. Mid-region of the heart is considerably dilated and bent to the right.
6. **Vitelline vein** and **vitelline artery** have developed.
7. **Proamnion, notochord, neural groove, neural fold, area vasculosa, area pellucida** and **Hensen's node** are also present.

27. Chick embryo-43 hours (Whole mount)

Comments:

1. The embryo of 43 hours incubation exhibits the **cranial flexion** and **torsion.**
2. The cephalic region of the embryo is twisted in such a manner that the left side comes to lie next to the yolk and right side away from the yolk.
3. Brain region is differentiated into **prosencephalon, mesencephalon, metencephalon** and **myelencephalon.**
4. **Auditory pits** arc also appcarcd in this stagc.
5. **Heart** shows its division into **ventricular region, arterial region** and **sinus region. Truncus arteriosus** also develops.
6. **Omphalomesenteric veins, omphalomesenteric arteries** and **extra-embryonic vascular plexus** are well developed.
7. 19 pairs of somites are formed.

8. **Primitive streak** diminishes to a small primitive plate.
9. **Neural plate** is very well developed.
10. Development of lateral body folds also takes place.

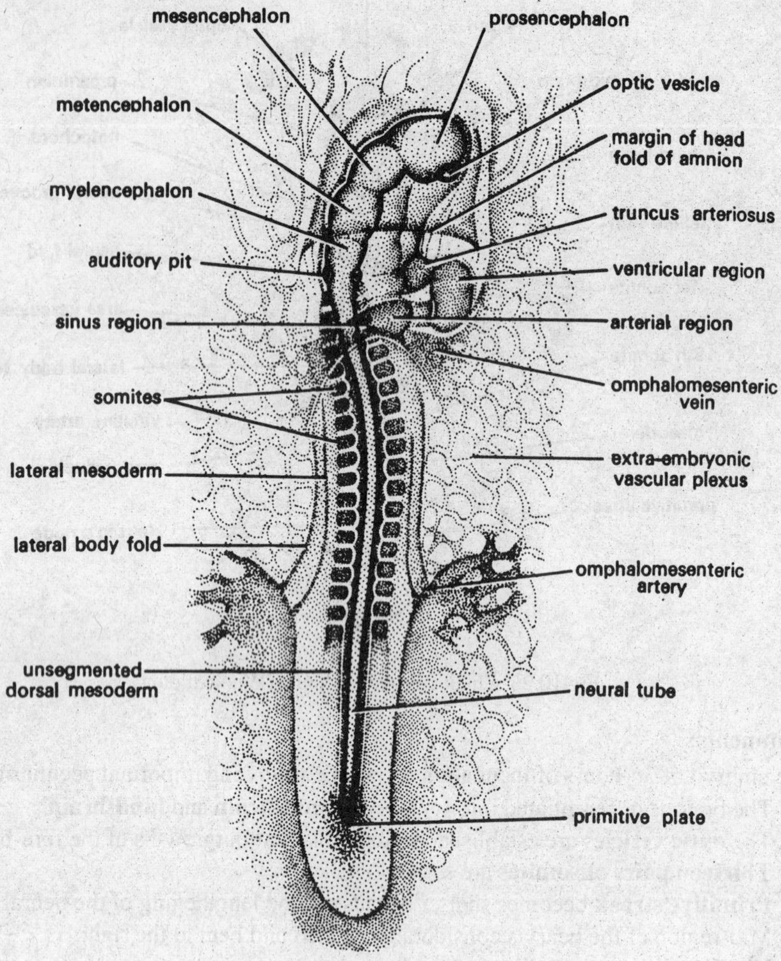

Fig.10.27. Chick embryo-43 hours (Whole mount).

28. Chick embryo-55 hours (Whole mount)

Comments:

1. In embryo of 55-hours of incubation, the entire head becomes free from the yolk by the progress towards the caudal end of the subcephalic fold.

2. Torsion takes place in the whole anterior half of the embryo and is completed in the cephalic region, so that the head now lies left side down on the yolk.

3. The posterior half of the embryo is still in its original position, ventral surface prone on the yolk.

4. At the extreme posterior end, the beginning of the candal fold marks off the tail region of the embryo from the extraembryonic membranes.

5. Due to **cranial flexure** the brain is bent nearly double on itself.

6. **Optic vesicles** and **auditory vesicles** are brought opposite each other at nearly the same antero-posterior level.

7. In the optic vesicle the **optic cup** ahd **lens** are differentiated.

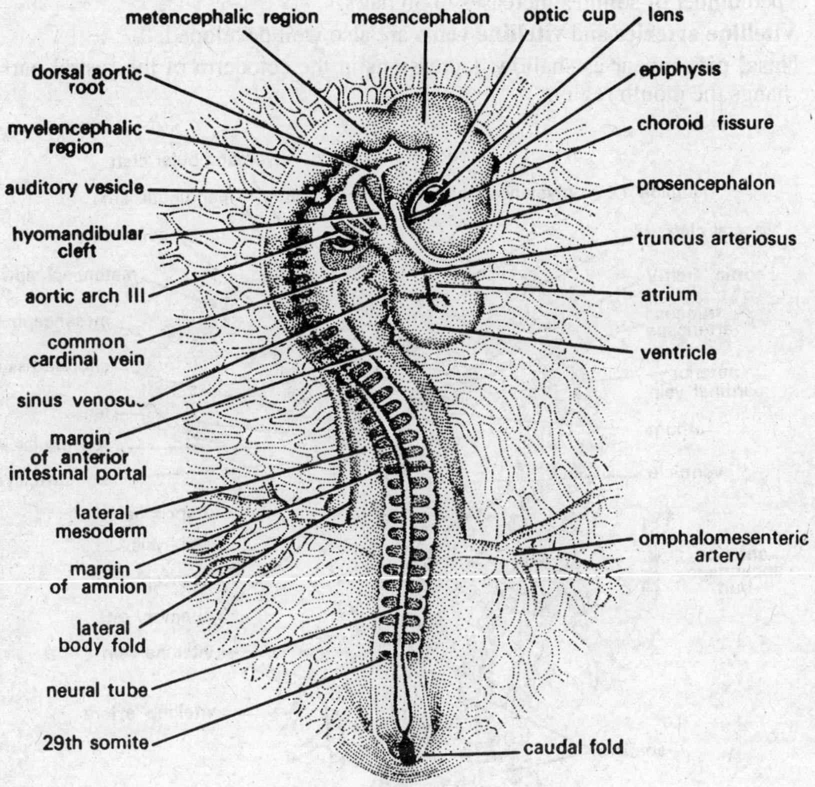

Fig.10.28. Chick embryo-55 hours (Whole mount).

8. In the mid-dorsal wall of the diencephalic region a small evagination has appeared. This evagination is the **epiphysis.**

9. **Branchial arches, branchial clefts, dorsal aorta** and **aortic arches** have appeared.

10. **Heart** is differentiated into **atrium** and **ventricle. Truncus arteriosus** and **sinus venosus** have also developed.

11. The number of somites increases to 29 pairs.

12. Besides the above mentioned structures, **common cardinal vein, anterior intestinal portal, lateral body folds, neural tube** and **omphalomesenteric artery** are also present.

29. Chick embryo-72 hours (Whole mount)

Comments:

1. The chick embryo of 72-hours of incubation has been affected by **torsion** throughout its entire length.

2. The torsion is complete well posterior to the level of heart but the caudal portion of the embryo is not turned on its side.

3. Due to the **cranial** and **cervical flexures,** the long axis of the embryo shows nearly right angled bends in the mid brain and in the neck region.

4. The mid-body region is slightly concave dorsally because of the fact that the embryo is still broadly attached to the yolk in this region.

5. The **visceral arches** are thicker and more conspicuous than in the anterior embryo.

6. Both the anterior and posterior **appendage buds** have appeared in the embryo.

7. **Telencephalon** is also formed.

8. In the eye, **lens, sensory** and **pigment layers** are developed.
9. The number of somites increases to 36 pairs.
10. **Vitelline arteries** and **vitelline veins** are also well developed.
11. **Nasal pits** appear as shallow depressions in the ectoderm of the rostral part of the head which overhangs the mouth region.

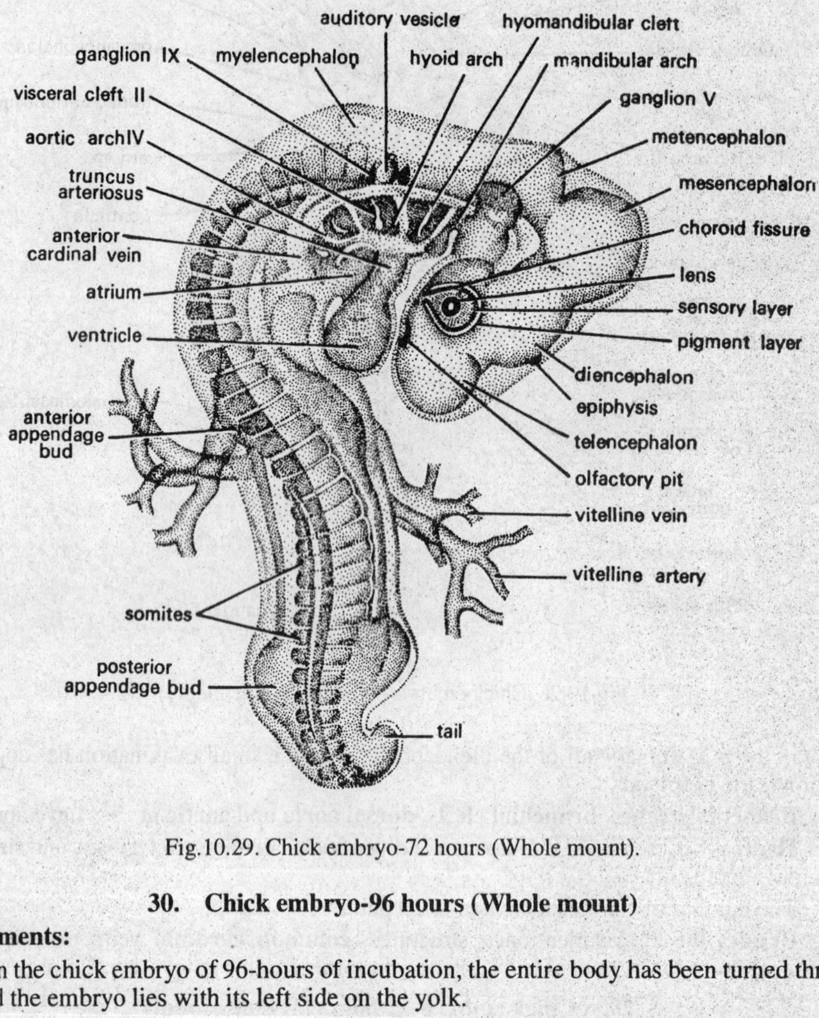

Fig.10.29. Chick embryo-72 hours (Whole mount).

30. Chick embryo-96 hours (Whole mount)

Comments:

1. In the chick embryo of 96-hours of incubation, the entire body has been turned through 90 degrees and the embryo lies with its left side on the yolk.
2. At the end of 96 hours the body folds have undercut the embryo so that it remains attached to the yolk only by a slender stalk.
3. The yolk stalk soon becomes elongated, allowing the embryo to become first straight in the mid-dorsal region and then convex dorsally.
4. The progressive increase in the cranial, cervical, dorsal and caudal flexures results in the bending of the embryo on itself so that its originally straight long axis becomes C-shaped and its head and tail lie close together.
5. **Optic cup** shows the more developed **lens.**
6. **Endolymphatic duct** arises from the auditory vesicle.
7. **Visceral arches** have become very much thickened.
8. **Appendage buds** increase rapidly in size and become elongated.
9. The number of somites increases to 41 pairs.

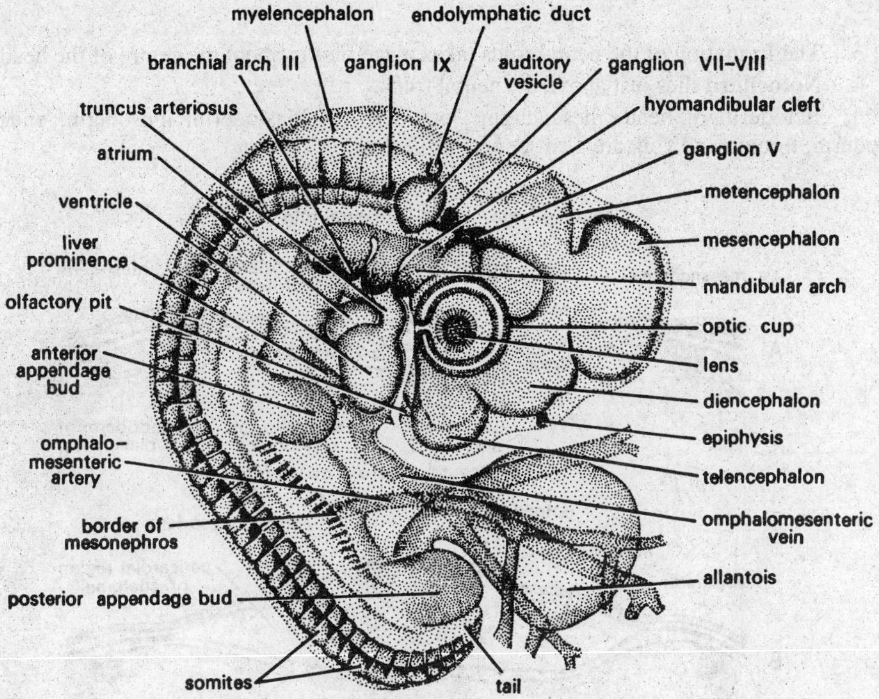

Fig.10.30. Chick embryo-96 hours (Whole mount).

10. **Allantois** has also appeared.
11. **Omphalomesenteric artery** and **omphalomesenteric vein** are also developed.

31. T.S. of 18-hours chick embryo through Hensen's node

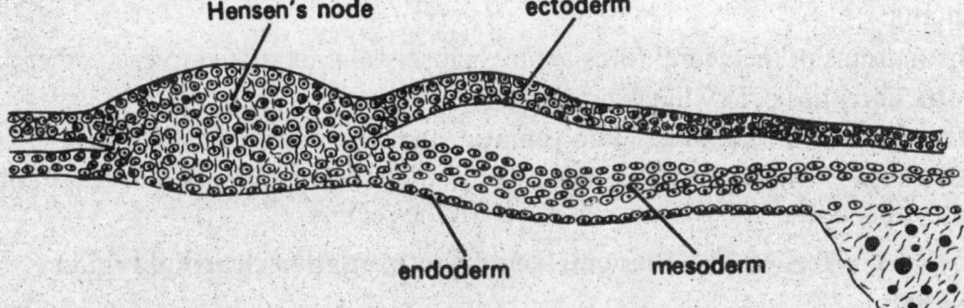

Fig.10.31. T.S. of 18-hours chick embryo through Hensen's node.

Comments:

1. The section shows **ectoderm, mesoderm, endoderm** and **Hensen's node.**
2. **Hensen's node** is a thickening of cells.
3. **Ectoderm** lies above and **endoderm** below.
4. **Mesoderm** lies in between the ectoderm and endoderm.

32. T.S. of 24-hours chick embryo

(A) T.S. through head

Comments:

1. The **neural plate** is folded and forms a nearly complete tube.
2. Dorsally the margins of the **neural folds** of either side have approached each other and lie almost in contact.

3. The formation of the neural folds takes place first in about the centre of the head region.

4. **Notochord** lies just above the neural folds.

5. Ectoderm of head, mesenchyme, ectoderm of blastoderm, mesoderm, endoderm of blastoderm, foregut and yolk are also seen in the section.

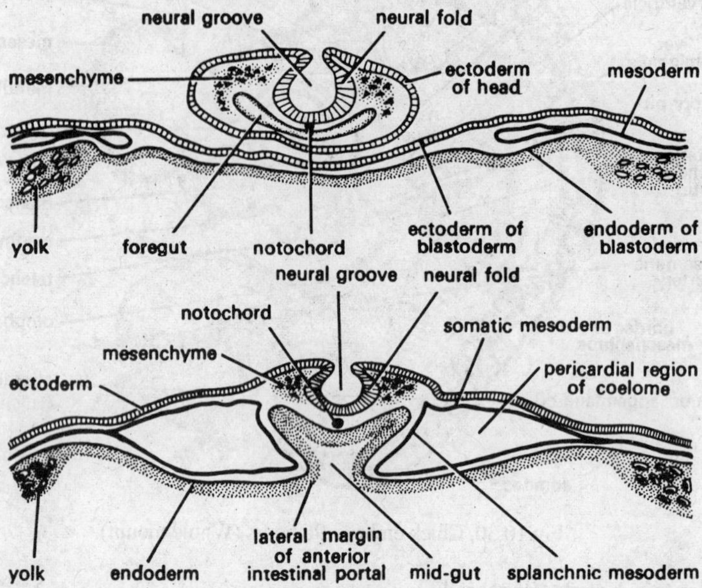

Fig.10.32. T.S. of 24-hours chick embryo. A–T.S. through head; B–T.S. through mid-body.

(B) T.S. through mid-body

Comments:

1. The margins of the neural folds are less closely approximated to each other.

2. **Notochord** lies below the neural folds.

3. **Mesoderm** is differentiated into **somatic** and **splanchinc mesoderm.**

4. Ectoderm, endoderm, mesenchyme, pericardial region of coelom, midgut, lateral margin of anterior intestinal portal and yolk are also seen in the section.

33. T.S. of 28-hours chick embryo through pericardial region

Comments:

1. This section (Fig. 10.33) shows formation of heart from mesoderm.

2. **Endocardial primordia** are approximated to each other.

3. The **epimyocardial areas** of the mesoderm are brought together first ventrally and then dorsally to the endocardium.

4. **Splanchnic mesodermal layers** of the opposite side of the body become opposed to each other, dorsal and ventral to the heart, they form double-layered supporting membranes called respectively, the **dorsal mesocardium** and **ventral mesocardium.**

5. The section also exhibits somatopleure and splanchnopleure, closed neural canal, notochord, foregut and pericardial coelom.

34. T.S. of 33-hours chick embryo through mid-brain region

Comments:

1. **Somatopleure** comprising of ectoderm and somatic mesoderm, lies on the dorsal surface.

2. **Splanchnopleure** composed of endoderm and splanchinic mesoderm, lies on the ventral surface.

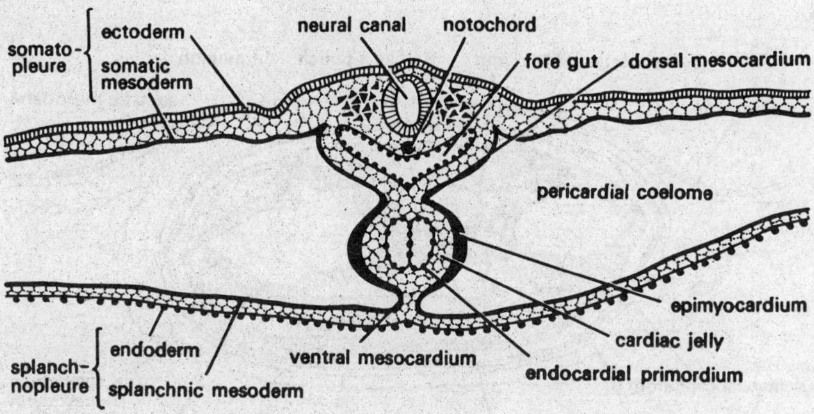

Fig.10.33. T.S. of 28-hours chick embryo through pericardial region.

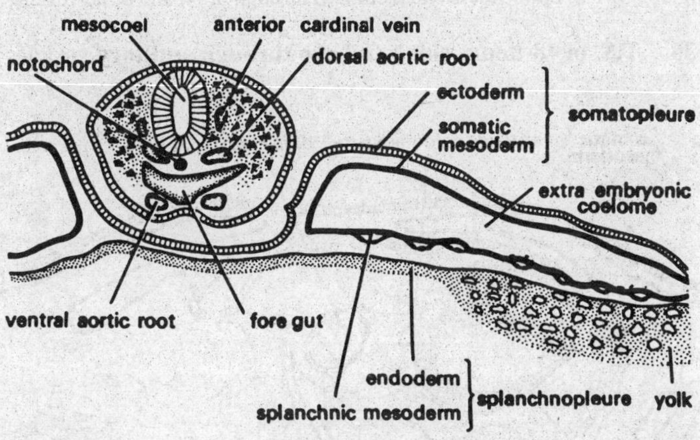

Fig.10.34. T.S. of 33-hours chick embryo through mid-brain region.

3. In the median line tubular **mesocoel, anterior cardinal veins, dorsal aortic roots, notochord, forgut** and **ventral aortic roots** are present.

4. **Extra-embryonic coelom** lies between the somatopleure and splanchnopleure.

5. **Yolk** is present below the splanchnopleure on either side.

35. T.S. of 48-hours chick embryo through brain and optic vesicles

Comments:

1. **Ectoderm** is single-celled thick layer lying on the dorsal side and endoderm lies on the ventral side.

2. Below the ectoderm is present **somatic mesoderm** and **splanchnic mesoderm** lies above the endoderm.

3. Head portion of the embryo is enclosed within amnion.

4. Brain is represented by a large cavity **prosencephalon** (fore-brain) and small cavity **myelencephalon** (hind-brain).

5. In between the two cavities of brain are present **Rathke's pouch, gut, dorsal** and **ventral aortic roots.**

6. **Optic vesicles** and **lens** are present on either side of the prosencephalon.

7. On either side of the myelencephalon lies the **anterior cardinal vein.**

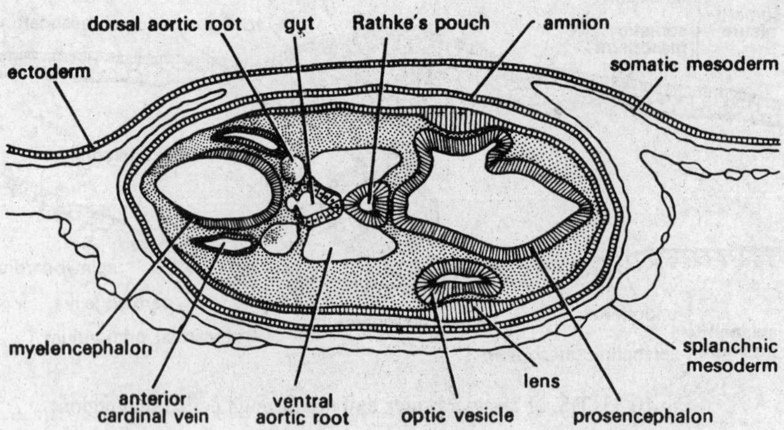

Fig.10.35. T.S. of 48-hours chick embryo through brain and optic vesicles.

36. T.S. of 48-hours chick embryo through auditory vesicles

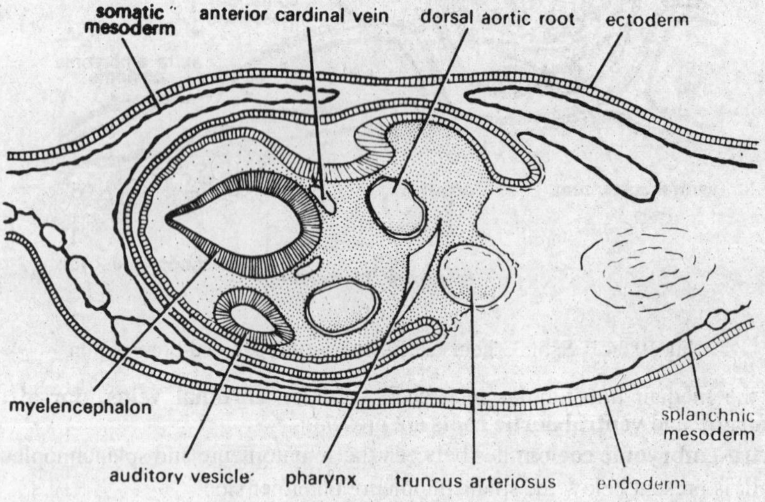

Fig.10.36.. T.S. of 48-hours chick embryo through auditory vesicles vesicles.

Comments:

1. **Ectoderm** lies on the dorsal side and **endoderm** lies on the ventral side. Both the layers are single-celled thick.
2. Below the ectoderm is present **somatic mesoderm** and **splanchnic mesoderm** lies above the endoderm.
3. Head part of the embryo is enclosed within the amnion.
4. Brain is represented by **myelencephalon.**
5. **Auditory vesicles** are in the form of small cavities, one cavity lying on either side of the myelencephalon.
6. Sections of anterior cardinal veins, dorsal aortic roots, pharynx and turncus arteriosus are also seen.

37. T.S. of 48-hours chick embryo through heart

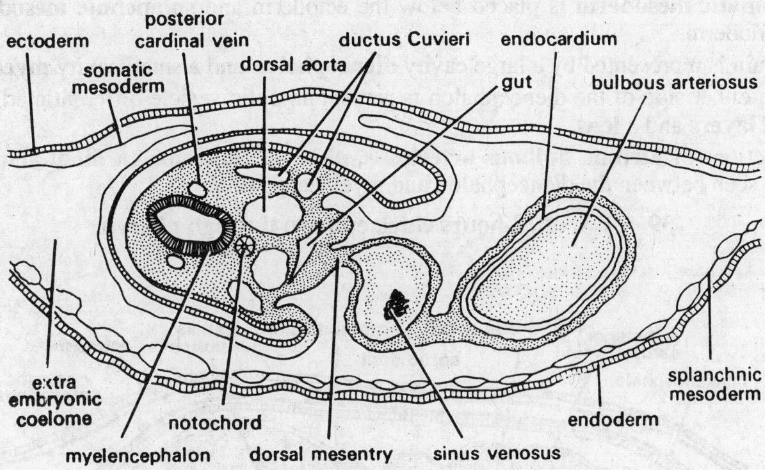

Fig. 10.37. T.S. of 48-hours chick embryo through heart.

Comments:

1. **Ectoderm** lies on the dorsal side and endoderm lies on the ventral side. Both the layers are single-celled thick.
2. **Somatic mesoderm** is lying below the ectoderm, while **splanchnic mesoderm** is lying above the endoderm.
3. **Heart** is represented by **sinus venosus** and **bulbous arteriosus** lying in the extra-embryonic coelom between two layers of mesoderm.
4. Sections of **myelencephalon, posterior cardinal veins, notochord, dorsal aorta,ductus Cuvieri** and **gut** are also seen.
5. **Dorsal mesentery** is connecting the heart with the head portion.

38. T.S. of 72-hours chick embryo through brain and eye

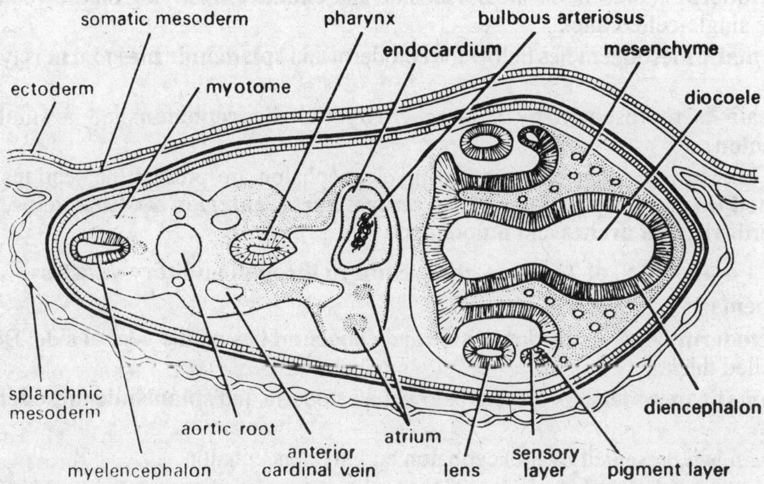

Fig. 10.38. T.S. of 72-hours chick embryo through brain and eye.

Comments:

1. **Ectoderm** lies on the dorsal side and **endoderm** is lying on the ventral side. Both the layers are single-celled thick.

2. **Somatic mesoderm** is placed below the ectoderm and **splanchnic mesoderm** is lying above the endoderm.

3. Brain is represented by a large cavity **diencephalon** and a small cavity **myelencephalon.**

4. On either side of the diencephalon is present an optic vesicle differentiated into **sensory** and **pigment layers** and a **lens.**

5. Sections of **atrium, bulbous arteriosus, pharyax, anterior cardinal veins** and **aortic root** are also seen between the diencephalon and myelencephalon.

39. T.S. of 72-hours chick embryo through pharynx

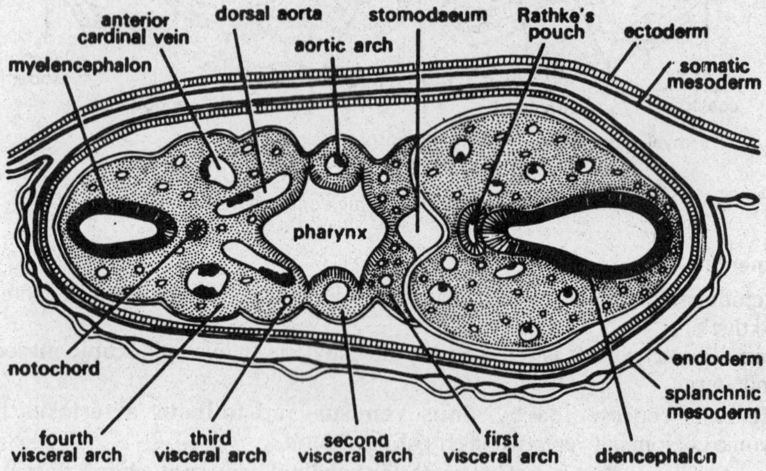

Fig. 10.39. T.S. of 72-hours chick embryo through pharynx.

Comments:

1. **Ectoderm** is present on the dorsal side and **endoderm** is lying on the ventral side. Both the layers are single-celled thick.

2. **Somatic mesoderm** lies below the ectoderm and **splanchnic mesoderm** is lying above the endoderm.

3. Brain is represented by a large cavity the **diencephalon** and a small cavity the **myelencephalon.**

4. In between the diencephalon and myelencephalon are present the sections of **Rathke's pouch, stomodaeum, pharynx, dorsal** and **ventral aorta, anterior cardinal veins**, first, second, third and fourth **visceral arches** and **notochord.**

40. T.S. of 72-hours chick embryo through auditory vesicles

Comments:

1. **Ectoderm** lying on the dorsal side and endoderm lies on the ventral side. Both the layers are single-celled thick.

2. **Somatic mesoderm** is present below the ectoderm and **splanchnic mesoderm** lies above the endoderm.

3. Brain is represented by diencephalon and myelencephalon.

4. The diencephalon and myelencephalon are almost continuous.

5. Auditory vesicles lie on either side of the myelencephalon.
6. **Amnion** surrounds the auditory vesicles and regions of brain.

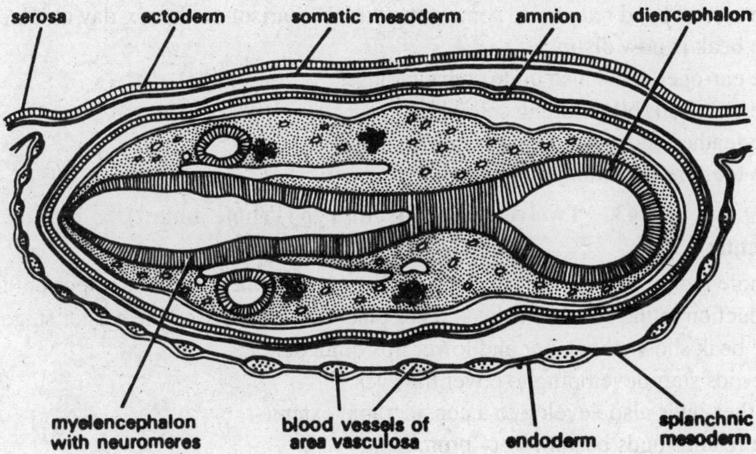

Fig.10.40. T.S. of 72-hours of chick embryo through auditory vesicles.

41. Six-day chick embryo (Whole mount)

Comments:

The whole mount of **six-day chick embryo** shows the following noteworthy features :

1. The head is very large in comparison to the rest of the body.
2. Eyes are readily distinguishable.
3. The ear also begins to appear.
4. The beak begins to appear near the lower edge of the head.
5. Neck and trunk are also distinguishable.
6. Feather buds start appearing.
7. Developing toes are also marked out on the feet.

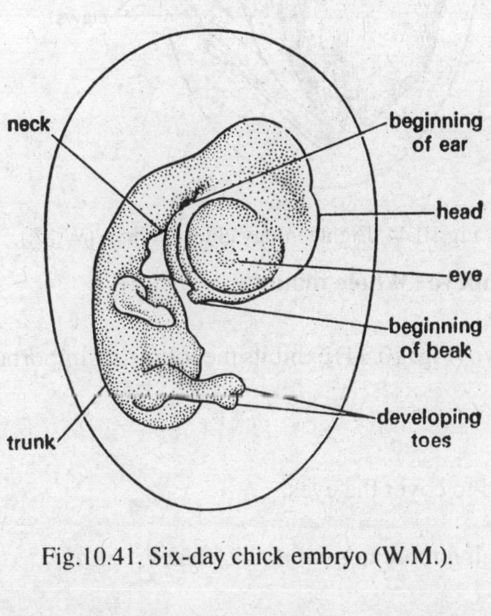

Fig.10.41. Six-day chick embryo (W.M.).

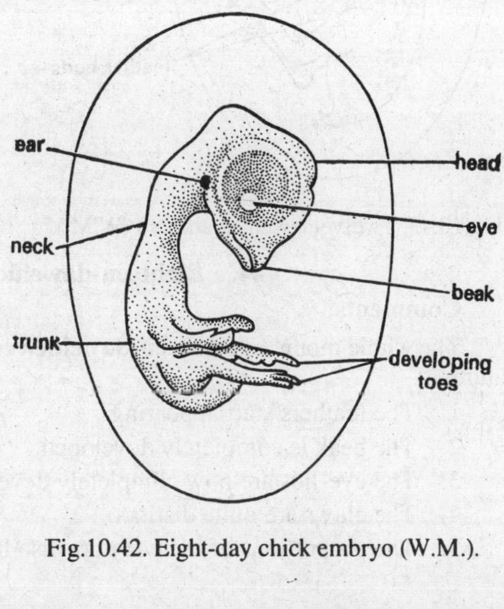

Fig.10.42. Eight-day chick embryo (W.M.).

42. Eight-day chick embryo (Whole mount)

Comments:

The whole mount of **eight-day chick embryo** (Fig. 10.42) exhibits the following important features :

1. The size of head reduces to some extent in comparison to the six-day chick embryo.
2. The beak is now distinct.
3. The ear-opening is also quite conspicuous.
4. The eye is slightly more developed.
5. The feather buds become quite conspicuous.
6. The toes are becoming more clear.

43. Twelve-day chick embryo (Whole mount)

Comments:

The whole mount of **twelve-day chick embryo** shows the following important features :

1. Reduction in the size of the head takes place as compared to the earlier stage.
2. The beak shows the upper and lower lips quite distinct.
3. Eye-lids start developing to cover the eyes.
4. Feather-buds also develop to a considerable extent.
5. The feather-buds become very prominent.
6. The legs are more or less formed.
7. Toes have claws.
8. The elongation in the size of feather buds gives the embryo a more bird-like appearance.

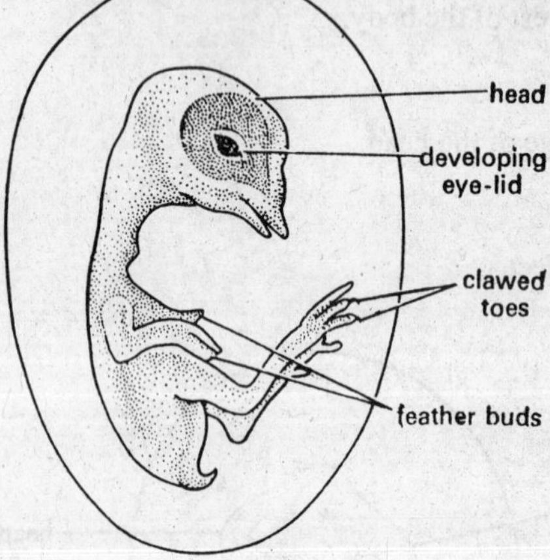

Fig.10.43. Twelve-day chick embryo (W.M.).

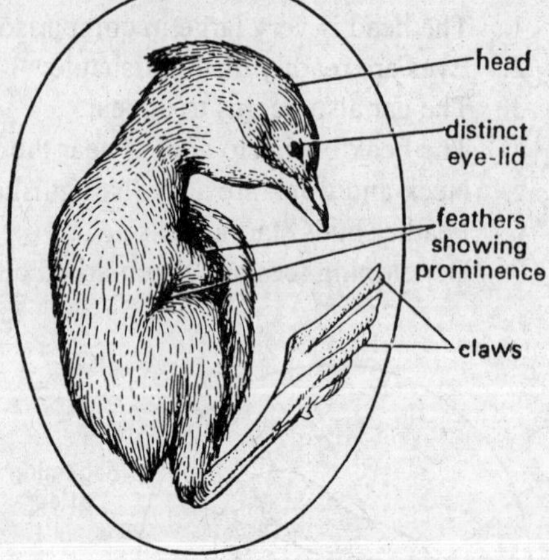

Fig.10.44: Eighteen-day chick embryo (W.M.).

44. Eighteen-day chick embryo (Whole mount)

Comments:

The whole moun. of **eighteen-day chick embryo** (Fig. 10.44) exhibits the following important features :

1. The feathers start appearing.
2. The beak is completely developed.
3. The eye-lids are now completely developed to cover the eyes.
4. The claws are quite distinct.
5. The proportions of the body are also becoming more normal in this stage.

45. Twenty one-day chick embryo (Whole mount)

Comments:

The whole mount of **twenty one day chick embryo** shows the following noteworthy features :

1. The body proportions are now completely distinct.
2. The feathers have become almost normal.
3. The embryo has grown up so much that it almost completely fills the shell.
4. It is now almost ready to hatch.
5. Being ready for hatching the yolk sac has been drawn out half-way into the abdominal cavity.

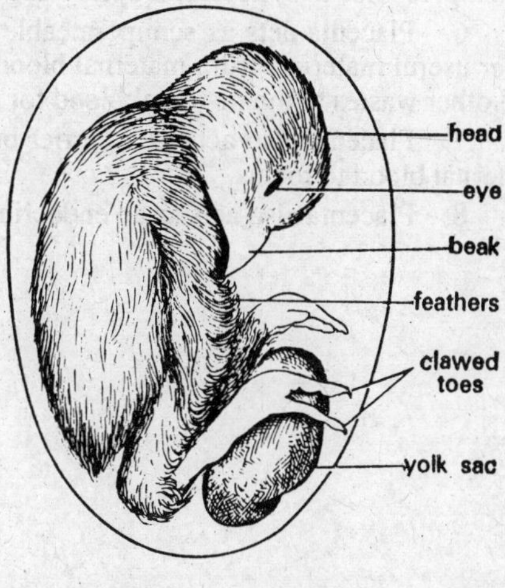

10.45. Twenty one-day chick embryo (W.M.).

46. T.S. Chorio-alloantoic Placenta

Comments:

1. Placenta is a characteristic structure which develops during the embryonic development of viviparous mammals.
2. Allantois of the foetus develops to join with chorion to form chorio-allantoic membrane.
3. The chorio-allantoic membrane develops chorionic villi which fit into similar structures formed in uterine mucosa. This connected structure between foetal and uterine tissues is called **placenta.**
4. The early villi are compact, bush-like tufts. Their main stem arises from the chorionic membrane and all the ends of anchoring villi attach and mingle with the eroded surface of endometrium (decidua basalis).

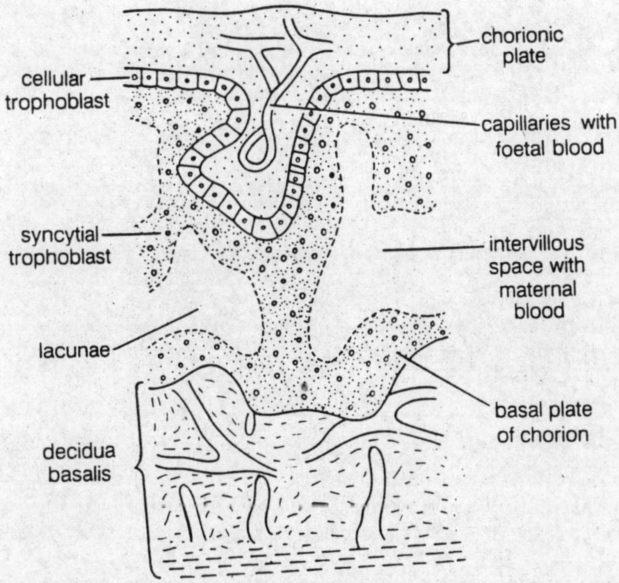

Fig.10.46. T.S.chorio-allantoic placenta of a mammal.

5. The villus in its centre bears a connective tissue core in which blood vessels are embedded. This core is covered by a single layer of **cellular trophoblast,** also called **layer of Langerhans,** which gives rise to **syncytial trophoblast** that marks the external boundary of the villus.

6. Placenta acts as semipermeable membrane allowing diffusion of oxygen, nutrients and other useful materials from maternal blood to the foetus and foetal metabolites like carbon dioxide and other wastes to the maternal blood for removal.

7. Placenta also acts as a barrier but certain pathogens and drugs can pass through it from maternal blood to foetus.

8. Placenta also acts as an endocrine gland.

CHAPTER 11

ECONOMIC (APPLIED) ZOOLOGY: SPECIMENS AND SLIDES

Economic zoology also referred to as applied zoology deals with the useful and harmful affairs of various animals occurring in nature. Although all organisms have their own importance in nature but some animals are very useful to mankind, while some others cause great loss to our economy. This chapter, however, deals with some of those animals which are economically important to human beings.

PROTOZOA

1. *Entamoeba histolytica*

Classification:

Phylum	...	**Protozoa**	→	Microscopic and acellular.
Subphylum	...	**Plasmodroma**	→	Cilia absent; locomotion either by pseudopodia or flagella.
Class	...	**Rhizopoda**	→	Pseudopodia chief organs of locomotion and food catching.
Order	...	**Lobosa**	→	Pseudopodia short, blunt lobose type.
Suborder	...	**Nuda**	→	Body naked or enclosed in a thin membrane.
Genus	...	*Entamoeba*		
Species	...	*histolytica*		

Comments:

1. *Entamoeba histolytica* is widespread in tropical and subtropical countries.
2. It occurs in the colon of man and feeds on the mucous memberane destroying the tissues by an enzyme which it secretes.
3. The size varies from 0.06 mm to 0.05 mm in diameter.
4. It is **dimorphic** occurring in two forms: (i) **trophic** or **trophozoite**, (ii) **precystic** or **minuta**. Both these forms occur in man.Trophozoite is pathogenic, while minuta is non-pathogenic.
5. Cytoplasm is differentiated into ectoplasm and endoplasm.
6. Endoplasm contains single spherical **nucleus** and **food vacuoles**.
7. Contractile vacuole is absent.
8. One or two pseudopodia are present.
9. Reproduction by binary fission and spore formation.
10. *E. histolytica* causes **amoebic dysentery**. When it infects liver, it causes **amoebic hepatitis.**

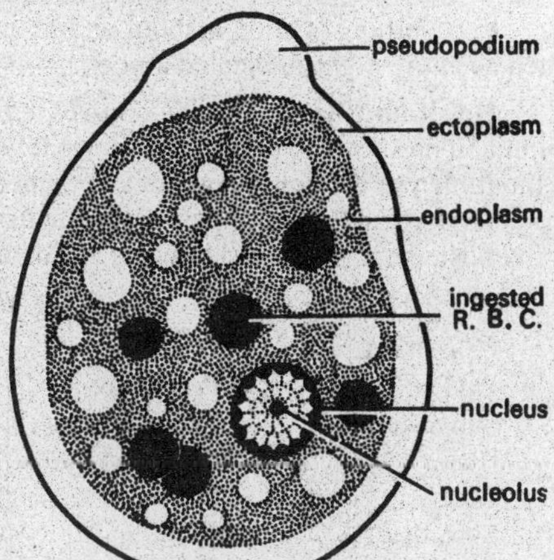

Fig. 11.1. *Entamoeba histolytica.*

11. Mode of infection is oral. Man is infected by taking contaminated food or water with cysts of *E. histolytica.*

Habit and habitat: *Entamoeba hystolytica* is found as a parasite in the digestive tract of man, frog and cockroaches in the tropical and subtropical regions.

2. *Entamoeba coli*

Classification:

Phylum	...	**Protozoa**	Characters same as those of *Entamoeba*.
Subphylum	...	**Plasmodroma**	
Class	...	**Rhizopoda**	
Order	...	**Lobosa**	
Suborder	...	**Nuda**	
Genus	...	*Entamoeba*	
Species	...	*coli*	

Comments:

1. The trophozoite of *Entamoeba coli* measures 15 to 40 microns (average individual 20 to 35 microns) in diameter.

2. The cytoplasm is not well differenetiated into ectoplasm and endoplasm.

3. The endoplasm is granular and contains bacteria, faecal debris of various sizes in food vacuoles.

4. Nucleus is 5 to 8 microns in diameter containing a comparatively larger nucleus which is not placed in the centre.

5. The cyst is spherical or often ovoid, highly refractile; 10 to 30 microns in diameter.

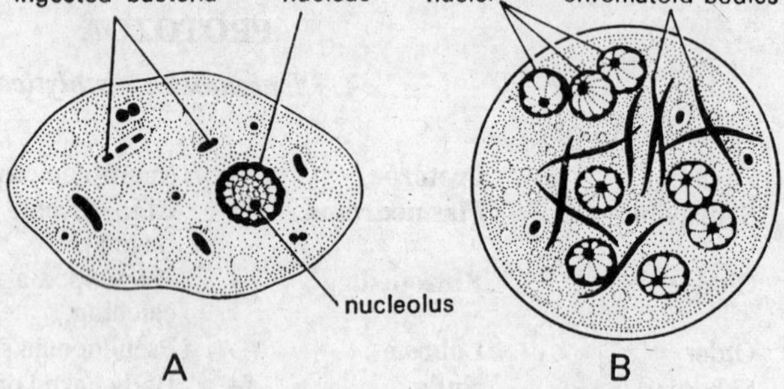

Fig. 11.2. *Entamoeba coli.* A–Mature trophozoite; B–Octonucleate cyst.

6. Immature cyst contains 1, 2 or 4 nuclei, one or more large glycogen bodies and a small number of filamentous chromatoid bodies with sharply pointed ends.

7. Mature cyst contains 8 nuclei and a few or no chromatoid bodies.

8. Nothing is known about its life cycle in human intestine. According to **Hegner,** the cysts hatch as entire eight-nucleated amoeba.

Habit and habitat : *Entamoeba coli* is the commonest species of *Entamoeba* found in the colon and has been stated to occur probably in 50 per cent of human population. This amoeba lives in the lumen of the colon and does not enter the tissues of the wall. It is a harmless species feeding on bacteria, particles of indigested food and other debris but never on blood cells or other living tissues of the host, therefore, considered as **endocommensal**.

3. *Entamoeba gingivalis*

Classification:

Phylum	...	**Protozoa**	Characters same as those of *Entamoeba*.
Subphylum	...	**Plasmodroma**	
Class	...	**Rhizopoda**	
Order	...	**Lobosa**	
Suborder	...	**Nuda**	
Genus	...	*Entamoeba*	
Species	...	*gingivalis*	

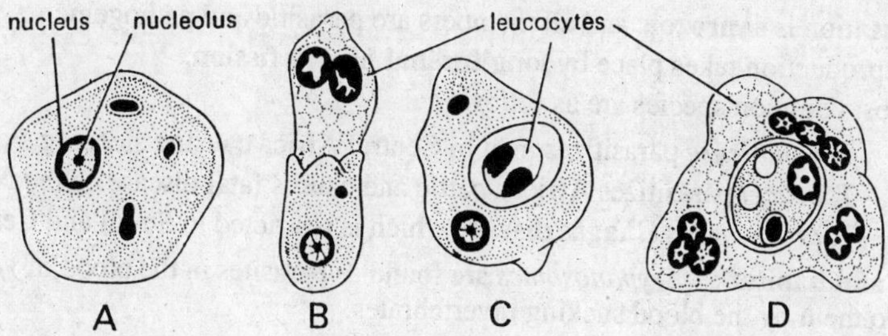

Fig. 11.3. *Entamoeba gingivalis.* A–Matur trophozoite; B–*Entamoeba* engulfing a leucocyte; C–Leucocyte engulfed in a food vacuole; D–Three leucocytes are engulfing a trophozoite.

Comments:

1. *Entamoeba gingivalis* is commonly known as **mouth amoeba**.
2. The trophozoite is as active as that of *Entamoeba histolytica* and measures 8 to 30 microns (average 10 to 20 microns) in diameter.
3. The cytoplasm is well differentiated, endoplasm hyaline but vacuolated and contains a large number of leucocytes and bacteria in food vacuoles.
4. Nucleus is 2 to 4 microns in diameter and contains a small central nucleolus.
5. Monopodial movements in some individuals have also been observed.
6. This species does not form cyst since it is directly transmitted from one human mouth to another by contact.
7. Although *Entamoeba gingivalis* seems to occur more commonly in diseased mouths, there is no convincing evidence that the organism causes oral disease.

Habit and habitat : *Entamoeba gingivalis* inhabits the mouth of human beings. It lives in carious teeth, in tar tar and debris accumulated around the roots of teeth and in abscesses of gums, tonsils, etc. According to **Kofoid** 75 per cent or more people over 40 years of age harbour this amoeba.

4. *Trypanosoma*

Classification:

Phylum	...	**Protozoa**	→	Microscopic and acellular.
Subphylum	...	**Plasmodroma**	→	Cilia absent; locomotion either by pseudopodia or flagella.
Class	...	**Mastigophora**	→	Locomotory organs flagella.
Order	...	**Protomonadina**	→	Amoeboid or euglenoid flagellates; flagella usually one or two; mouth and gullet absent; mostly parasitic.
Genus	...	*Trypanosoma*		

Comments:

1. *Trypansoma* (Fig. 11.4) is polymorphic.
2. It has a colourless, slender, elongated, spindle-shaped body tapering at both ends, ranging in length from 15 to 16 microns.
3. Body is covered by the periplast or **pellicle** somewhat compressed laterally and twisted spirally.
4. The anterior end is more pointed than the posterior end and it bears the **flagellum**.
5. The body is convex on one edge and concave on the other, the convex edge of the body is thrown into irregular folds and is called **undulating membrane**.
6. Along the edge of the undulating membrane runs a stout thread or **axoneme** which is attached with the **basal granule** (blepharoplast) at the posterior end. Just posterior to the basal granule is a small rod-shaped body, the **parabasal body** (kinetoplast).
7. **Nucleus** is rounded and central in position.

8. Contractile vacuoles are absent.
9. Nutrition is **saprozoic** and all members are parasitic and pathogenic.
10. Reproduction takes place by **longitudinal binary fission**.
11. Most common species are as follows :
    (i) *T. gambiense* parasitizes man in Central Africa causing sleeping sickness disease.
    (ii) *T. brucei* parasitizes African cattle and causes fatal disease called **Nagana**.
    (iii) *T. cruzi* causes **Chagas disease** which is restricted to South and Central America.

Habit and habitat : *Trypanosomes* are found as parasites in the blood of vertebrates and are transmitted to them by the blood sucking invertebrates.

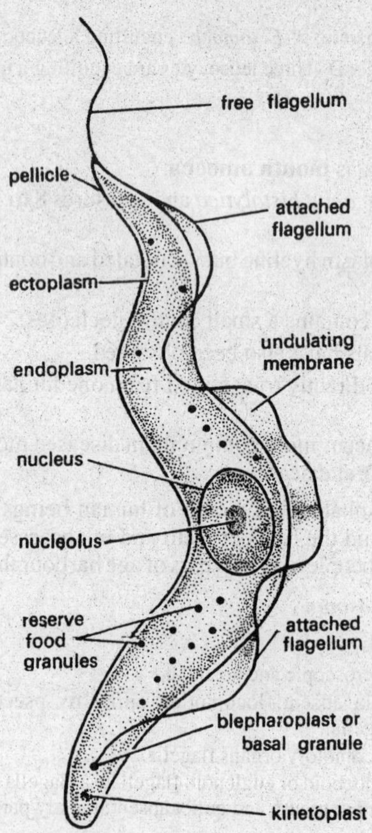

Fig. 11.4. *Trypanosoma.*

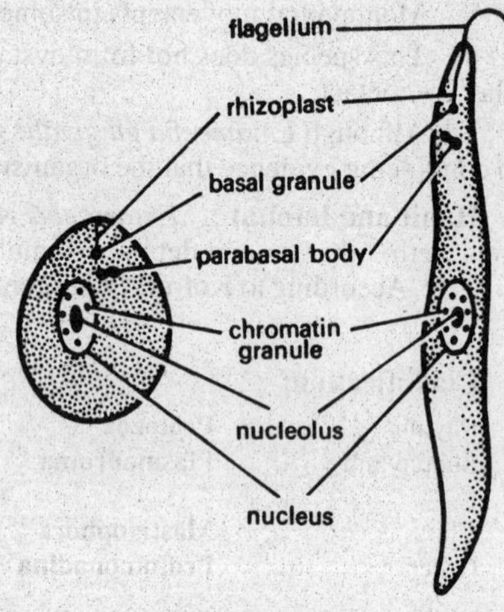

Fig. 11.5. *Leishmania.*

5. *Leishmania*

Classification:

Phylum	...	**Protozoa**	Characters same as those of *Trypanosoma*.
Subphylum	...	**Plasmodroma**	
Class	...	**Mastigophora**	
Order	...	**Protomonadina**	
Genus	...	*Leishmania*	

Comments:

The genus *Leishmania* (Fig. 11.5) includes three species which are common parasites of ma*viz.*, (i) *Leishmania donovoni* causing **kala-azar**, (ii) *Leishmania tropica* causing **oriental sor** and (iii) *Leishmania brassiliensis* causing **American leishmaniasis.**

Leishmania donovoni has two phases in the life cycle, one flagellar *Leishmania form* in the reticulo-endothelial cells of man and other *Leishmania form* in the gut of blood sucking fly *Phlebotomus*. *Leishmania form* is intracellular and is found in the cells of liver, spleen, bone marrow, intestine and lymph glands in the reticulo-endothelial system. This form is oval or spherical 1–3 μ in diameter with a limiting membrane, the periplast or pellicle. The cytoplasm contains an oval **nucleus**, rod-shaped or dot-like kinetoplast and a **parabasal body**. Reproduction by **binary fission**.

6. *Trichomonas*

Classification:

Phylum	...	**Protozoa**	→	Microscopic and acellular.
Subphylum	...	**Plasmodroma**	→	Cilia absent; locomotion either by pseudopodia or flagella.
Class	...	**Mastigophora**	→	Locomotory organs flagella.
Order	...	**Polymastigina**	→	Minute, oval or elongated covered with a delicate pellicle; flagella 3–8 but generally 4; mouth elongated slit-like; intestinal parasites.
Genus	...	*Trichomonas*		

Comments:

1. The body is pear-shaped tapering posteriorly, provided with four **flagella** of which one is directed backward united to the body by an **undulating membrane.**

2. **Cytostome** is present antero-ventrally and is used for the ingestion of food.

3. An **axostyle** which is stout rod-like extends from the posterior end medially through the body.

4. **Blepharoplast** is minute structure situated at the anterior end, while **parabasal body** is club-shaped extends posteriorly from the basal granule.

5. Single large **nucleus** is situated anteriorly provided with several chromatin granules.

6. Feeding is saprozoic but most species ingest bacteria, yeast and other solid food.

7. Reproduction is **asexual** by longitudinal binary fission.

8. *Trichomonas* are parasites in many vertebrates and in few invertebrates. Three species are found in man which are as follows :

(i) *T. hominis* in the colon; (ii) *T. buccalis* in the mouth; and (iii) *T. vaginalis* in the human vagina and urethra.

9. The disease caused by this parasite is called **trichomoniasis.**

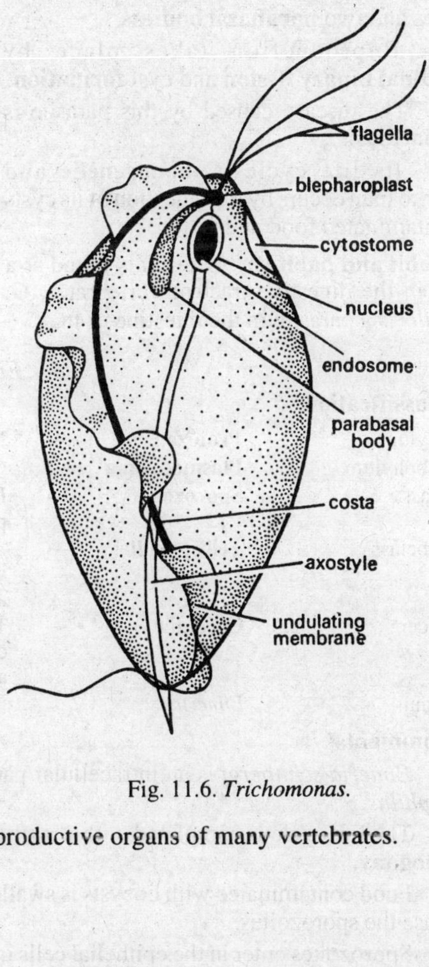

Fig. 11.6. *Trichomonas.*

Habit and habitat : *Trichomonas* is found as a parasite in the alimentary canal and reproductive organs of many vertebrates.

7. *Giardia*

Classification:

Phylum	...	**Protozoa**	Characters same as those of *Trichomonas*.
Subphylum	...	**Plasmodroma**	
Class	...	**Mastigophora**	
Order	...	**Polymastigina**	
Genus	...	*Giardia*	

Comments:

1. The body is bilaterally symmetrical and pear-shaped in appearance with the anterior end rounded and posterior end tapering.
2. The dorsal surface is convex and the ventral surface is flattened provided with a concave sucker.
3. Axostyles form the median langitudinal axis of the body.
4. Four pairs of **flagella** arise from four pairs of **basal granules**.
5. Protoplasm contains two **nuclei**, two **axostyles** and two **parabasal bodies**.
6. Reproduction takes place by **longitudinal binary fission** and **cyst formation**.
7. The disease caused by this parasite is called **giardiasis**.
8. Its life cycle is **monogenetic** and infection to man occurs by the ingestion of its cysts with contaminated food and water.

Habit and habitat: *Giardia* is found as a parasite in the digestive tract of vertebrates. *G. intestinalis* is a parasite in the intestine of man.

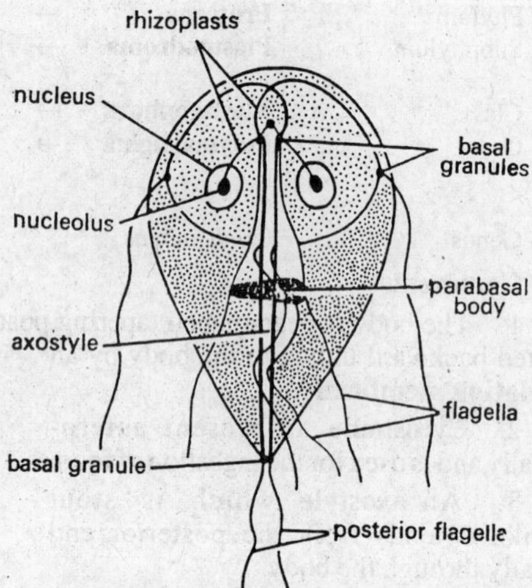

Fig. 11.7. *Giardia*.

8. *Eimeria*

Classification:

Phylum	...	**Protozoa**	→	Microscopic and acellular.
Subphylum	...	**Plasmodroma**	→	Cilia absent.
Class	...	**Sporozoa**	→	Locomotory organs absent in adult; exclusively parasitic forms.
Subclass	...	**Telosporidia**	→	Adult trophozoite uninucleate; sporozoites elongated and without polar capsules in spores; spore cases simple and each containing several sporozoites.
Order	...	**Coccidia**	→	Trophozoite minute intra-cellular parasites of epithelial cells of invertebrates and vertebrates; young sporozoites and merozoites elongated and motile.
Genus	...	*Eimeria*		

Comments:

1. *Eimeria schubergi* is an intracellular parasite in the epithelial cells of gut of *Scolopendra* and *Geophilus*.
2. The life cycle is completed within a single host passing through the phases of **schizogony** and **sporogony**.
3. Food contaminated with oocysts is swallowed, the host's digestive juices dissolve the cysts and release the sporozoites.
4. Sporozoites enter in the epithelial cells in the bile duct and enlarge as feeding trophozoites.

5. Trophozoites undergo rapid schizogony to produce many merozoites.

6. The **merozoites** develop into gametocytes, also within epithelial cells; some become large macrogametes and other divide to yield many microgametes.

7. Macrogamete and microgamete form a zygote that secretes an elliptical covering oocyst.

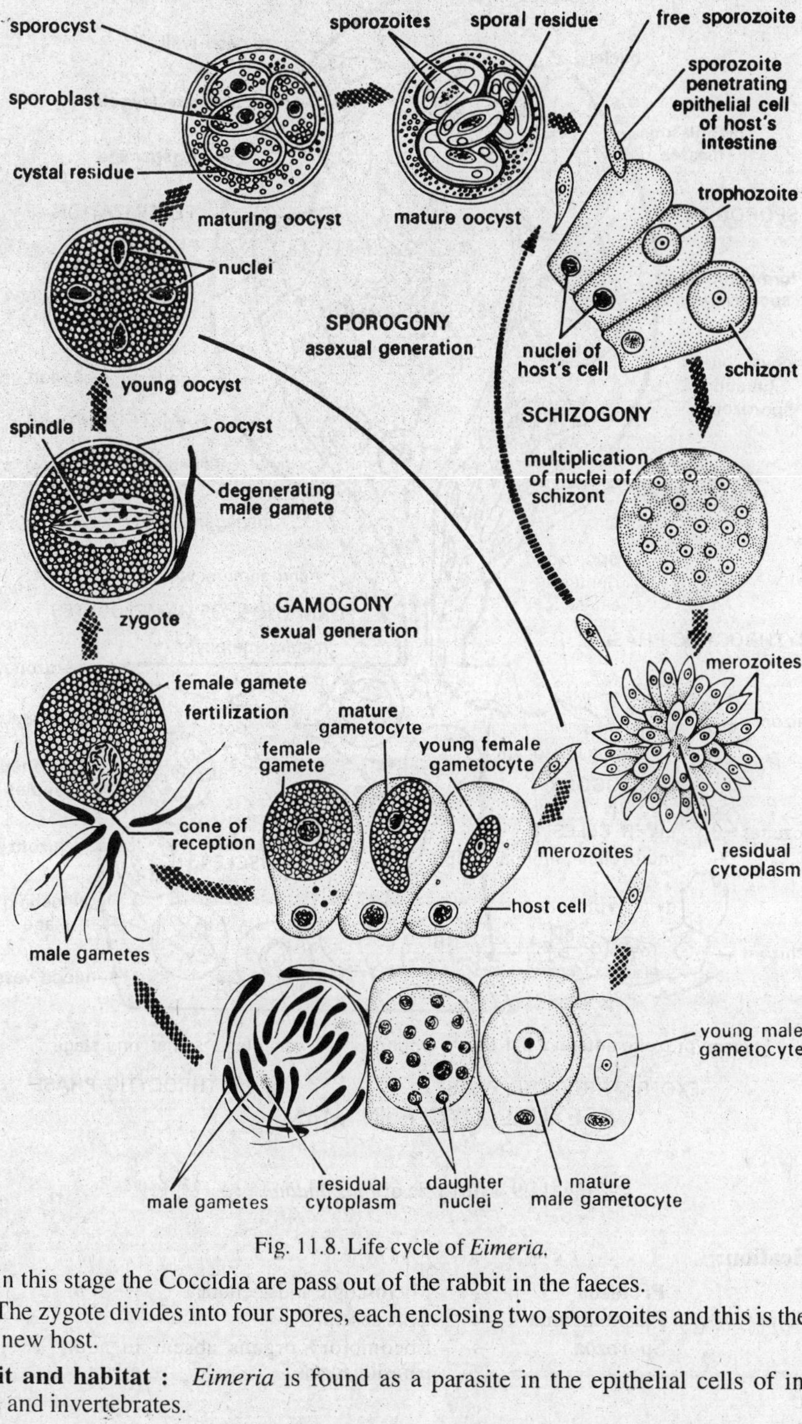

Fig. 11.8. Life cycle of *Eimeria*.

8. In this stage the Coccidia are pass out of the rabbit in the faeces.

9. The zygote divides into four spores, each enclosing two sporozoites and this is the infective stage for a new host.

Habit and habitat : *Eimeria* is found as a parasite in the epithelial cells of intestine of vertebrates and invertebrates.

9. *Plasmodium* (Malarial Parasite)

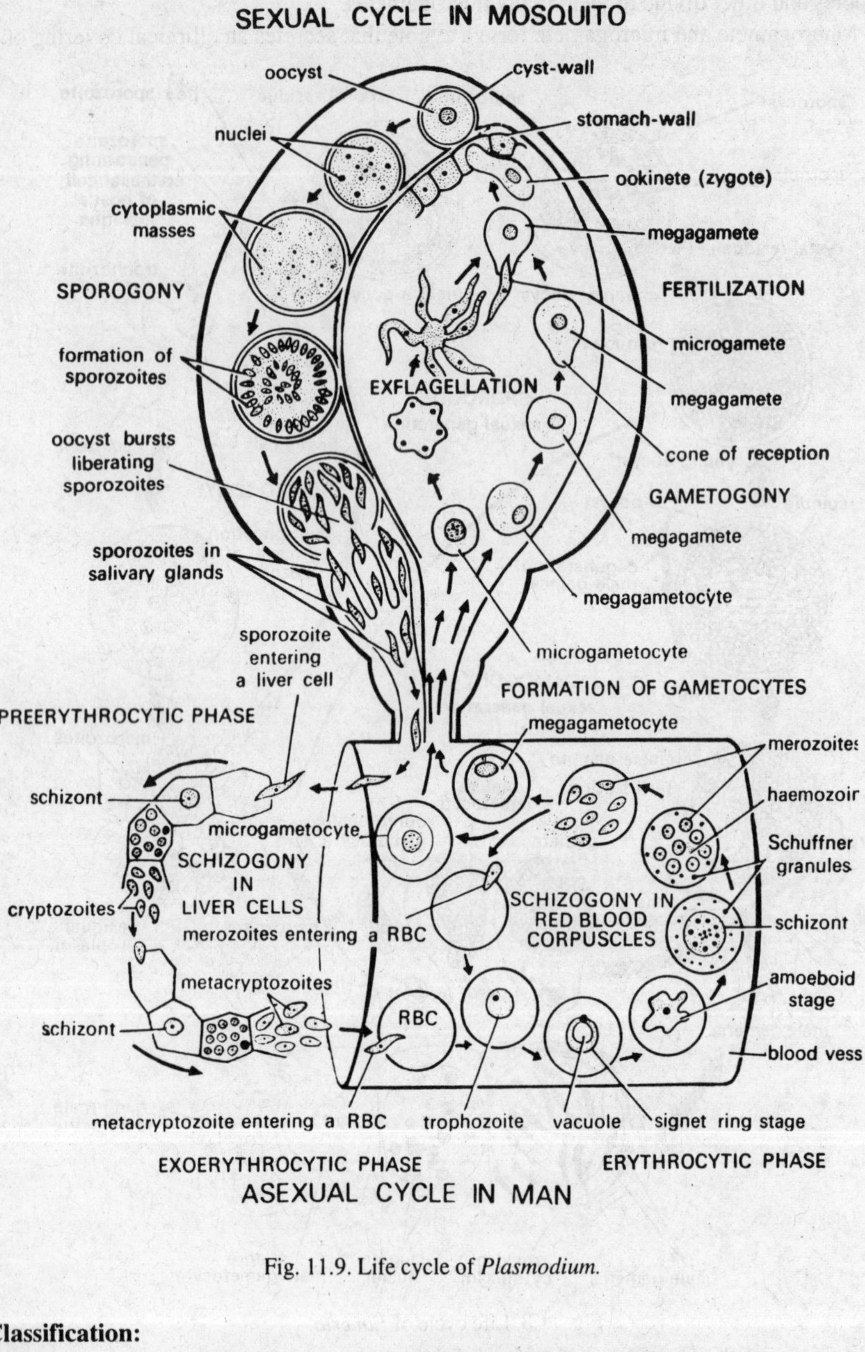

Fig. 11.9. Life cycle of *Plasmodium.*

Classification:

Phylum	...	**Protozoa**	→	Microscopic and acellular.
Subphylum	...	**Plasmodroma**	→	Cilia absent.
Class	...	**Sporozoa**	→	Locomotory organs absent in adult; exclusively parasitic forms.

Subclass	...	**Telosporidia**	→	Adult trophozoite uninucleate; sporozoites elongated and without polar capsules in spores; spore cases simple and each containing several sporozoites.
Order	...	**Haemosporidia**	→	Trophozoite small, amoeboid and intracellular; sporozoites naked; zygote motile, parasites of blood of vertebrates.
Genus	...	*Plasmodium*		

Comments:

1. *Plasmodium* (Fig. 11.9) is an intracellular blood parasite of man and other vertebrates.
2. The life history of *Plasmodium* is completed in two hosts, *viz.*, partly in definitive host, the man and partly in intermediate host, the female *Anopheles* mosquito.
3. When an infected female *Anopheles* mosquito bites a man, **sporozoites** are introduced in the blood from where they reach in liver cells through blood streams and multiply to form merozoites.
4. After few schizogonic cycles in the liver, the merozoites enter the red blood corpuscles (R. B. Cs) and feed on the contents of R. B. Cs.
5. After 2–8 schizogonic changes in the main blood stream, the merozoites assume different shapes and known as **gametocytes**.
6. Gametocytes cannot develop further in the blood of man, therefore, they wait for female *Anopheles* mosquito to suck them with the blood.
7. When gametocytes are sucked in by the female *Anopheles* with the blood of man, they undergo sporogony for further development.
8. There are four species of *Plasmodium* causing different types of fever :
   (i) *Plasmodium vivax* causes benign tertian fever.
   (ii) *P. falciparum* causes malignant tertain fever.
   (iii) *P. malariae* causes quratan fever.
   (iv) *P. ovale* causes ovale or mild tertian fever.

Habit and habitat : *Plasmodium* is found as parasite in the blood of vertebrates.

10. *Sarcocystis*

Classification:

Phylum	...	**Protozoa**	→	Microscopic and acellular.
Subphylum	...	**Plasmodroma**	→	Cilia absent.
Class	...	**Sporozoa**	→	Locomotory organs absent in adult; exclusively parasitic forms.
Subclass	...	**Neosporidia**	→	Adult trophozoite multinucleate; sporozoites amoeboid with polar capsules in spores; spore cases complex, each having single sporozoite.
Order	...	**Sarcosporidia**	→	Trophozoites elongated, spore cases absent, parasites in the muscles of reptiles, birds and mammals.
Genus	...	*Sarcocystis*		

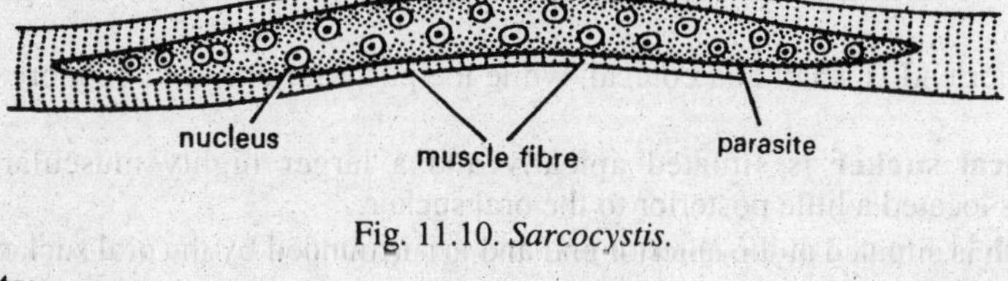

Fig. 11.10. *Sarcocystis*.

Comments:

1. *Sarcocystis* forms a long spindle-shaped multinucleate cyst which is found embedded in the striped muscle fibres of the host.
2. The masses of spindle-shaped parasites form long slender, cylindrical bodies with pointed ends known as "**Miescher's** or **Rainey's tubes**"

3. Within each tube a large number of sickle-shaped spores are formed. These spores are known as **Rainey's corpuscles**.

4. Very little is known regarding the mode of transmission of this parasite.

5. *Sarcocystis* causes a disease known as **sarcosporidiosis**.

Habit and habitat : *Sarcocystis* is a parasite of vertebrates especially in sheep, rabbit, pig, rat, cattle, horses and also in reptiles and birds. It is occasionally found in man.

PLATYHELMINTHES

11. *Polystoma*

Classification:

Phylum	...	**Platyhelminthes**	→	Acoelomate, organ grade and flatworm.
Class	...	**Trematoda**	→	Ecto-or endoparasitic; body wall without epidermis and cilia; well-developed suckers present.
Order	...	**Monogenea**	→	Oral sucker weak or absent; anterior end with a pair of adhesive structures; posterior end with an adhesive disc with hooks. Life cycle is completed in one host only.
Genus	...	*Polystoma*		

Comments:

1. *Polystoma* (Fig. 11.11) is leaf-like and dorso-ventrally flattened.

2. Anterior end is provided with a weak oral sucker.

3. **Mouth** is surrounded by the oral sucker.

4. **Pharynx** is muscular and leads into the bifurcated **intestine** which finally unites posteriorly. Both the branches of intestine also join with each other by transverse connections.

5. Posterior disc or opisthaptor is expanded and contains three pairs of large suckers and 2 or 3 hooks.

6. **Hermaphroditic.** Male reproductive organs consist of single testis, a vas deferens and a penis. Female system comprises single ovary, oviduct, two vaginae and longitudinal vitelline ducts.

7. Breeding season starts in spring

Habit and habitat : *Polystoma* is found in the urinary bladder of frogs and turtles.

12. *Fasciola*

Classification:

Phylum	...	**Platyhelminthes**	→	Acoelomate, organ grade and flatworms.
Class	...	**Trematoda**	→	Ecto-or endoparasitic; body wall without epidermis and cilia; well developed suckers present.
Order	...	**Digenea**	→	Endoparasitic, mostly with two suckers without hooks. Life cycle is completed in two hosts.
Genus	...	*Fasciola*		
Species	...	*hepatica*		

Comments:

1. *Fasciola hepatica* (Fig. 11.12) is commonly known as **liver fluke**.

2. Body is leaf-like, dorso-ventrally flattened, measures 25–30 mm in length and 4–5 mm in breadth.

3. Anterior end is small and conical, while the posterior end is large more rounded in front than behind.

4. An **oral sucker** is situated apically and a larger highly muscular ventral sucker (**acetabulum**) is located a little posterior to the oral sucker.

5. **Mouth** is situated at the anterior end and is surrounded by the oral sucker.

6. Digestive system is simple, **pharynx** is muscular, oesophagus short and branched and diverticulated intestine.

7. Between the oral and ventral sucker is a median genital pore through which pass eggs to the exterior.

8. **Excretory pore** lies at the extreme posterior end of the body.

9. **Hermaphroditic.** Male system consists of testes, vasa deferentia, seminal vesicle, ejaculatory duct and penis, while female system comprises ovary, uterus and vitelline glands.

10. Life cycle is complicated includes an intermediate host. *Limnaea* (a mollusc– freshwater snail).

11. Liver fluke causes a disease known as **liver rot**.

Habit and habitat : *Fasciola hepatica* is found as an endoparasite in the bile ducts of liver of sheep.

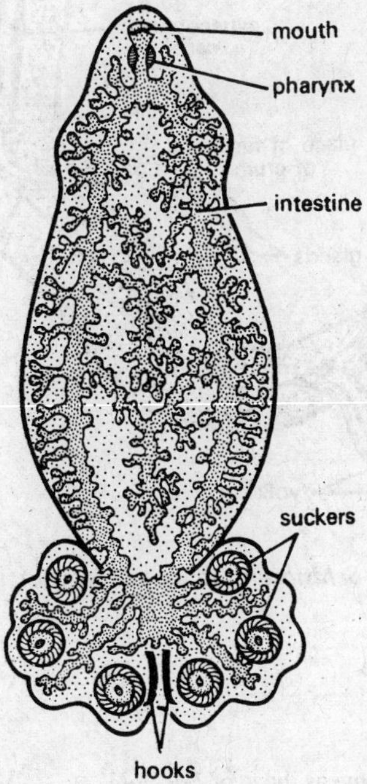

Fig. 11.11. *Polystoma*

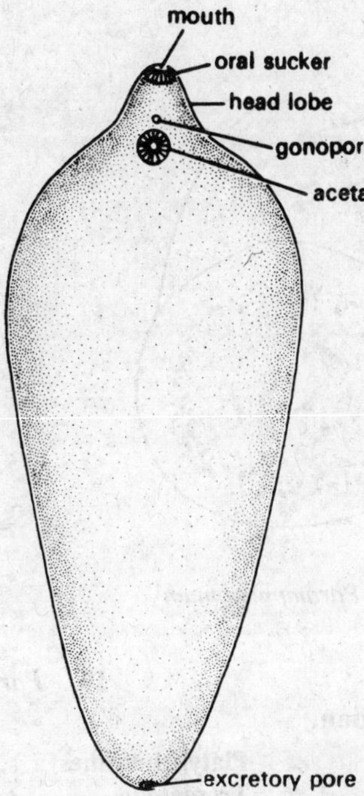

Fig. 11.12. *Fasciola hepatica*.

13. *Schistosoma*

Classification:

Phylum	...	**Platyhelminthes**	Characters same as those of *Fasciola*.
Class	...	**Trematoda**	
Order	...	**Digenea**	
Genus	...	*Schistosoma*	

Charaters:

1. *Schistosoma* (Fig. 11.13) is commonly called the **blood fluke**.
2. Body long slender in form and greyish or pinkish in colour.
3. Sexes are completely separate.
4. **Male** is usually 8 to 16 mm in length, has a cylindrical stout and flattened body.
5. **Female** is longer than male, usually 15 to 20 mm in length, has a more slender delicate cylindrical body.
6. Males carry females permanently in their **gynecophoric canals** formed by the infolding of the ventral body wall.
7. Life cycle includes a secondary host which is *Bulinus* or *Planorbis* (a mollusc, fresh water snail)

8. Its infection causes a disease called **schistosomiasis**.

9. Its 3 species are: *S. haematobium*, *S. mansoni* and *S. japonicum*.

Distribution : Schistosomes have been reported from Africa (particularly from Egypt), West Indies, South America, Japan, China, Celebes and Philippines.

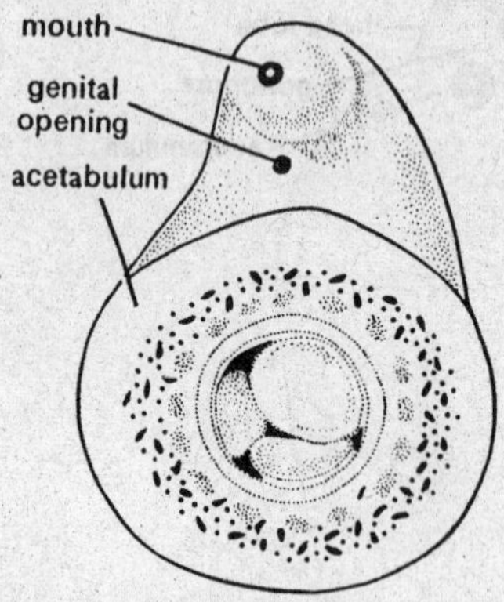

Fig. 11.14. *Paramphistomum*

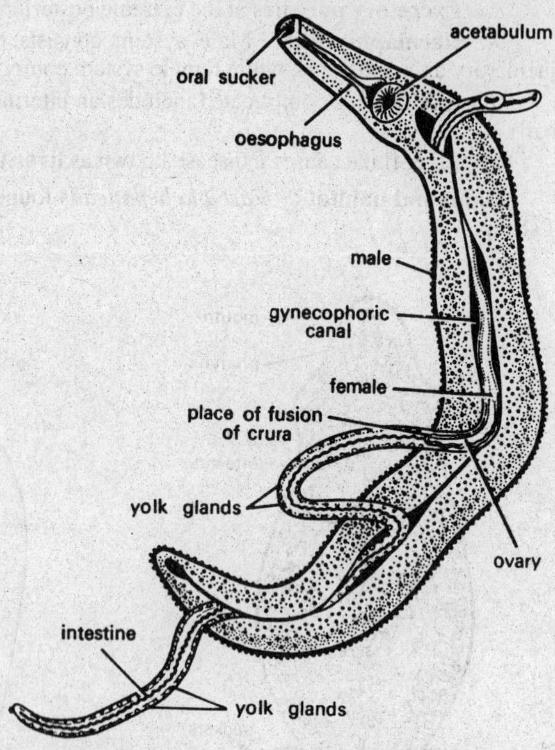

Fig. 11.13. *Schistosoma* (male and female).

14. *Paramphistomum*

Classification:

Phylum	...	**Platyhelminthes**	Characters same as those of *Fasciola.*
Class	...	**Trematoda**	
Order	...	**Digenea**	
Genus	...	*Paramphistomum*	

Comments:

1. *Paramphistomum* (Fig. 11.14) is small, triangular in shape and dorsoventrally flattened.
2. The oral sucker is absent.
3. The ventral sucker is large, muscular and situated posteriorly. It acts as an adhesive organ.
4. It is called amphistome parasite, because acetabulum is large and found near posterior end of the body.
5. The intestine is forked and unbranched.
6. The male system comprises two testes with tandem arrangement, sperm ducts, seminal vesicle and cirrus.
7. The female system comprises ovary, ootype, vitellaria and folded uterus.
8. The genital antrum lies just beneath the fork of intestine.
9. The eggs are large.
10. The lymphatic system is present.

Habit and habitat: *Paramphistomum* is an endoparasite in the rumen of sheep, goat, deer, cattle, etc.

Distribution: It is reported from India, Sri Lanka, Burma, other parts of Asia, Europe and U.S.A.

15. *Diphyllobothrium*

Classification:

Phylum	...	**Platyhelminthes**	→	Acoelomate, organ grade and flatworms.
Class	...	**Cestoda**	→	Endoparasitic in the intestines of vetrebrates; body divided into few to many segments (proglottids); anterior end with hooks and suckers.
Subclass	...	**Eucestoda**	→	Body elongated and ribbon-like; anterior end bears an expanded scolex bearing adhesive structure; each proglottid with more than one set of reproductive organs.
Order	...	**Pseudophyllidea**	→	Endoparasitic in intestine of vertebrates; scolex with two to six suckers rarely without adhesive organs; yolky glands numerous.
Genus	...	*Diphyllobothrium*		
Species	...	*latum*		

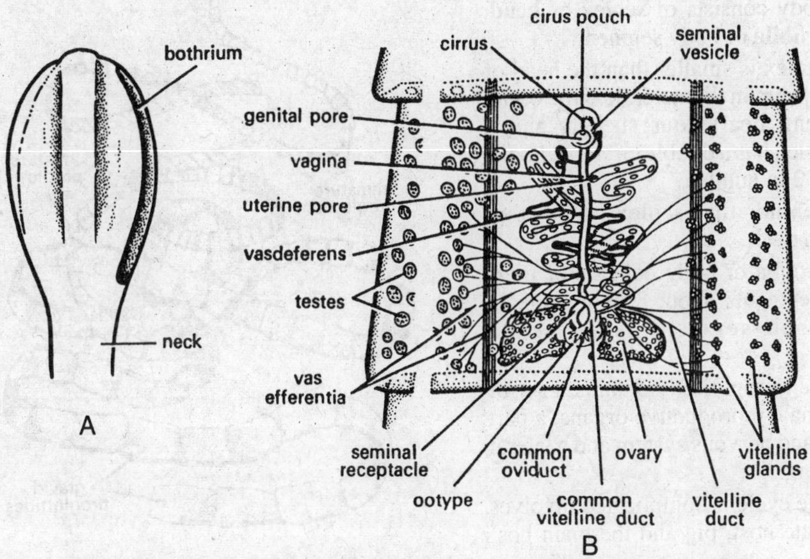

Fig. 11.15. *Diphyllobothrium latum.* A–Scolex; B–A mature proglottid.

Comments:

1. *Diphyllobothrium latum*, the **broad fish tapeworm**, is the largest and most injurious parasite in the intestine of man.
2. Body length reaches upto 20 metres have 3,000 to 4,000 proglottids
3. **Scolex** is fusiform, bears two slit-like bothria, followed by the long slender neck.
4. Mature proglottid is broader than long.
5. Excretory system protonephridial type.
6. **Hermaphroditic.** Male system consists of numerous testes, vas deferens, seminal vesicle and cirrus. Female system consists of a bilobed ovary, oviduct and a coiled uterus. Each segment has a single set of reproductive system.
7. Life cycle involves two intermediate hosts, one is copepod (*Cyclops*) and another fish.
8. Infection in man results by eating insufficiently cooked infected fish.

Habit and habitat: *Diphyllobothrium latum* is found as a parasite in the intestine of man. It is worldwide in distribution.

16. *Taenia solium*

Classification:

Phylum	...	**Platyhelminthes**	Characters same as those of *Diphyllobothrium.*
Class	...	**Cestoda**	
Subclass	...	**Eucestoda**	
Order	...	**Taenioidea**	→ Endoparasitic in the intestines of birds and mammals, scolex with four suckers often with an apical rostellum armed with hooks; yolk gland single and compact.
Family	..	**Taeniidae**	→ Rostellum permanently extruded, armed with two circlets of hooks; gravid uterus with main stem and lateral branches.
Genus	...	*Taenia*	
Species	...	*solium*	

Comments:

1. *Taenia solium* is commonly known as **pork tapeworm.**
2. Body consists of **scolex** or **head, neck** and **strobila** or body segment.
3. **Scolex** is smaller than the head of a pin, about 1 mm in diameter. It is the organ of attachment, bears four suckers and a rostellum which has a double circlet of hooks about 28 to 32 in number.
4. Behind the scolex is a thin unsegmented **neck.**
5. Strobila or body consists of large number of segments about 800 or more in number. Each segment is termed a **proglottid**.
6. Each progloittid contains a set of male and female reproductive organs, a part of excretory and nervous system and a lateral genital opening.
7. Life cycle is complicated involves an intermediate host, pig and the main host being man.

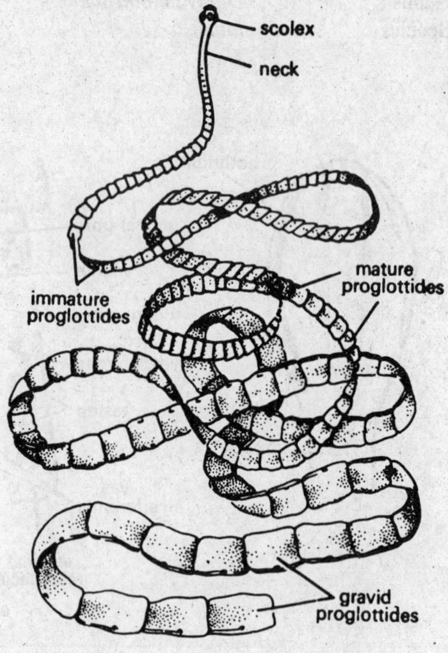

Fig. 11.16. *Taenia solium.*

8. The intermediate host of *Taenia solium* is pig in which infective cysticercus larva or the bladder worm is found encysted.
9. Pig is infected by onchospheres or hexacanths after eating contaminated human faeces.
10. Man in turn gets infection by consuming measly pork.
11. The disease caused by *Taenia solium* is called **taeniasis.**

Habit and habitat: *Taenia solium* is commonly found in the intestine of man in places where pork is eaten as food.

17. *Taenia saginata*

Classification:

Phylum	...	**Platyhelminthes**	Characters same as those of *Taenia solium.*
Class	...	**Cestoda**	
Subclass	...	**Eucestoda**	
Order	...	**Taenioidea**	
Family	...	**Taeniidae**	
Genus	...	*Taenia*	
Species	...	*saginata*	

Comments:

1. *Taenia saginata* is commonly called **beef tapeworm**.
2. The body of the worm is dorso-ventrally flattened and much larger than *Taenia solium*, reaches a length of 5 to 7 meters.

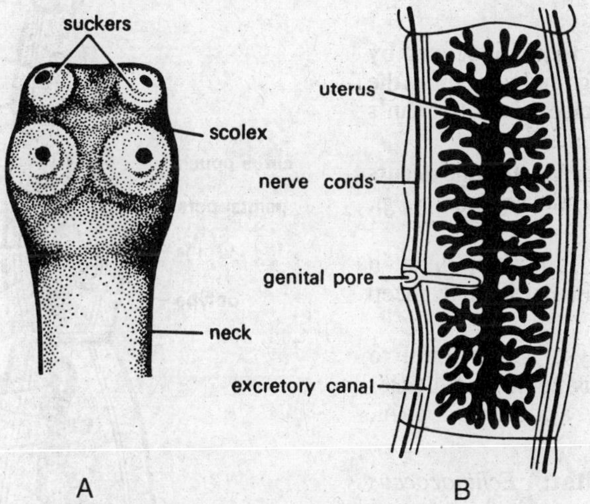

Fig. 11.17. *Taenia saginatta.* A–Scolex; B–Proglottid.

3. Body is divisible into **scolex**, **neck** and **strobila**.
4. **Scolex** is 1.5 to 2 mm in diameter, bears four large suckers for adhesion and devoid of rostellar hooks.
5. Both mature and ripe proglottids are larger than those of *T. solium*.
6. Life history is similar to that of *T. solium*.
7. Intermediate hosts are cattle, usually cow and buffalo.
8. Infection takes place by ingestion of improperly cooked beef.

Habit and habitat: *Taenia saginata* is found in the intestine of man. It is cosmopolitan in distribution.

18. *Echinococcus granulosus*

Classification:

Phylum	...	**Platyhelminthes**	Characters same as those of *Taenia solium*.
Class	...	**Cestoda**	
Subclass	...	**Eucestoda**	
Order	...	**Taenioidea**	
Family	...	**Taeniidae**	
Genus	...	*Echinococcus*	
Species	...	*granulosus*	

Comments:

1. *Echinococcus granulosus* is commonly called **dog tapeworm**.
2. The worm is only 2–8 mm in length and consists of a **scolex**, **neck** and three or four successive large **segments** (proglottids).
3. The **scolex** or head bears four **suckers** and a protrusible **rostellum** provided with double rows of 30 to 36 hooks.
4. **Proglottids** are usually three, first proglottis is generally immature, second proglottis is mature, while the third is large and gravid.

5. It is **hermaphrodite**. Male reproductive system consists of spherical testes, vas deferens and cirrus. Female reproductive system comprises ovaries, oviduct, vitellaria, ootype, uterus and vagina.

6. Its gravid proglottis is elongated containing branched uterus with onchospheres.

7. Its infection to man occurs by playing with infected dogs and sometimes the infective onchosphere also reach to man's body along with food or drink.

8. Hydatid cyst or larval stage occurs in man and other domestic animals, *e. g.*, monkey and cattle.

9. *Echinococcus* is cosmopolitan in distribution specially in cattle and sheep raising areas.

10. Its hydatid cysts are harmful to man; the cystic fluid is toxic and causes vomiting, diarrhoea, abdominal pain, eosinophilia, etc.

Habit and habitat: *Echinococcus granulosus* is an endoparasite in the intestine of dogs, cats and foxes (primary hosts) etc. Its secondary hosts are man, rabbits, kangaroos, sheep and cattle.

Distribution: Cosmopolitan, especially found in sheep and cattle raising areas. Its infection has been reported from Australia, Japan, India, New Zealand, China and U.S.A.

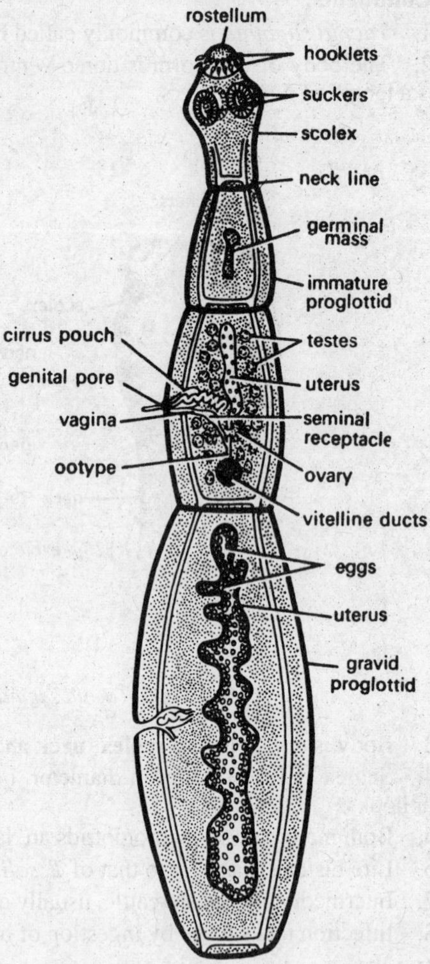

Fig. 11.18. *Echinococcus granulosus*.

19. Hydatid Cyst

Comments:

1. The hydatid cyst of *Echinococcus* is found in man or herbivorous animals like rabbits, kangaroos, sheep and cattle.

2. When the embryonated egg capsules are passed out with faeces from the host and reach somehow to an intermediate host which is man or herbivorous animals (named above), the egg-shell dissolve and hexacanth released.

3. The hexacanth makes its way generally to the liver or lung and rarely to the kidneys, spleen, heart and brain.

4. The hexacanth then changes into a hollow **bladder** around which the host secretes an enveloping fibrous **cyst wall**. This is now called a **hydatid cyst** which is filled with colourless fluid.

5. The bladder wall within the cyst buds off exogenously or endogenously a number of daughter **brood capsules** which become free or attached to bladder wall.

6. The scolices are finally formed in the brood capsules.

7. When such brood capsules reach the main host (dog, cat or wolf), the scolices develop into individual *Echinococcus*.

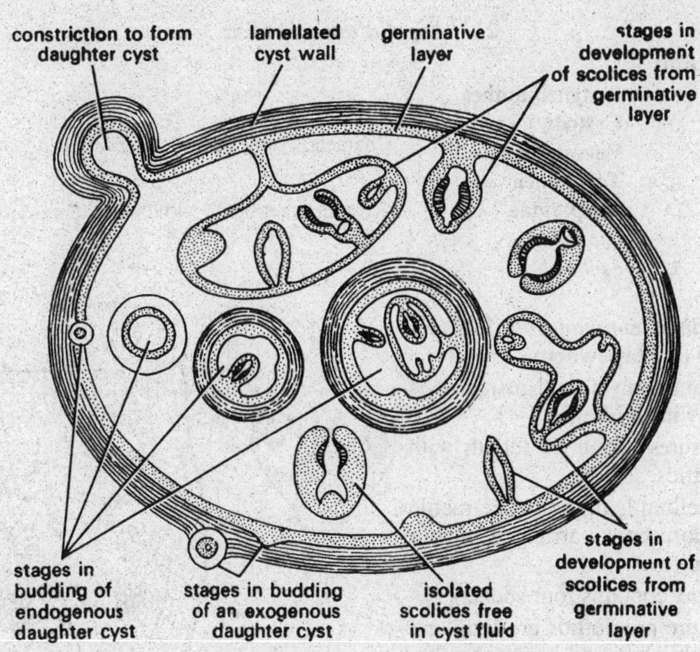

Fig. 11.19. Hybrid cyst of *Echinococcus.*

20. *Hymenolepis nana*

Classification:

Phylum	...	**Platyhelminthes**	Characters same as those of *Taenia solium.*
Class	...	**Cestoda**	
Subclass	...	**Eucestoda**	
Order	...	**Taenioidea**	
Family	...	**Taeniidae**	
Genus	...	*Hymenolepis*	
Species	...	*nana*	

Comments:

1. *Hymenolepis nana* (Fig. 11.20) is commonly known as **dwarf tapeworm**.

2. It is the smallest tapeworm measures 7 to 10 mm in length.

3. Body consists of a **scolex, neck** and **proglottids**.

4. **Scolex** bears a well developed retractile **rostellum** with a crown of 20–30 hooks.

5. **Neck** is slender and long.

6. **Proglottids** are usually broader than long.

7. Life cycle is completed in a single host. No intermediate host is required.

8. *Hymenolepis nana* causes severe toxic symptoms including abdominal pain and diarrhoca, etc.

9. It is world wide in distribution. It is the commonest tapeworm in Southern United States.

Habit and habitat: *Hymenolepis nana* is an endoparasite in the intestine of man.

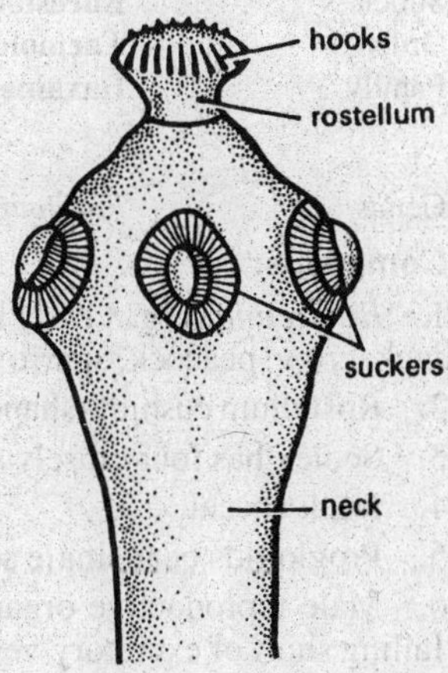

Fig. 11.20. Scolex of *Hymenolepis nana*

21. *Dipylidium caninum*

Classification:

Phylum	...	**Platyhelminthes**	Characters same as those of *Taenia solium.*
Class	...	**Cestoda**	
Subclass	...	**Eucestoda**	
Order	...	**Taenioidea**	
Family	...	**Dilepididae**	→ Rostellum retractile with one or more circlets of hooks.
Genus	...	*Dipylidium*	
Species	...	*caninum*	

Comments.

1. *Dipylidium caninum* (Fig. 11.21) is commonly called **dog tapeworm**.
2. It is commonly found in pet dogs and cats but rarely in children.
3. It measures 25 cm in length with about 150 proglottids.
4. Its rostellum is muscular, retractile and bears rose-thorn hooks arranged in four transverse circular rows.
5. Its scolex contains four suckers.
6. Its mature proglottids are long and slender with double sets of hermaphrodite reproductive organs and a genital pore on each lateral margin.
7. Testes numerous scattered throughout proglottid.
8. Uterus reticulate which breaks into uterine capsules containing about 30 embryos or less.

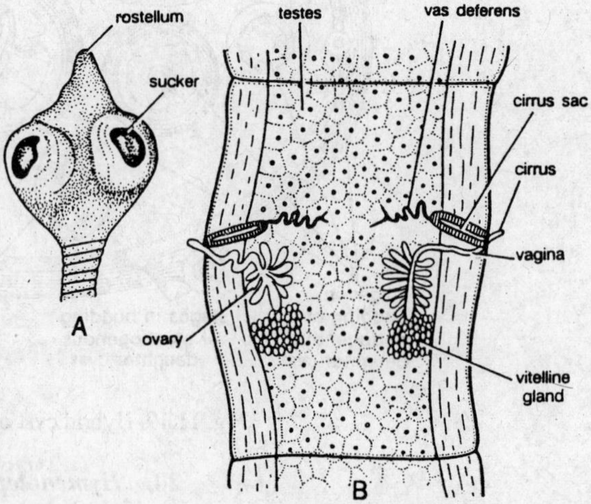

Fig. 11.21. *Dipylidium caninum.*
A–Scolex; B–Mature proglottid.

9. Dog-louse and dog-flea are secondary hosts in which cysticercoids develop.

22. *Raillietina*

Classification:

Phylum	...	**Platyhelminthes**	Characters same as those as *Taenia solium.*
Class	...	**Cestoda**	
Subclass	...	**Eucestoda**	
Order	...	**Taenioidea**	
Family	...	**Davaineidae**	→ Rostellum simple surrounded by a large number of hammer-shaped hooks; suckers spinose; parasites in most of the avian orders.
Genus	...	*Raillietina*	

Comments:

1. *Raillietina* (Fig. 11.22) is a large-sized tapeworm parasitizing pigeon and other birds like fowl, duck, crow, peacock, sparrow, etc.
2. Rostellum cushion-shaped with double set of hammer-shaped hooks arranged in two rows.
3. Scolex has four closely set suckers which are spinose.
4. Neck absent.
5. Proglottids bear single set of hermaphrodite reproductive organs.
6. Male reproductive organs consist of many testes, sperm duct coiled and cirrus sac with cirrus falling short of excretory vessels.
7. Female reproductive organs consist of somewhat median ovary, post ovarian compact vitelline gland, receptaculum seminis and uterus.
8. Uterus breaks down into egg capsules containing one to several capsules.

9. Genital pores alternate irregularly.
10. Secondary hosts are houseflies and ants in which cysticercoids develop.

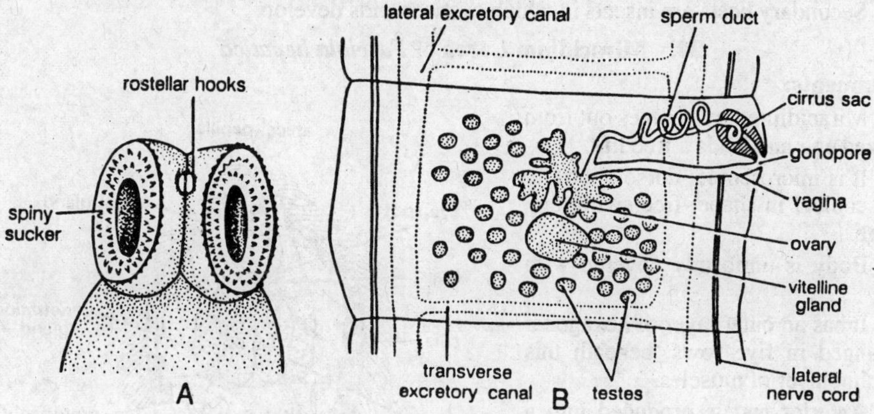

Fig. 11.22. *Raillietina.* A–Scolex; B–Mature proglottid.

23. *Cotugnia*

Classification

Phylum	...	**Platyhelminthes**	Characters same as those of *Raillietina*.
Class	...	**Cestoda**	
Subclass	...	**Eucestoda**	
Order	...	**Taenioidea**	
Family	...	**Davaineidae**	
Genus	...	*Cotugnia*	

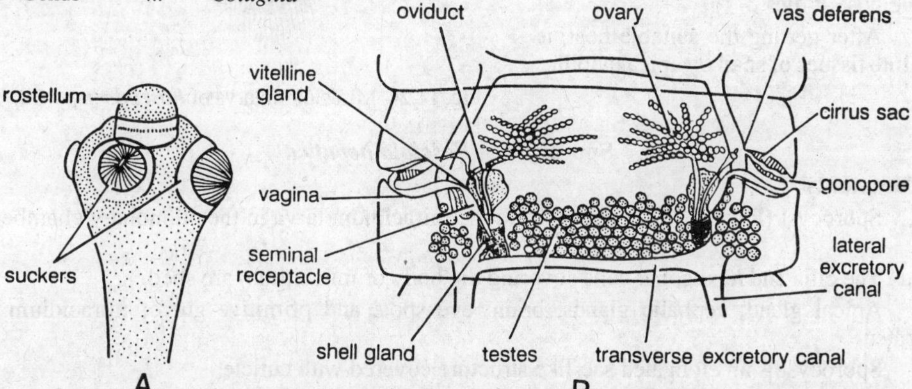

Fig. 11.23. *Cotugnia.* A–Scolex; B–Mature proglottid.

Comments:

1. *Cotugnia* is medium-sized davaineid tapeworm parasitizing pigeon, fowl, duck, peacock, parrot, crow, kite, etc.
2. Rostellum simple with a large number of hammer-shaped hooks arranged in two rows.
3. Scolex bears four spinose suckers.
4. Proglottids bear double set of hermaphrodite reproductive organs (the only davaineid with double set of reproductive organs).
5. Proglottids very short, broader than long except the last ones which are longer than broad.
6. Testes numerous, lie in extravascular field, may or may not cross excretory vessels.
7. Ovaries medially placed.
8. Genital ducts dorsal to excretory vessels and nerve stem.

9. Cirrus sac subcylindrical.
10. Genital pores bilateral.
11. Secondary hosts are insects in which cysticercoids develop.

24. Miracidium Larva of *Fasciola hepatica*

Comments:

1. Miracidium larva comes out from the fertilized egg and leads a free life.
2. It is microscopic, dorso-ventrally flattened, conical in shape, free-swimming larval stage.
3. Body is uniformly covered with **cilia**.
4. It has an outer layer of hexagonal cells, arranged in five rows, beneath this layer is a thin layer of muscles.
5. Anterior end is produced into a conical lobe, the **apical papilla**.
6. Internal structures, **apical gland**, **cephalic** or **penetration glands, brain,** two **eye spots,** two **flame cells**, **rudimentary gut** and **germ cells** are seen.
7. Miracidium larva swims in search of an intermediate host which is *Limnaea truncatula* for about 4–30 hours.
8. If it does not come in contact with a suitable host, it dies.
9. After getting the suitable host, it pierces into tissues of snail by apical papilla.

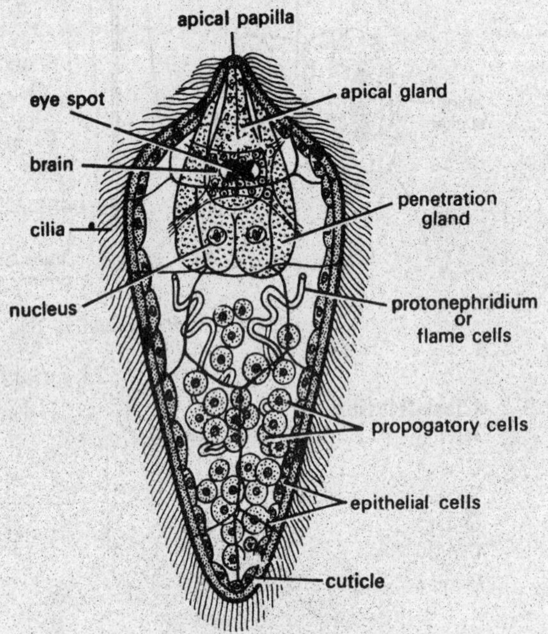

Fig. 11.24. Miracidium larva of *Fasciola hepatica.*

25. Sporocyst of *Fasciola hepatica*

Comments:

1. Sporocyst (Fig. 11.25) develops from the miracidium larva in the pulmonary chamber of snail.
2. The cilia and hexagonal cells covering the body of miracidium are shed.
3. Apical gland, cephalic glands, brain, eye spots and primitive gut of miracidium are degenerated.
4. Sporocyst is an elongated sac-like structure covered with cuticle.
5. Body wall consists of sub-epithelial cells, muscles and mesenchyme.
6. Body sac has **flame cells** and **germ cells**.
7. Germ cells multiply and give rise to next larval stage known as **redia larva**.

26. Redia Larva of *Fasciola hepatica*

Comments:

1. Redia larva (Fig. 11.26) develops from the germ cells of the sporocyst.
2. The body of redia is an elongated sac-like.
3. Anterior end bears the **mouth** leading into muscular pharynx, which finally leads into sac-like intestine.
4. Just behind the pharynx is a muscular ring-like swelling known as **collar**, which helps in locomotion.
5. Posterior region is also provided with two stumpy processes known as **lappets** helpful in locomotion.

6. Just posterior to collar a permanent aperture, the **birth pore** is seen.
7. The space between the body wall and intestine contains few germ cells.
8. Germ cells often give rise to second generation the **daughter rediae**.
9. Redia gives rise to a new type of larva known as **cercaria larva** from the germ cells.
10. **Cercaria larva** comes out from redia through birth pore.

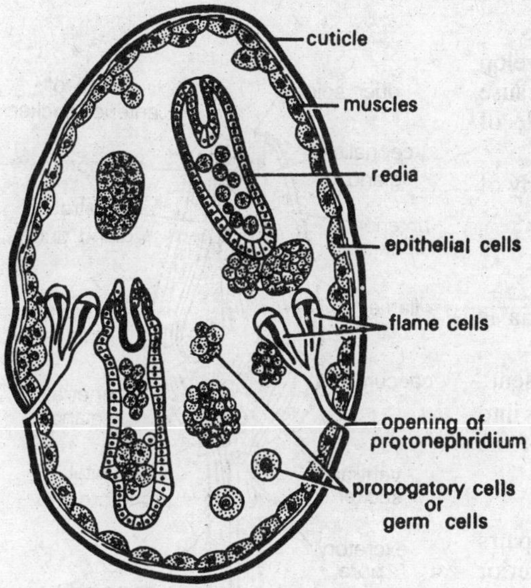

Fig. 11.25. Sporocyst of *Fasciola hepatica.*

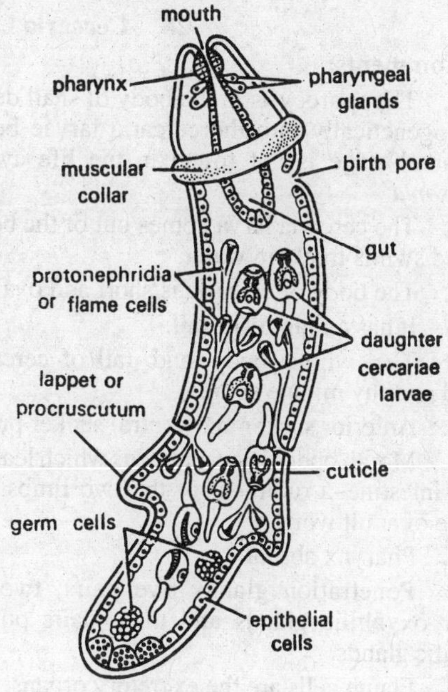

Fig. 11.26. Redia larva of Fasciola hepatica.

27. Cercaria Larva of *Fasciola hepatica*

Comments:

1. Body of cercaria (Fig. 11.27) is flat and oval bearing a tail.
2. Body is covered by cuticle and body wall consists of cuticle, muscles and mesenchyme.
3. It has two suckers, an **anterior oral sucker** surrounding the mouth and a **ventral sucker** situated in the middle of body.
4. Digestive system comprises mouth, muscular pharynx, oesophagus and inverted U "∩" shaped intestine.
5. Excretory organs of the flame cell type are present.
6. Body space is filled with parenchyma and contains a few **cystogenous glands** on each side which form the cyst of the future larva.
7. Rudimentary reproductive organs are also seen.

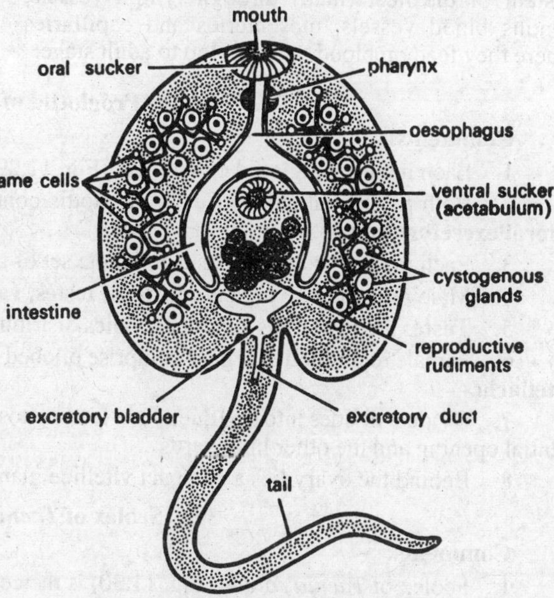

Fig. 11.27. Cercaria larva of *Fasciola hepatica.*

8. Cercaria larva comes out from the redia through the birth pore and also from the body of snail.

9. It is a free-swimming larva and after swimming for a short period it attaches to the aquatic plants.

10. Finally cercaria larva undergoes encystment and the encysted larva is known as **metacercaria** which is swallowed by the final host, sheep.

28. Cercaria Larva of *Schistosoma*

Comments:

1. The sporocysts in the body of snail develop parthenogenetically into the cercaria larvae because redia larval stage is not found in the life cycle of *Schistosoma.*

2. The cercaria larva comes out of the body of snail and swims freely in water.

3. The body of cercaria is short and oval.

4. It has a long **bifid tail**.

5. The whole body and tail of cercaria is covered over by minute spines.

6. Anterior sucker and ventral sucker present.

7. Mouth opens in oesophagus which leads into a trifid intestine–a rudiment of the two limbs of the intestine of adult worm.

8. Pharynx absent.

9. Penetration glands five pairs; two pairs anterior oxyphilic glands and three pairs posterior basophilic glands.

10. Flame cells are the excretory organs.

11. It infects man either by penetrating through the skin or by drinking infected water.

12. Finally cercaria larvae reach the portal system of the host (man) through lymph vessels, venous blood vessels, mesenteries and capillaries where they feed on blood and develop to adult stage.

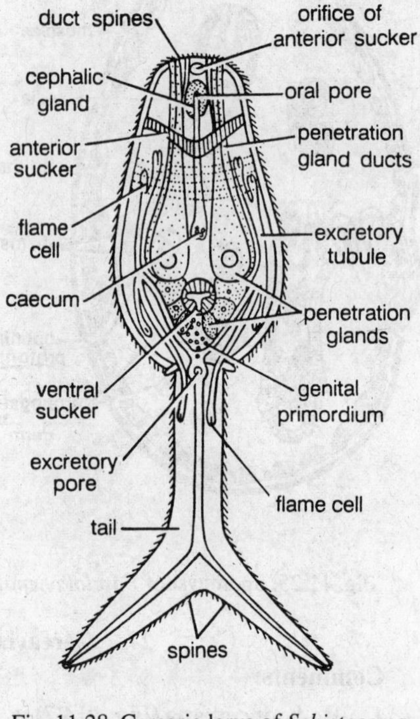

Fig. 11.28. Cercaria larva of *Schistosoma.*

29. Mature Proglottid of *Taenia solium*

Comments:

1. Each mature proglottid or segment (Fig. 11.29) is square-like in shape.

2. Both the lateral sides of each proglottis contain **lateral longitudinal nerve cords** and **lateral excretory canals**.

3. Each mature proglottis has a complete set of male and female reproductive organs.

4. Male reproductive organs consist of **testes, vasa efferentia**, **vas deferens** and **cirrus**.

5. **Testes** are numerous spherical bodies distributed throughout the proglottid.

6. Female reproductive organs comprise bilobed **ovary**, **oviduct**, **ootype**, **vagina**, **uterus** and **vitellaria**.

7. Oviduct divides into two ducts, one leading to vagina opening into genital atrium by female gential opening and the other into uterus.

8. Behind the ovary lies a compact vitelline gland.

30. Scolex of *Taenia solium*

Comments:

1. Scolex of *Taenia solium* (Fig. 11.30) is the cephalic portion of the body.

2. It is koob-like, smaller than the head of a pin measuring about 1 mm in diameter.

3. It is the organ of attachment and also contains the brain and nephridial canals.
4. It is roughly quadrate in structure and has two distinct parts, the proximal and distal.
5. The proximal part bears four adhesive **suckers**.
6. The distal part contains a small retractile prominence in the middle which is known as **rostellum**.
7. Rostellum has double circle of hooks about 28–32 in number.

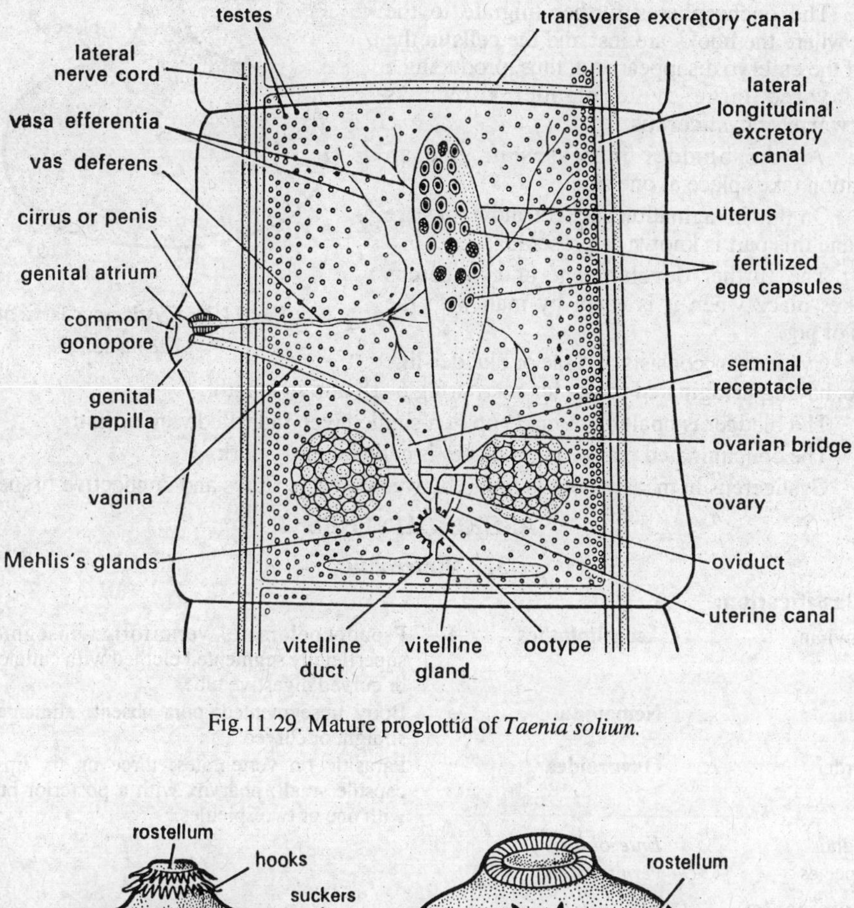

Fig.11.29. Mature proglottid of *Taenia solium*.

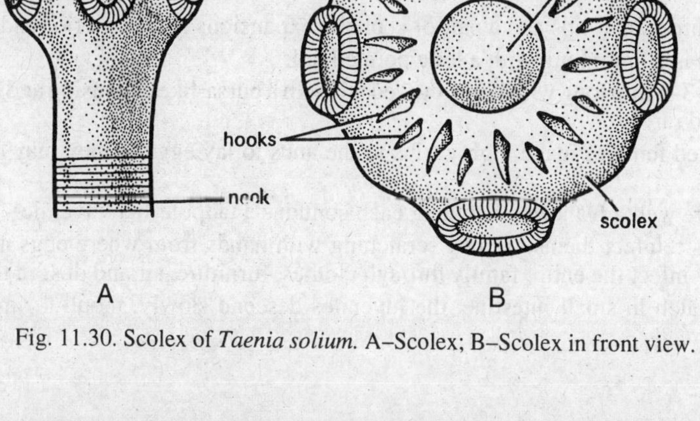

Fig. 11.30. Scolex of *Taenia solium*. A–Scolex; B–Scolex in front view.

31. Cysticercus (Bladderworm) Larva of *Taenia solium*

Comments:

1. Cysticercus larva is also known as bladderworm and develops in the muscles of pig, the intermediate host.

2. The onchospheres first reach the stomach of pig with faeces of man.

3. The onchospheres further migrate to the muscles where the hooks are lost and the cells in the centre of the embryo disappear and, thus, producing a single layered large ovoid bladder known as **bladderworm** or **cysticercus**.

4. As the bladder increases in size an invagination takes place at one side.

5. On this invagination suckers and hooks are formed and this part is known as proscolex.

6. The further development of cysticercus larva takes place when it is eaten by man with the muscles of pig.

7. Cysticercus consists of a sac or bladder-like structure, having invaginated proscolex measuring 6–18 mm in length.

8. The bladder is opalescent made up of a single layer and filled with a fluid.

9. The contaminated part of pig muscles is called measly pork.

10. Cysticercus in measly pork is found between muscle fibres and connective tissue.

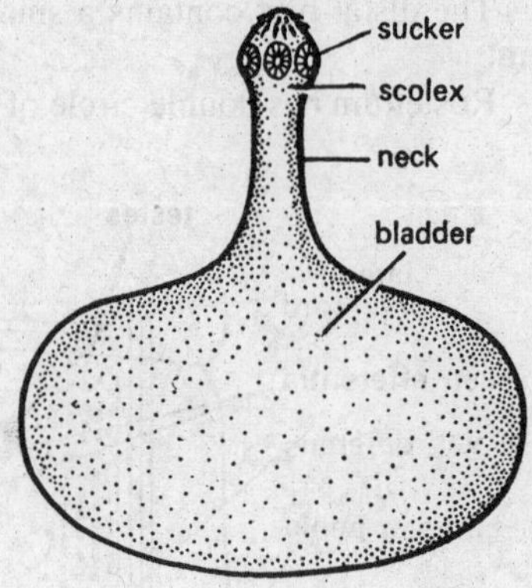

Fig. 11.31. Cysticercus larva of *Taenia solium.*

ASCHELMINTHES

32. *Enterobius*

Classification:

Phylum	...	**Aschelminthes**	→	Pseudocoelomate, vermiform; unsegmented or superficially segmented clothed with cuticle; straight or curved digestive tube.
Class	...	**Nematoda**	→	Body unsegmented; cilia absent; alimentary canal straight or curved.
Order	...	**Oxyuroidea**	→	Parasitic on vertebrates; three or six lips, buccal capsule small; pharynx with a posterior bulb; male with one or two spicules.
Genus	...	*Enterobius*		
Species	...	*vermicularis*		

Comments:

1. *Enterobius vermicularis* (Fig. 11.32) is commonly known as **pin-worm** or **seat-worm** or **thread-worm**.

2. It has three small lips and a pair of cephalic expansions at the anterior end.

3. Female is 10 mm long with a long pointed tail.

4. Male is 3–5 mm long with blunt, curved tail with a bursa-like expansion and a sinlge spicule. Males are few and rare.

5. Fertilized females make nighty trips to the anus to lay eggs or they may creep out of the anus and lay eggs.

6. Eggs are well advanced when laid, each contains a tadpole-like juvenile.

7. Persons reinfect themselves by scratching with hands from where eggs may get into the mouth. Eggs may infect the entire family through clothes, furniture, air and dust in the room.

8. Eggs hatch in small intestine, the juveniles descend slowly, moult 4 times and become adults.

9. *Enterobius* causes pinworm disease or **enterobiasis**, their movements cause intense itching of the anus, inflammation of mucous membrane of colon and appendix and often insomnia and lack of appetite.

10. Enteric-coated capusles of gentian violet of one grain given for 8 days remove most of the worms.

Habit and habitat: *Enterobius vermicularis* is cosmopolitan, but more common in Europe and America. In some communities 40–100% of the population may be infected. It is a parasite in the human caecum, colon and appendix.

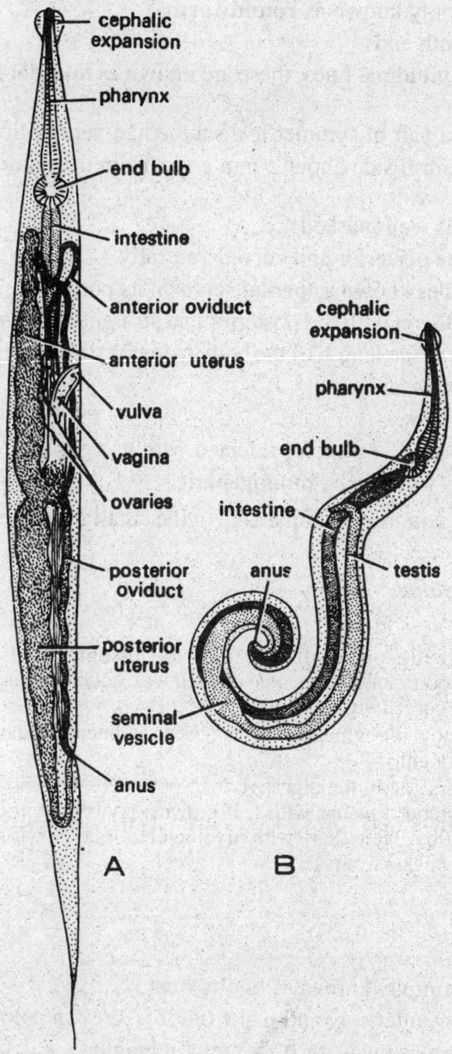

Fig. 11.32. *Enterobius vermicularis*.
A–Female; B–Male.

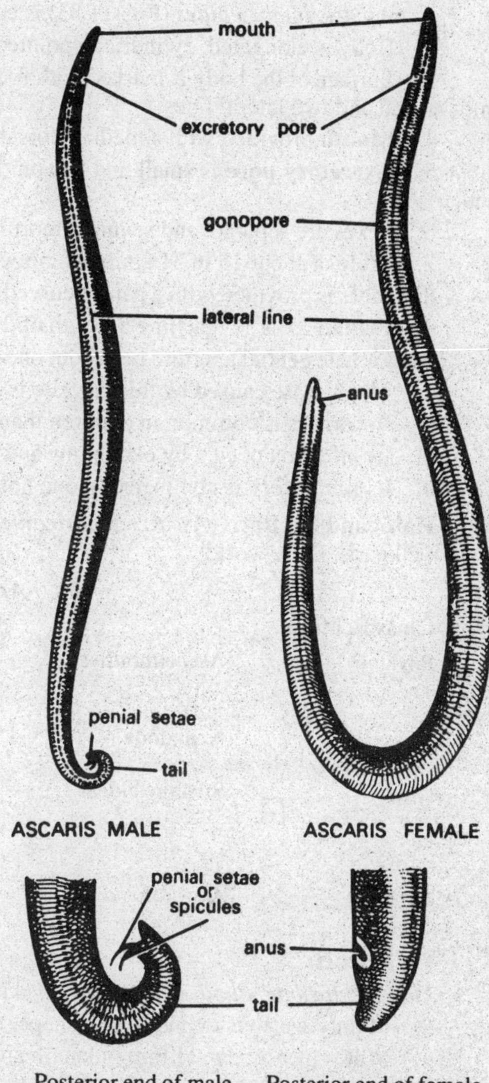

Posterior end of male. Posterior end of female.
Fig. 11.33. *Ascaris*.

33. *Ascaris*

Classification:

Phylum ... **Aschelminthes** → Pseudococlomate, vermiform; unsegmented or superficially segmented clothed with cuticle; straight or curved digestive tract.

Class	...	**Nematoda**	→ Body unsegmented; cilia absent; alimentary canal straight.
Order	...	**Ascaroidea**	→ Parasitic in the intestine of vertebrates; three prominent lips; pharynx without a posterior bulb; tail in female straight and blunt tail in male with two spicules
Genus	...	*Ascaris*	
Species	...	*lumbricoides*	

Comments:

1. *Ascaris lumbricoides* (Fig. 11.33) is commonly known as **roundworm**.
2. Body is elongated, cylindrical, pointed at both ends.
3. Surface of the body is marked with four longitudinal lines, these are known as mid-dorsal, mid-ventral and two lateral lines.
4. **Mouth** provided with a median dorsal and a pair of symmetrical submedian ventral lips.
5. **Excretory pore** is small and lies on the ventral side about 2 mm away from the anterior end.
6. Sexes are separate and sexual dimorphism is well marked.
7. **Male** is about 15 to 31 cm in length with the posterior end curved ventrally.
8. Male is provided with a pair of curved spicules known as **penial setae** on its posterior end.
9. **Female** is large reaching a length of 20 to 35 cm with the posterior end straight and blunt.
10. Female gential aperture lies about one-third of the length of the body from the anterior end.
11. The disease caused by this parasite is called **ascariasis**.
12. Ascariasis is common in children than in adults.
13. Its infection occurs by eating raw and uncooked vegetables infected with its eggs.
14. Its life history is completed in one host only (man), *i.e.*, **monogenetic.**

Habit and habitat: *Ascaris lumbricoides* is a common endoparasite in the small intestine of man in all parts of the world.

34. *Ancylostoma*

Classification:

Phylum	...	**Aschelminthes**	→ Pseudocoelomate, vermiform; unsegmented or superficially segmented clothed with cuticle; straight or curved digestive tube.
Class	...	**Nematoda**	→ Body unsegmented; cilia absent; alimentary canal straight.
Order	...	**Strongyloidea**	→ Parasitic in the digestive tract of vertebrates; mouth without lips but with leaf crowns; pharynx without bulbs; females with ovejecters; males with a copulatory bursa.
Genus	...	*Ancylostoma*	
Species	...	*duodenale*	

Comments:

1. *Ancylostoma duodenale* (Fig. 11.34) is commonly known as **hookworm**.
2. Mature worm is cylindrical in shape, narrow anteriorly and white or ivory grey in colour.
3. **Males** measure 8–11 mm in length and **females** measure 10–13 mm in length.
4. Buccal cavity is oval and the buccal capsule is made of articulated grooved portion.
5. **Excretory pore** mid-ventral in position just behind the nerve ring.
6. Male is provided with a **campanulate bursa** which is broader than long and supported by fleshy rays.
7. **Vulva** of the female is behind the middle of the body and the tail is tipped by a minute spine.
8. Fertilization is internal and occurs in the intestine of host. Fertilized eggs passed out with the faeces and after some time develop into infective stage.

9. Infection takes place by penetration of the skin of host by the infective larval stage.

10. *Ancylostoma duodenale* causes severe injury to the intestine and other parts of the host collectively referred to as **ancylostomiasis**.

Habit and habitat: *Ancylostoma duodenale* is found as an endoparasite in the intestine of man. It is endemic in Western Europe, tropical and subtropical Asia, Africa and various Pacific Islands and South America.

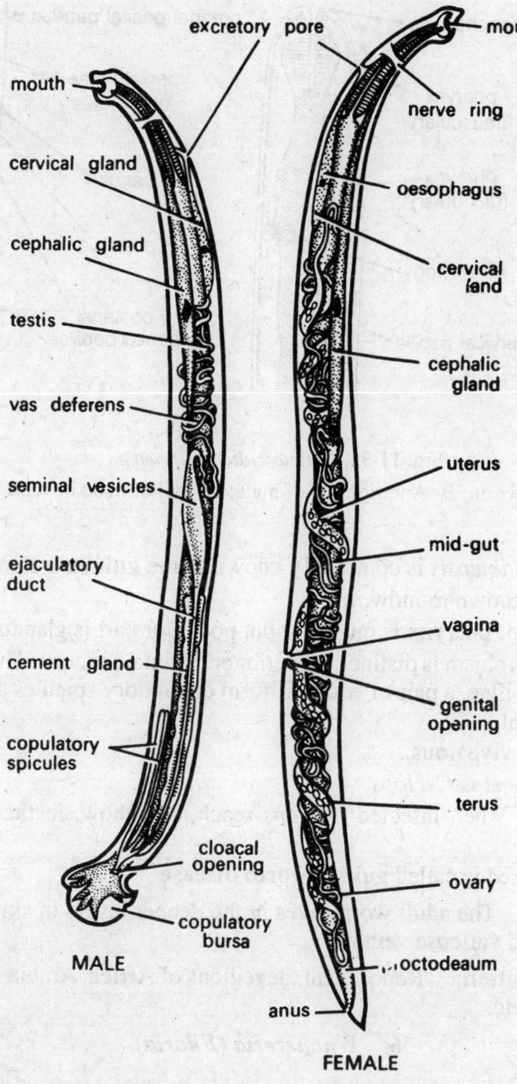

Fig. 11.34. *Ancylostoma duodenale*.

35. *Dracunculus medinensis*

Classification:

Phylum	...	**Aschelminthes**	→	Pseudocoelomate, vermiform; unsegmented or superficially segmented clothed with cuticle; straight or curved digestive tube.
Class	...	**Nematoda**	→	Body unsegmented; cilia absent; alimentary canal straight.

Order	...	**Dracunculoidea** →	Parasitic in vertebrates; mouth without lips, buccal capsule cuticularized; pharynx muscular and glandular without bulb; vulva in female in the middle of body; males with equal filiform spicules; bursa absent.
Genus	...	*Dracunculus*	
Species	...	*medinensis*	

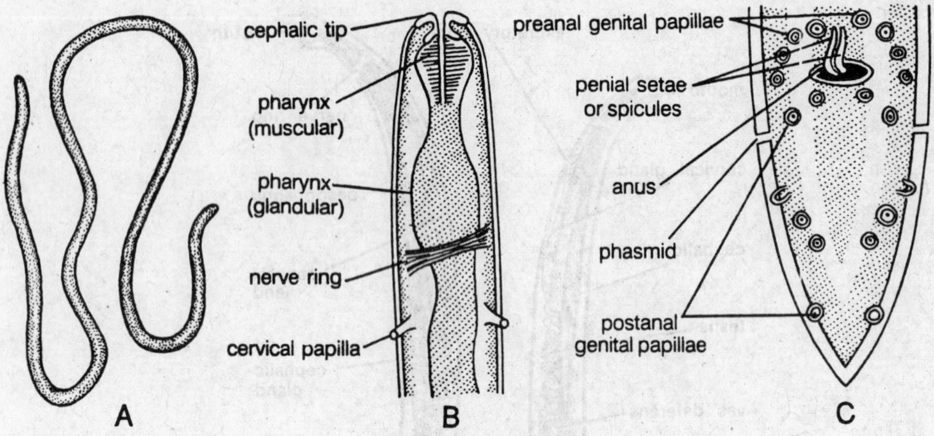

Fig. 11.35. *Dracunculus medinensis* .
A–Adult worm; B–Anterior end of male; C–Tail of male in ventral view.

Comments:

1. *Dracunculus medinensis* is commonly known as the **guinea worm** or **serpents worm**.
2. It is the largest known roundworm.
3. Its anterior part of pharynx is muscular but posterior part is glandular.
4. The sexual dimorphism is distinct; posterior end of male is curved having 4 pairs of preanal and 6 pairs of postanal papillae, a pair of equal filiform copulatory spicules and gubernaculum. The tail end of female is straight.
5. The females are vivparous.
6. The secondary host is *Cyclops*.
7. Man is infected when infected *Cyclops* reach, somehow, in the alimentary canal with contaminated water.
8. The disease caused is called **guinea worm disease**.

Habit and habitat: The adult worm lives in the deeper layers of skin of the arms, shoulder and legs much like a coiled varicose vein.

Geographic distribution:. Reported in the regions of Africa, Arabia, Persia, North America, Guinea, China and India, etc.

36. *Wuchereria (Filaria)*

Classification:

Phylum	...	**Aschelminthes** →	Pseudocoelomate, vermiform; unsegmented or superficially segmented, clothed with cuticle; straight or curved digestive tube.
Class	...	**Nematoda** →	Body unsegmented; cilia absent; alimentary canal straight,
Order	...	**Filarioidea** →	Mouth devoid of lips but with buccal capsule; bursa absent.
Genus	...	*Wuchereria* (*Filaria*)	
Species	...	*bancrofti*	

Comments:

1. *Wuchereria* (*Filaria*) is the most important human parasite in the lymph glands or ducts.

2. Adult worms are creamy white in colour, filiform and cylindrical in shape with both ends tapering and terminating bluntly.

3. Head slightly swollen and provided with two rows of small sessile papillae.

4. Mouth is unarmed and devoid of buccal capsule.

5. **Male** is about 40 mm in length with the caudal end curved like the tendril of vine.

6. **Female** is about 80 to 100 mm in length bearing the vulval opening just behind the anterior extremity of the body.

7. Female worms give birth to **microfilariae** which are surrounded by delicate membrane or sheath.

8. **Microfilaria** is long, cylindrical provided with a striated cuticle.

9. Microfilariae may either live in lymph or migrate into the blood capillaries.

10. *Culex* mosquito serves as an intermediate host which sucks the blood along with microfilariae from the infected host.

11. The further development of microfilariae takes place in the stomach and thoracic muscles of the mosquito.

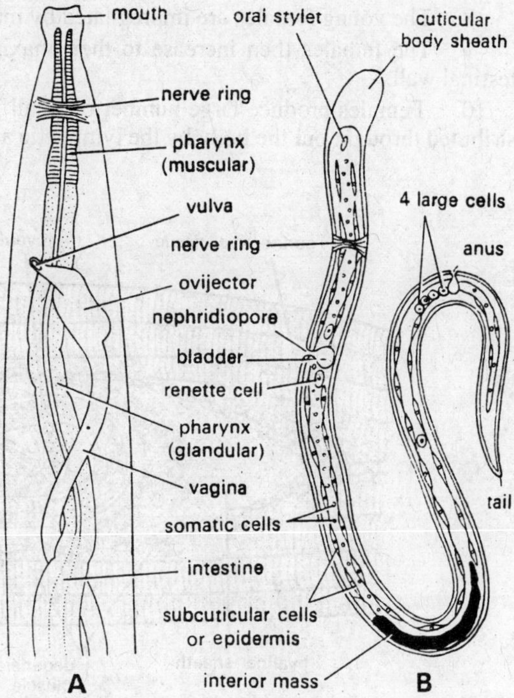

Fig. 11.36. A–Anterior part of female *Wuchereria bancrofti;* B–Microfilaria of *W. bancrofti.*

12. *Wuchereria* (=*Filaria*) *bancrofti* causes a serious disease known as **filariasis** or **wuchereriasis** or **elephantiasis**.

Habit and habitat: *Wuchereria bancrofti* is found as a parasite in the lympth glands or ducts of man. It is widely distributed in tropical and subtropical countries of the world.

37. *Trichinella*

Classification:

Phylum	...	**Aschelminthes**	→	Pseudocoelomate, vermiform; unsegmented or superficially segmented clothed with cuticle; straight or curved digestive tube.
Class	...	**Nematoda**	→	Body unsegmented; cilia absent; alimentary canal straight.
Order	...	**Trichuroidea**	→	Mouth without lips; pharynx with a stichosome, males with copulatory apparatus.
Genus	...	*Trichinella*		
Species	...	*spiralis*		

Comments:

1. **Male** worm (Fig.11.38 B) is 1.4–1.6 mm in length, has pointed anterior end and fleshy posterior end provided with a large conical papillae.
2. **Female** worm (Fig. 11.38 A) is 3–4 mm in length, pointed anteriorly and fleshy posteriorly.
3. The mouth leads into pharynx which contains large granular cells, called as stichosome.
4. Vulva is anterior.
5. The worms enter the digestive tract as larvae encysted in the muscle fibres of infected pork.

6. Cyst is dissolved in the stomach or intestine and the larvae penetrate into the mucosa of small intestine.

7. In the small intestine the larvae undergo a series of moults to beome adult.

8. The young females are impregnated by males which subsequently die and are eliminated.

9. The females then increase to their maximum size and penetrate more deeply into the intestinal wall.

10. Females produce large number of small juvenile worms in the intestinal wall which are distributed throught out the body by the lymphatic and blood streams.

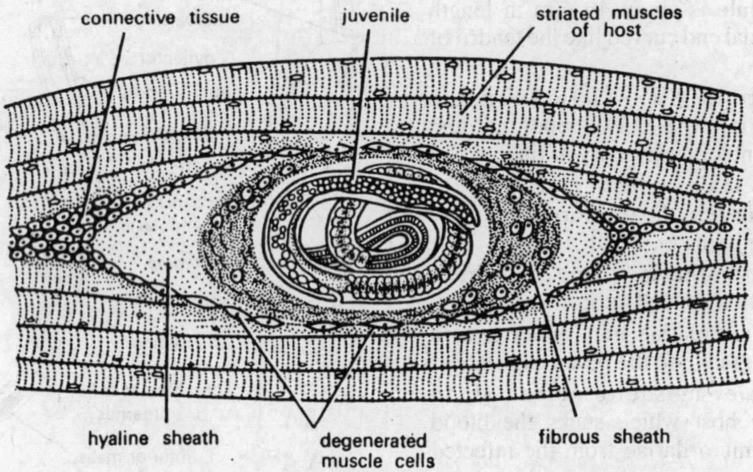

Fig. 11.37. Encysted larva of *Trichinella spiralis.*

11. The juvenile worms penetrate chiefly into the ends of skeletal muscles such as diaphragm and intercostals, etc., and form lemon-shaped cysts.

12. *Trichinella spiralis* causes a very serious and dreaded disease known as **trichinosis.**

Habit and habitat : *Trichinella spiralis* is an endoparasite in the intestine of man, pig, rat and other vertebrates and cosmopolitan in distribution.

38. Egg of *Ascaris*

Comments:

1. The egg of *Ascaris* (Fig. 11.39) is microscopic and oval in shape.

2. The eggs are yellowish or brownish in colour and surrounded by a tough, thick and transparent shell.

3. The shell is made of three layes: inner **lipoid layer**, middle **chitinous layer** and outer **protein layer**.

4. The protein layer is characteristically warty or coarsely mammillated.

5. The fertilized egg measures about 60–70 microns by 40–50 microns.

6. The fertilized eggs leave the females body through the gonopore and reach the host's intestine.

7. Due to high temperature and inadequate moisture and oxygen supply in the intestine, the eggs do not develop here and finally pass out with the faeces of the host.

8. Outside the host's body under suitable conditions of temperature, moisture and oxygen, the eggs develop into infective stage.

9. The optimum temperature for development is 85 °F.

10. The eggs may remain alive or viable in the soil for several years under advarse conditions.

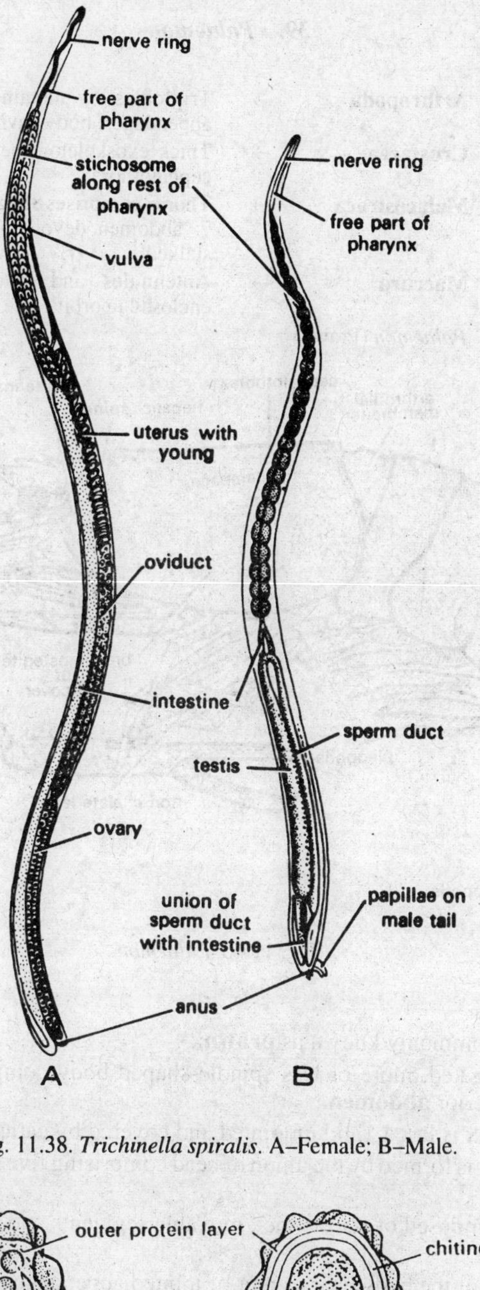

Fig. 11.38. *Trichinella spiralis*. A–Female; B–Male.

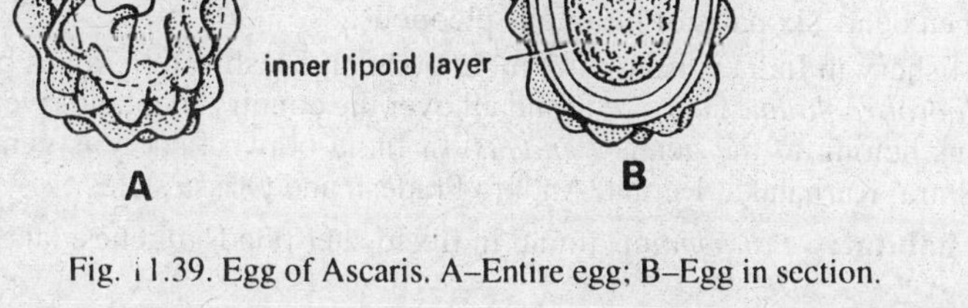

Fig. 11.39. Egg of Ascaris. A–Entire egg; B–Egg in section.

ARTHROPODA

39. *Palaemon*

Classification:

Phylum	...	**Arthropoda**	→	Triploblastic, metamerically segmented; jointed appendages: body cavity haemocoel.
Class	...	**Crustacea**	→	Thick exoskeleton; head fused with thorax to form cephalothorax.
Subclass	...	**Malacostraca**	→	Thorax comprises 8 segments and abdomen 6 or rarely 7; abdomen devoid of caudal styles: eyes usually stalked.
Suborder	...	**Macrura**	→	Antennules and antennae both large; eyes not enclosed in orbits.
Genus	...	*Palaemon* (Prawn)		

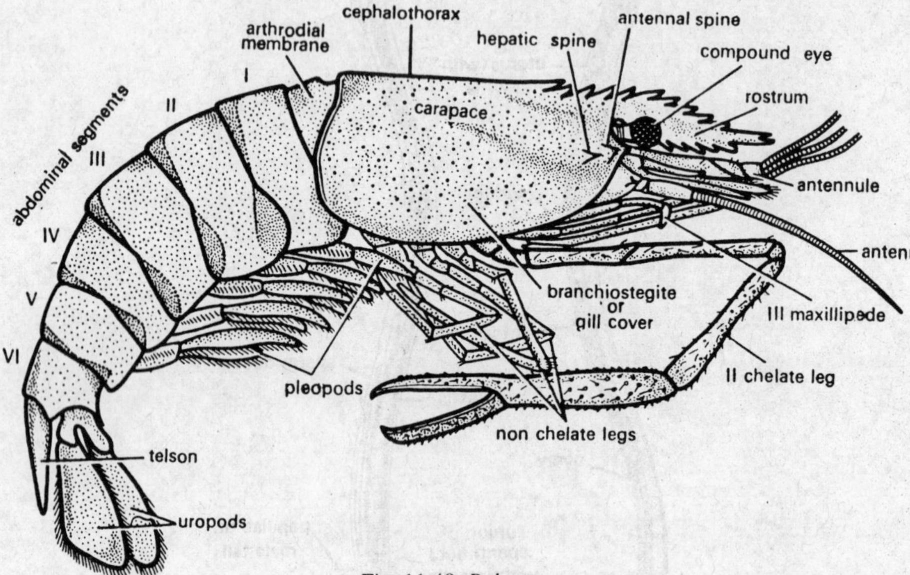

Fig. 11.40. *Palaemon* .

Comments:

1. *Palaemon* is commonly known as **prawn.**

2. It has an elongated, more or less spindle-shaped body composed of two parts, anterior **cephalothorax** and posterior **abdomen.**

3. **Cephalothorax** is large, rigid unjointed and covered by carapace.

4. Cephalothorax is formed by the union of head comprising five segments and eight segments of thorax.

5. **Abdomen** composed of six distinct, movable segments and a terminal conical piece, the **telson**.

6. Each segment of the body bears a pair of jointed appendages.

7. Thoracic appendages are three pairs of maxillipedes and five pairs of walking legs.

8. Abdomen bears six pairs of biramous pleopods.

9. Prawn fishery in India is becoming more popular. Freshwater prawns belong to the genus *Palaemon* and *Macrobrachium*. They are found all over the country in inland rivers and hill streams. All marine prawns belong to the genus *Penaeus*. In India prawn fishery is generally practiced in Gujarat, Maharashtra, Karnataka, Kerala, Andhra Pradesh and Orissa states.

Habit and habitat: *Palaemon* is found in freshwater ponds, ditches, lakes and rivers.

40. *Schistocerca gregaria* (Locust)

Classification:

Phylum	...	**Arthropoda**	→	Triploblastic; metamerrically segmented; jointed appendages; body cavity haemocoel.
Class	...	**Insecta**	→	Air breathing, terrestrial or aquatic; body divisible into head, thorax and abdomen; three pairs of legs; two pairs of wings.
Subclass	...	**Pterygota**	→	Wings usually present; abdomen devoid of appendages except genitalia and cerci; metamorphosis simple or complex.
Division	...	**Exopterygota**	→	Metamorphosis primitive or simple; young stages nymphs; wings develop externally.
Order	...	**Orthoptera**	→	Mouth parts strong biting and chewing; short or long: simple or segmented cerci.
Genus	...	*Schistocerca*		
Species	...	*gregaria*		

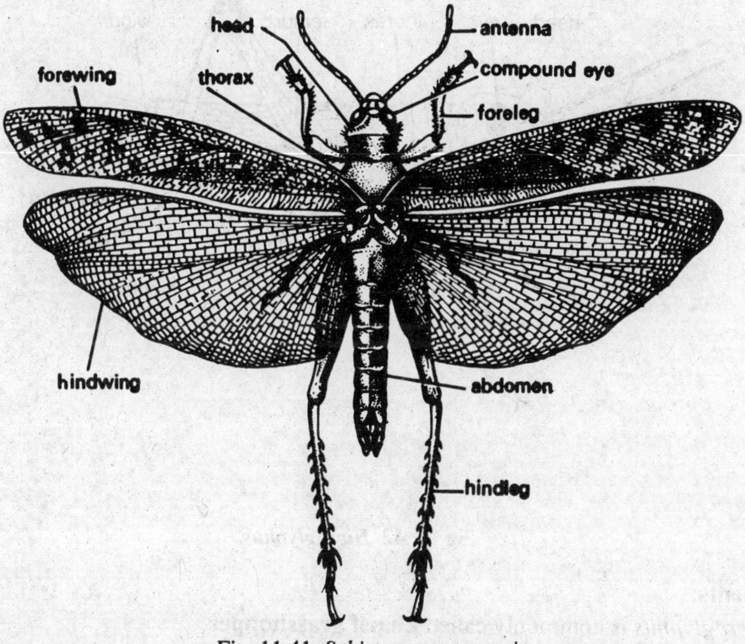

Fig. 11.41. *Schistocerca gregaria.*

Comments:

1. *Schistocerca gregaria* is commonly known as **locust.**
2. Body is divisible into **head, thorax** and **abdomen**.
3. **Head** bears short antennae and well developed compound eyes.
4. Mouth parts **biting** and **chewing type.**
5. **Thorax** is well developed and bears three pairs of legs and two pairs of dissimilar wings.
6. Fore-wings are elongated, thick, leathery and straight, while hind-wings are large, membranous.
7. **Abdomen** comprises ten segments with eleventh vestigial segment.
8. The locust is hervivorous in diet and gregarious in nature, migrating or swarming in great numbers
9. This locust is very destructive to crops, etc.

Habit and habitat: *Schistocerca gregaria* is found in the deserts in tropical regions of the world.

41. *Hieroglyphus*

Classification:

Phylum	...	**Arthropoda**	Characters same as those of *Schistocerca gregaria.*
Class	...	**Insecta**	
Subclass	...	**Pterygota**	
Division	...	**Exopterygota**	
Order	...	**Orthoptera**	
Family	...	**Acrididae**	→ Hind femora greatly swollen at base and elongated.
Genus	...	*Hieroglyphus*	

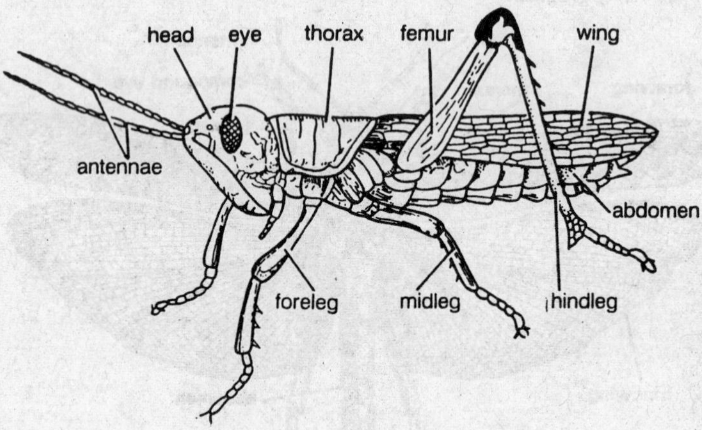

Fig. 11.42. *Hieroglyphus.*

Comments:

1. *Hieroglyphus* is commonly called **kharif grasshopper**.
2. Its two species *H. banian* and *H. nigrorepletus* are the major pests of poddy crops.
3. They are **polyphagous**, hence, also feed on sugarcane, maize, millets, jowar, bajra, sunhemp, arhar and grasses.
4. The adults are greenish, dry grass coloured, yellow hoppers.
5. *H. nigrorepletus* has three black lines running latrally on either side of the thorax, while *H. banian* possesses 2–3 black markings.
6. Female lays eggs in masses called **egg pod**, each containing 30–40 eggs, in the soil around rice fields from September to November.
7. Hatching of eggs takes place after few days of first shower in June or early July, attain adulthood within 8–10 weeks and start mating in September or early October.
8. Male dies after copulation, while female after laying eggs.

Nature of damage: The nymphs and adults cause enormous losses to the crop by chewing and cutting various parts of the plant like leaves, flowers, grain, etc., when their population is greatly high. They completely defoliate the plants leaving only the mid-ribs which affect the growth of the plant and finally resulting into low yield.

Control: Dusting and spraying of crop with BHC 5% @ 5 Kg per hectare; malathion 5% @ 20 Kg per hectare for dusting and @ 2.5 litre per hectare for spraying.

Distribution: Widely distributed in paddy growing areas of our country, especially in eastern Uttar Pradesh.

42. *Odontotermes* (Termite)

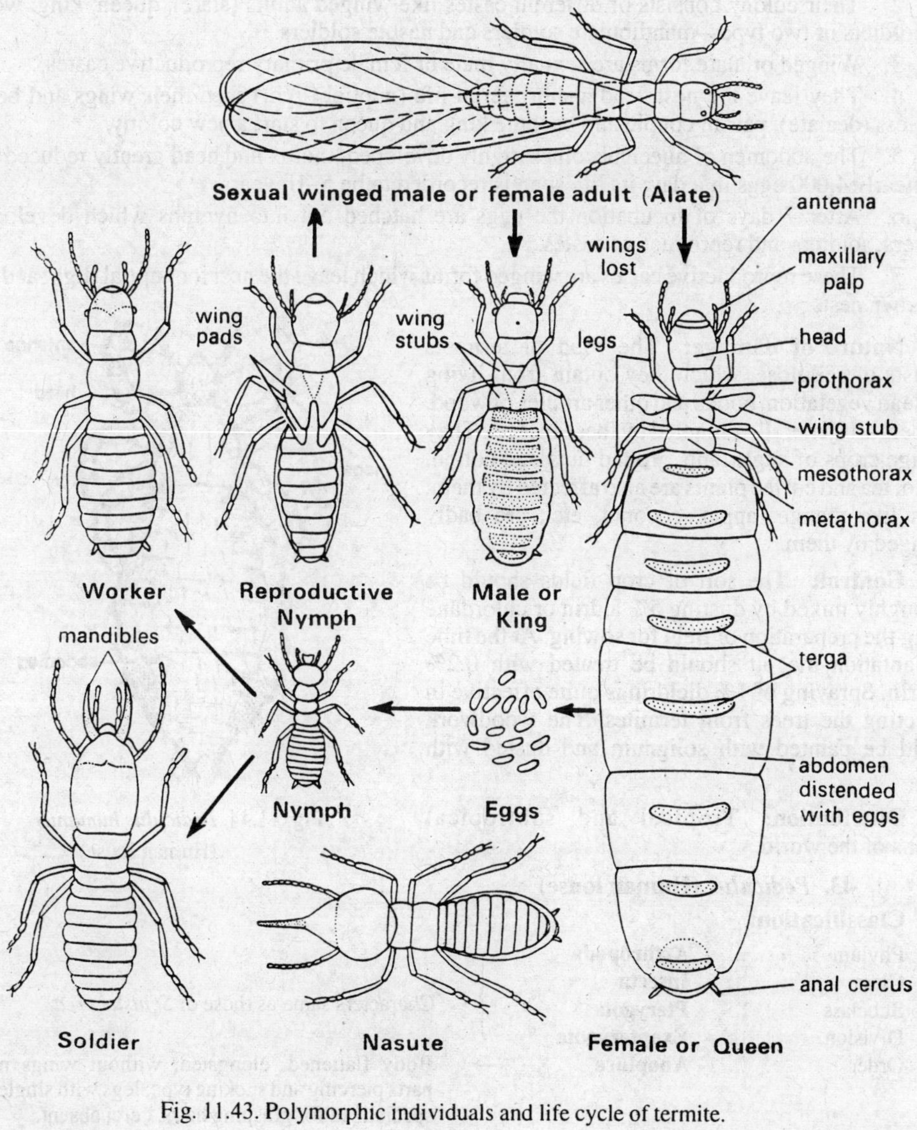

Fig. 11.43. Polymorphic individuals and life cycle of termite.

Classification:

Phylum	...	**Arthropoda**	Characters same as those of *Schistocerca gregaria.*
Class	...	**Insecta**	
Subclass	...	**Pterygot**	
Division	...	**Exopterygota**	
Order	...	**Isoptera**	→ Wings equal sized; social and polymorphic insects.

Family ... **Termitidae** → Wings more or less opaque, venation distinct at postrior portion, head either nasute or with toothed mandibles; workers present.

Genus ... *Odontotermes*

Comments:

1. Termites or while ants (Deemak) (Fig. 11.43) are social and polymorphic insects.
2. Their colony consists of different castes like winged adults (alate), queen, king, workers and soldiers of two types–mandibulate soldiers and nasute soldiers.
3. Winged or alate forms are sexually male or female primary reproductive castes.
4. They leave the nest, lead a short aerial life (nuptial flight) shed their wings and become wingless (dealate), pair in couple and become king and queen to start a new colony.
5. The abdomen of queen becomes highly developed, thorax and head greatly reduced and it lays nearly 4,000 eggs in a day; its life span is recorded to be 5–10 years.
6. After 7 days of incubation the eggs are hatched out into nymphs which develop into workers, soldiers and reproductive castes.
7. These reproductive castes are winged forms which leave the nest for nuptial flight and make their own nests.

Nature of damage: The food of termites consists of cellulose, which they obtain from living and dead vegetation, books and other articles of wood. Timbers are generally tunneled by these insects. They damage crops of sugarcane, ground nuts and cotton. Potato, tea and coffee plants are also affected by them. Trees like mango, apple, coconut, etc., are badly damaged by them.

Control: The soil of crop fields should be thoroughly mixed by dusting 5% aldrin or chlordane during the preparation of field for sowing. At the time of plantation the pit should be treated with 0.2% dieldrin. Spraying of 1% dieldrin is quite effective in protecting the trees from termites. The woodwork should be painted with solignum and dusted with BHC.

Distribution: Tropical and subtropical regions of the world.

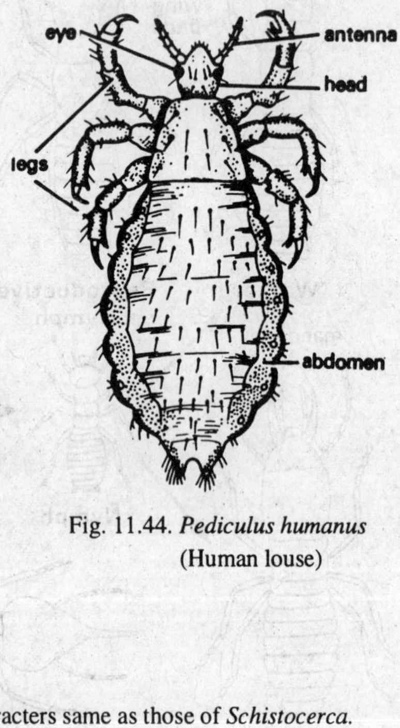

Fig. 11.44. *Pediculus humanus* (Human louse)

43. *Pediculus* (Human louse)

Classification:

Phylum ... **Arthropoda**
Class ... **Insecta**
Subclass ... **Pterygota**
Division ... **Exopterygota**
} Characters same as those of *Schistocerca*.

Order ... **Anoplura** → Body flattened, elongated, without wings mouth parts piercing and sucking type; legs with single claw specialized for grasping hairs, cerci absent.

Genus ... *Pediculus*

Species ... *humanus*

Comments:

1. *Pediculus* (Fig. 11.44) is commonly known as **human louse.**
2. Body is dorso-ventrally flattened and pale in colour with dark markings along the sides.
3. **Head** small bears a pair of compound eyes and a pair of five segmented antennae.
4. Mouth parts are **piercing** and **sucking type.**

5. Wings are absent.
6. Each leg bears a large curved claw adapted for clinging to the hairs of the host.
7. **Abdomen** is nine segmented.
8. In male, the posterior end is turned upwards and in female it does not turn upwards.
9. *Pediculus humanus capitis* is the head louse and *Pediculus humanus corporis* is the body louse.

Nature of damage: *Pediculus* is an ectoparasite of man. They suck the blood, cause annoyance, itching, irritation and transmit germs of certain diseases like relapsing fever, typhus fever, etc.

Control: Delousing should be done by using powder containing 2% malathion or 1% lindane. Delousing medicines like lycil or mediker should be used.

44. *Cimex*

Classification:

Phylum	...	**Arthropoda**	}	Characters same as those of *Schistocerca.*
Class	...	**Insecta**		
Subclass	...	**Pterygota**		
Division	...	**Exoptergota**		
Order	...	**Hemiptera**	→	Mouth parts piercing and sucking type; cerci absent; anal respiratory filaments present in some aquatic forms and anal filaments in male coccids.
Family	...	**Cimicidae**	→	Wings vestigial; clypeus triangular; Ocelli absent; ectoparasites of birds, bats, man.
Genus	...	*Cimex*		

Comments:

1. *Cimex* is commonly known as **bed bug.**
2. Body is dorso-ventrally flattened and oval, brownish in colour, measuring 5 mm in length.
3. **Head** is short and bears a pair of compound eyes and a pair of four segmented antennae.
4. Mouth parts are **piercing** and **sucking type**.
5. Mouth Parts in from of suctorial **rostrum** which lies bent under the head.
6. The rostrum consists of deeply grooved, jointed **labium** covered above by **labrum**. In the groove a pair of needle-like mandibles and maxillae are present.
7. Wings are absent.
8. **Abdomen** usually comprises seven segments.
9. Sexes are separate.
10. It is nocturnal and sucks human blood.

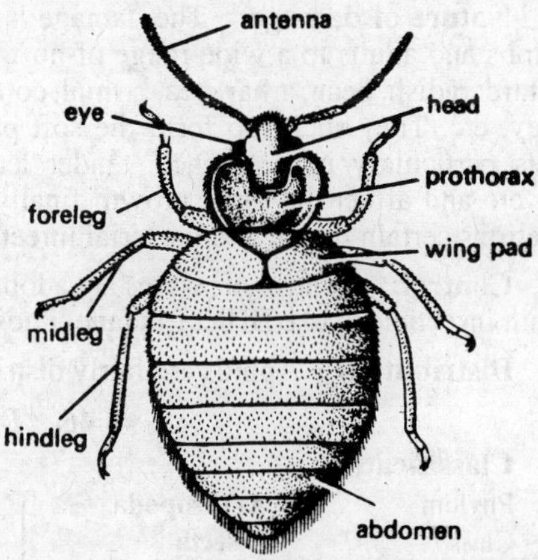

Fig. 11.45. *Cimex* (Bed bug).

Nature of damage: *Cimex* is an ectoparasite of man and prefers to live in various articles associated to human beings. They suck the blood, cause annoyance, irritation and transmit germs of various diseases like kala-azar, plague, typhoid, tuberculosis, etc., from man to man.

Control: The bug nuisance may be checked by pouring hot water on the furnitures, cots, etc. Spraying of kerosene oil, 5% toxaphene or 1 to 2% chlordane help in reducing bug poputation. Tick-20 is used for their effective control.

45. Aphid

Classification:

Phylum	...	**Arthropoda**	Characters same as those of *Cimex*.
Class	...	**Insecta**	
Subclass	...	**Pterygota**	
Division	...	**Exopterygota**	
Order	...	**Hemiptera**	
Family	...	**Aphididae**	→ Parthenogenetic oviparous females and usually also the males with functional mouth parts.
Type	...	Aphid	

Comments:

1. Aphid is commonly called as **plant lice**.
2. Body is short and stout, divisible into **head, thorax** and **abdomen**.
3. **Antennae** are quite large and reach about the middle of the body.
4. Compound eyes are well developed.
5. Mouth parts are **piercing** and **sucking type**.
6. Nine pairs of lateral abdominal spiracles are present.
7. Aphids are represented by winged or wingless forms.
8. They exhibit the phenomena of parthenogenesis.

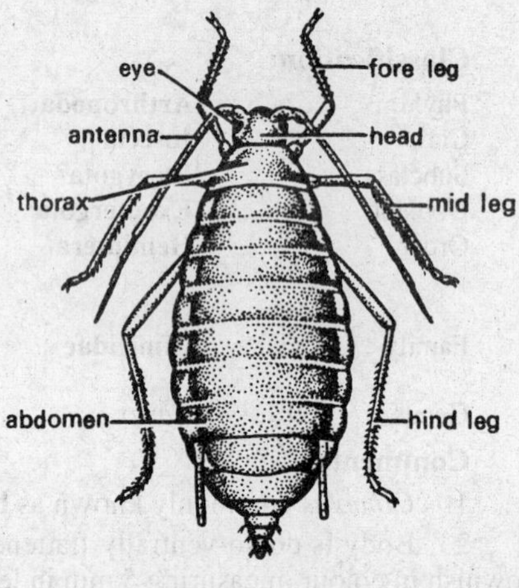

Fig. 11.46. Aphid.

Nature of damage: The damage is caused by nymphs and adults to a wide range of host plants like mustard, radish, bean, arhar, pea, brinjal, cotton, wheat, barley, etc. They suck sap form the soft parts of the plants particulary inflorescence, tender leaves, ears, pod, etc and affect the plant growth finally decreasing the yield of the crop. Due to their sticky secretions certain fungal and bacterial infections also occur.

Control: Control measures are adopted according the crop infested. Spraying and dusting of certain insecticides like BHC, Lindane, Phosphomidon, etc., help in controlling their nuisance.

Distribution. Aphids are fairly distributed throughout our country.

46. *Dysdercus koenigii*

Classification:

Phylum	...	**Arthropoda**	Characters same as those of *Cimex*.
Class	...	**Insecta**	
Subclass	...	**Perygota**	
Division	...	**Exopterygota**	
Order	...	**Hemiptera**	
Family	...	**Pyrrhocoridae**	→ Membrane with two large basal cells, emitting 7–8 branching veins.
Genus	...	*Dysdercus*	
Species	...	*koenigii*	

Comments:

1. *Dysdercus* (Fig. 11.47) is commonly known as **red cotton bug** or **red cotton stainer**.
2. It is a major pest of cotton but also infests lady's finger, maize and hollyhock.
3. It possesses **piercing** and **sucking** mouth parts.
4. It is conspicuous on account of its red colour with prominent dark brown or black marks on the forewings and white bands on abdomen.

5. Females lay eggs in clusters in the cracks of the soil near the plant which hatch out into nymphs within 7 days.

6. Nymphs feed often gregariously on full exposed bolls and undergo five moults to become adult.

Nature of damage: The damage is caused by both adults and nymphs by sucking the sap of the plant and bolls. Their excreta stain the tint; hence, the name cotton stainer. Thus, the cotton fibres are spoiled. The attacked seeds become worthless for sowing and oil extraction.

Control: Fields should be ploughed to expose the eggs to sun light. Dusting the infested crop with 10% BHC @ 25 kg per hectare is effective in controlling the pest. Spraying of 0.05% malathion is effective in controlling the pest.

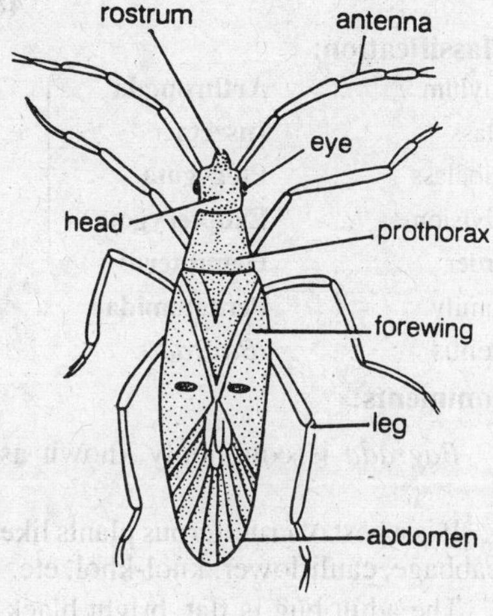

Fig. 11.47. *Dysdercus koenigii.*

Distribution: In our country it is a serious pest of cotton in U. P., Bihar, Andhra Pradesh, Tamil Nadu and Maharashtra.

47. *Leptocorisa varicornis*

Classification:

Phylum	...	**Arthropoda**	Characters same as those of *Cimex*.
Class	...	**Insecta**	
Subclass	...	**Pterygota**	
Division	...	**Exopterygota**	
Order	...	**Hemiptera**	
Family	...	**Coreidae**	→ Head short and narrower than the prothorax.
Genus	...	*Leptocorisa*	
Species	...	*varicornis*	

Comments:

1. *Leptocorisa* is commonly called **rice bug** or **gundhi bug** due to foul odour it emits.

2. It is a major pest of paddy but it also feeds on millets, jowar, maize, bajra, sugarcane and some grasses.

3. It is green or light brown in colour. It has long legs, 4 jointed antennae and 3 jointed tarsi.

4. The female bug lays oval eggs on the leaf blade, 10–20 in number in a single row. Eggs hatch into nymphs within 6–7 days which moult five times within 15–18 days to become adult.

Nature of damage: Both nymphs and adults suck the sap, with their piercing sucking mouth parts, when the ears are in milky stage and cause great loss to the yield.

Control: Dusting of 10% BHC or carbaryl at the flowering stage of crop gives effective control from this pest.

Distribution: Uttar Pradesh and Bihar are the worst affected regions of our country by this pest.

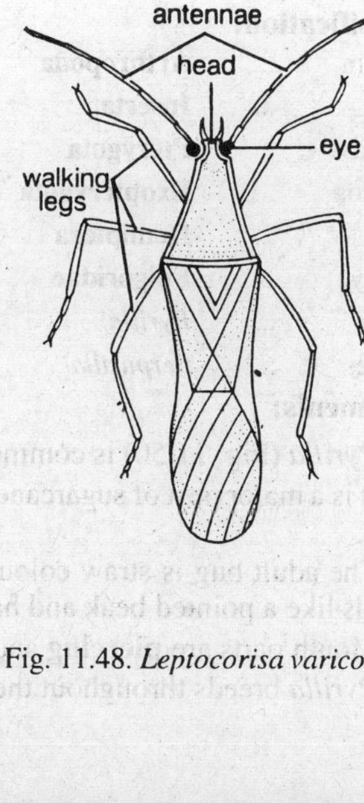

Fig. 11.48. *Leptocorisa varicornis.*

48. *Bagrada*

Classification:

Phylum	...	**Arthropoda**	Characters same as those of *Cimex*.
Class	...	**Insecta**	
Subclass	...	**Pterygota**	
Division	...	**Exopterygota**	
Order	...	**Hemiptera**	
Family	...	**Pentatomidae**	→ Scutellum very large; proboscis sheath 4-jointed.
Genus	...	*Bagrada*.	

Comments:

1. *Bagrada* is commonly known as **painted bug**.

2. It is a pest of cruciferous plants like mustard, turnip, cabbage, cauliflower, knol-khol, etc.

3. The adult bug is flat, bright black coloured with reddish yellow spots and dots.

4. The mouth parts are **piercing** and **sucking** type

5. The female bug lays 90–220 eggs, usually singly and they hatch into nymphs within 7 days. The nymphs moult four times within 20–22 days and become adult.

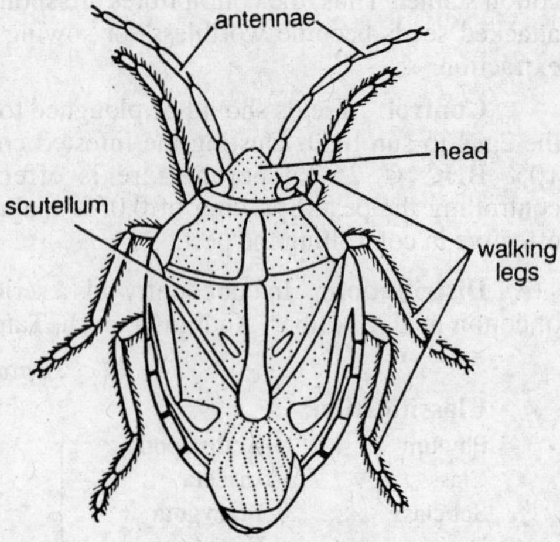

Fig. 11.49. *Bagrada*.

Nature of damage: The damage is caused by nymphs and adults both by sucking sap from the plant upon which they are seen clustering.

Control: The bugs may be dislodged from the plants and killed in kerosene water. Dusting of 5% malathion or 10% carbaryl dust @ 25 kg per hectare to the infested crop is an effective control measure.

Distribution: It is most active from February to April in Uttar Pradesh and widely distributed in Arabia, Burma, Iraq, Sri Lanka and East Africa.

49. *Pyrilla perpusilla*

Classification:

Phylum	...	**Arthropoda**	Characters same to those of *Cimex*.
Class	...	**Insecta**	
Subclass	...	**Pterygota**	
Division	...	**Exopterygota**	
Order	...	**Hemiptera**	
Family	...	**Fulgoridae**	→ Anal area of hind wing reticulate with many cross veins.
Genus	...	*Pyrilla*	
Species	...	*perpusilla*	

Comments:

1. *Pyrilla* (Fig. 11.50) is commonly called **sugarcane leafhopper**.

2. It is a major pest of sugarcane but it also attacks maize, wheat, barley, bajra, oats, sorghum, etc.

3. The adult bug is straw coloured, wings two pairs folded roof-shaped on the back and its head extends like a pointed beak and has red eyes.

4. Mouth parts are **piercing** and **sucking** type.

5. *Pyrilla* breeds throughout the year.

6. The female lays eggs in clusters, each cluster having 300– 500 eggs on the lower surface of the leaves or inside the leaf-sheaths. The eggs hatch in 7 to 22 days into dirty white coloured nymphs with a pair of anal tufts. Nymphs moult 5 times to become adult.

Nature of damage: The damage is caused by the nymphs and adults both by sucking the sap from the leaves. As a result, the leaves turn pale yellow and dry up. They secrete sweat sticky honeydew on which black shooty mould grows. Due to this, the plant suffers in growth, quantity and quality of juice deteriorates and the yield of the crop is affected adversly.

Control: Dusting of 5% BHC or spraying of 0.03% monocrotophos, dimethoate, phosphamidon or methyl demeton is effective in the control of this pest.

Distribution: It is distributed throughout India especially in U. P., Madhya Pradesh, Bihar and Maharashtra.

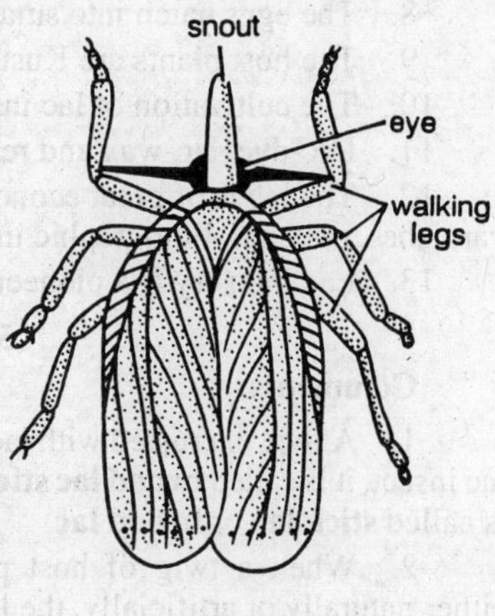

Fig. 11.50. *Pyrilla perpusilla.*

50. *Tachardia lacca*

Classification:

Phylum	...	**Arthropoda**	Characters same as those of *Cimex*.
Class	...	**Insecta**	
Subclass	...	**Pterygota**	
Division	...	**Exopterygota**	
Order	...	**Hemiptera**	
Family	...	**Lacciferidae** →	Body enclosed in a mass of resinous cells, each cell with 3 openings; adult female legless.
Genus	...	*Tachardia*	
Species	...	*lacca*	

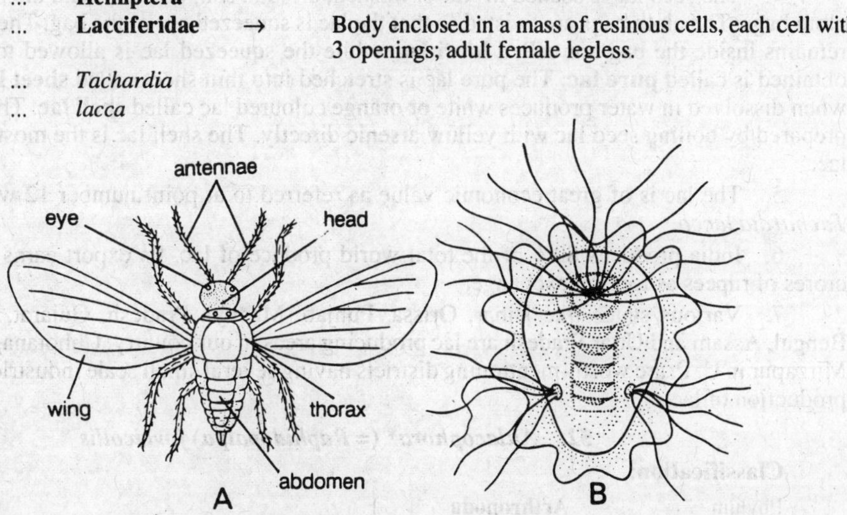

Fig. 11.51. *Tachardia lacca.* A–Male; B–Female.

Comments:

1. *Tachardia* is a small **scale insect** commonly called **lac insect**.
2. It is a useful insect secreting **lac or lakh**.
3. Sexual dimorphism well distinct.
4. Male is red coloured, small-sized bearing 10 segmented abdomen.
5. Female is larger having pyriform body and enclosed in resinous cells. Head, thorax and abdomen indistinct.
6. Mouth parts **piercing** and **sucking** type.
7. A mature female lays 200–500 eggs and dies after oviposition.

8. The eggs hatch into small active nymphs having a sap-sucking proboscis.
9. The host plants are Kusum, Ber, Babul, Sal, Palas, Peepal, Gular, etc.
10. The cultivation of lac insect for obtaining lac is called **lac culture**.
11. Lac-dye, lac-wax and resinous lac are obtained from this insect.
12. The lac is of great economic value as it is used in making toys, bracelets, bangles, paints, varnishes, polishes, lithographic ink, sealing wax, gramophone records and insulating material, etc.
13. Lac industry is a pioneer industry of India.

51. A twig encrusted with lac

Comments:

1. A twig encrusted with lac also bears eggs of lac insect, it is called **brood lac stick** and the lac alone is called **stick lac** or **brood lac**.
2. When a twig of host plant is inoculated either naturally or artificially, the larvae start sucking the sap of the twig and secreting lac which gets encrusted over the twig.
3. The encrusted lac from the twig is scrapped. The scrapped lac is then processed for removing various impurities like dead parts of lac insect, eggs colouring matter, etc., then crushed by mortars and obtained granules are air dried. It is then called **seed lac**.

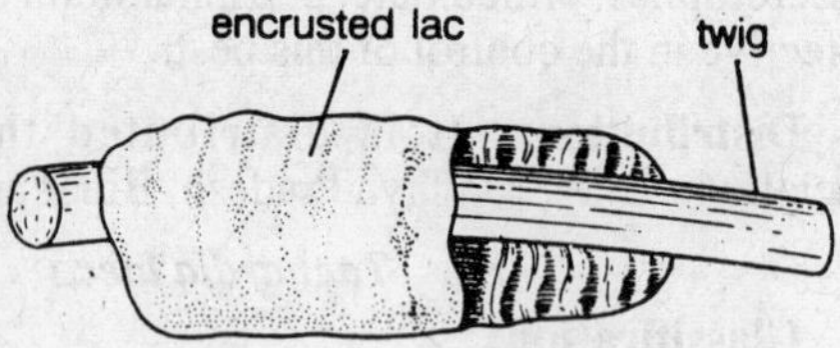

Fig. 11.52. A twig encrusted with lac.

4. The seed lac is soaked in water, washed, dried in sun, bleached and then heated to melt in cloth bags. The cloth bags are twisted so that the lac is squeezed out of the bag. The substance which remains inside the bag are called **kirri lac,** while the squeezed lac is allowed to cool. Lac, thus, obtained is called **pure lac**. The pure lac is stretched into thin sheet called **sheet lac**. The sheet lac when dissolved in water produces white or orange coloured lac called **shell lac**. The shell lac is also prepared by boiling seed lac with yellow arsenic directly. The shell lac is the most purified form of lac.
5. The lac is of great economic value as referred to at point number 12 while commenting *Tachardia lacca.*
6. India produces 80% of the total world produce of lac. Its export earns us more than 10 crores of rupees as foreign exchange.
7. Various states like Bihar, Orissa, Punjab, Madhya Pradesh, Gujarat, Rajasthan, West Bengal, Assam and Uttar Pradesh are lac producing areas in our country. Chhotanagpur in Bihar and Mirzapur in U. P. are worth-mentioning districts having several small scale industries involved in the production of lac.

52. *Aulacophora** (= *Raphidopalpa*) *foveicollis*

Classification:

Phylum	...	**Arthropoda**	}	Characters same as those of *Schistocerca gregaria.*
Class	...	**Insecta**		
Subclass	...	**Pterygota**		
Division	...	**Endopterygota**	→	Metamorphosis complete; wings develop internally.
Order	...	**Coleoptera**	→	Leathery integument; mouth parts biting and chewing type; wings two pairs, fore-wings or elytra is useless in flight; cerci absent.
Family	...	**Galerucidae**	→	Hind femora not swollen; tibiae never grooved; cucumber beetles.
Genus	...	*Aulacophora*		
Species	...	*foveicollis*		

* *Aulacophora* has been described under differrent generic names but *Raphidopalpa* is the most common one.

Comments:

1. *Aulacophora foveicollis* is commonly called **red pumpkin beetle**.
2. It is the most common pest of cucurbitaceous vegetables like pumpkin, ghia, tori, tinda, cucumber, melon, etc.
3. It is orange-red coloured oblong beetle.
4. It bears **biting** and **chewing** type of mouth parts.
5. Female lays orange coloured oval or long eggs nearly 60 to 300 in number either singly or in a group of 8–9 eggs near roots.
6. The eggs hatch into small **grubs** within 6–15 days which feed on the roots and undergo 4 moults within 13–25 days to become full grown grub which pupate in soil. After 7–17 days, adult beetle emerges from pupa and begins to feed on the host plant.

Nature of damage: The damage is caused both by beetles and grubs. The beetles feed on leaves, flowers and buds by cutting, chewing and making holes. In case of severe infestation plants are completely defoliated. The grubs bore into the roots, underground portion of stem and fruits that come in contact with the ground. Young seedlings are the worst effected.

Control: Use of 5% aldrin against grubs in the soil and dusting of 4% carbaryl or spraying of 0.05% malathion on the foliage part of the plants are effective.

Distribution: The north-western part of our country is badly affected by this pest. It is widely distributed throughout Asia, Australia, Africa and Southern Europe.

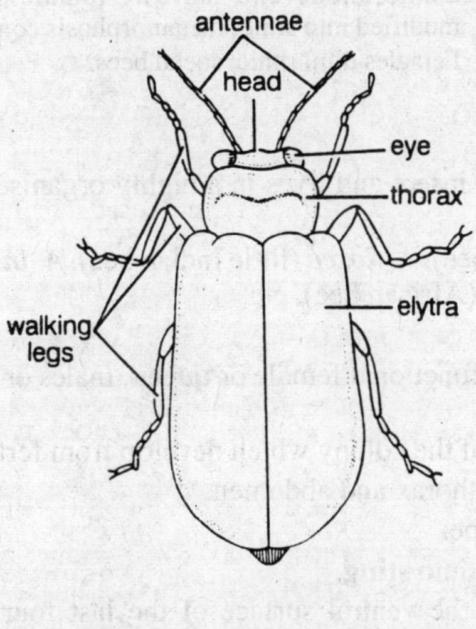

Fig. 11.53. *Aulacophora foveicollis.*

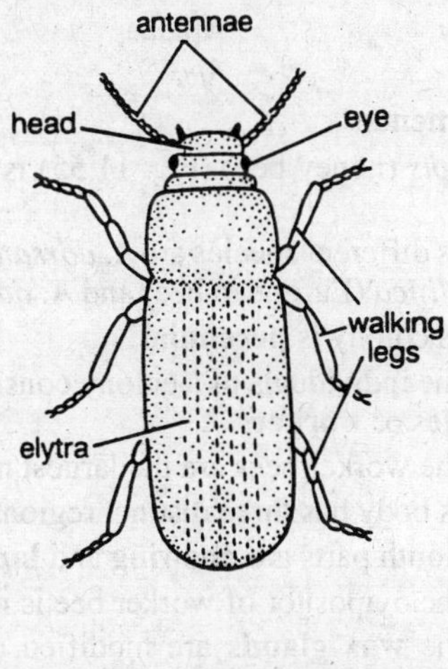

Fig. 11.54. *Tribolium.*

53. *Tribolium*

Classification:

Phylum	...	**Arthropoda**	Characters same as those of *Aulacophora*.
Class	...	**Insecta**	
Subclass	...	**Pterygota**	
Division	...	**Endopterygota**	
Order	...	**Coleoptera**	
Family	...	**Tenebrionidae** →	Antennae simple, clavate, moniliform, short; tarsi heteromerous; wings vestigial or absent; elytra immovably soldered along the middle.
Genus	...	*Tribolium*	

Comments:

1. *Tribolium* (Fig. 11.54) is commonly called **red flour beetle**.
2. It commonly attacks wheat flour but flour of all starchy materials are attacked.
3. Adults are very active. A fertilized female lays nearly 1000 eggs in the flour or grains. The eggs hatch into larvae in about 10 days which moult 6 or 7 times and then pupate. The adults emerge within 5–9 days.

Nature of damage: Adult and larval both stages are damaging. They damage flour in flour mills. Sometimes grain in store house are also attacked but they attack only those grains which are already damaged partly by other pest infestation. When severely damaged the flour changes into greyish colour which becomes unfit for human consumption.

Control: Preventive measures are the only way so that the pest does not reach to the flour mills, store houses or godowns. For this, the storing places should be clean and dry, and the infested flours should be immediately removed.

54. *Apis* (Honey bee)

Classification:

Phylum	...	**Arthropoda**	}	Characters same as those of *Aulacophora.*
Class	...	**Insecta**		
Subclass	...	**Pterygota**		
Division	...	**Endopterygota**		
Order	...	**Hymenoptera**	→	Social or parasitic insects; mouth parts biting, chewing, lapping and sucking; wings two pairs, membranous and narrow; female's ovipositor modified into sting; metamorphosis complex.
Family	...	**Apidae**	→	Females dimorphic; social bees.
Genus	...	*Apis*		

Comments:

1. *Apis* (honey bee) (Fig. 11.55) is a social insect and lives in a highly organised colony in bee hive.
2. Its different species are *A. dorsata* (rock bee), *A. florea* (little Indian bee), *A. indica* (Indian bee), *A. mellifica* (European bee) and *A. adamsoni* (African bee), etc.
3. Its colony is **perennial**.
4. The individuals of a colony consists of a functional female or **queen**, males or **drones** and sterile females or **workers**.
5. The worker bees are the largest number of the colony which develop from fertilized eggs.
6. Its body has three distinct regions; head, thorax and abdomen.
7. Mouth parts are **chewing** and **lapping** type.
8. The ovipositor of worker bee is modified into **sting**.
9. The **wax glands** are modified cells on the ventral surface of the last four abdominal segments.
10. They do all the duties of the colony relating to food collection, bringing nectar, secreting wax, looking after the youngs, building and cleaning the hive, etc.
11. The **queen** is larger in size having longer abdomen and its duty is to lay eggs only.
12. The **drones** are the male members of the colony, without stings and develop from unfertilized eggs. Their duty is to copulate with the queen.
13. The honey bees are economically important insects as they are the source of honey and bees wax.
14. Bee keeping is called **apiculture**.

55. Honey bee–Stages of life cycle

Comments:

1. The various stages in the life cycle of honey bee (Fig. 11.56) are egg, larva, pupa and adults which include worker, drone and queen.

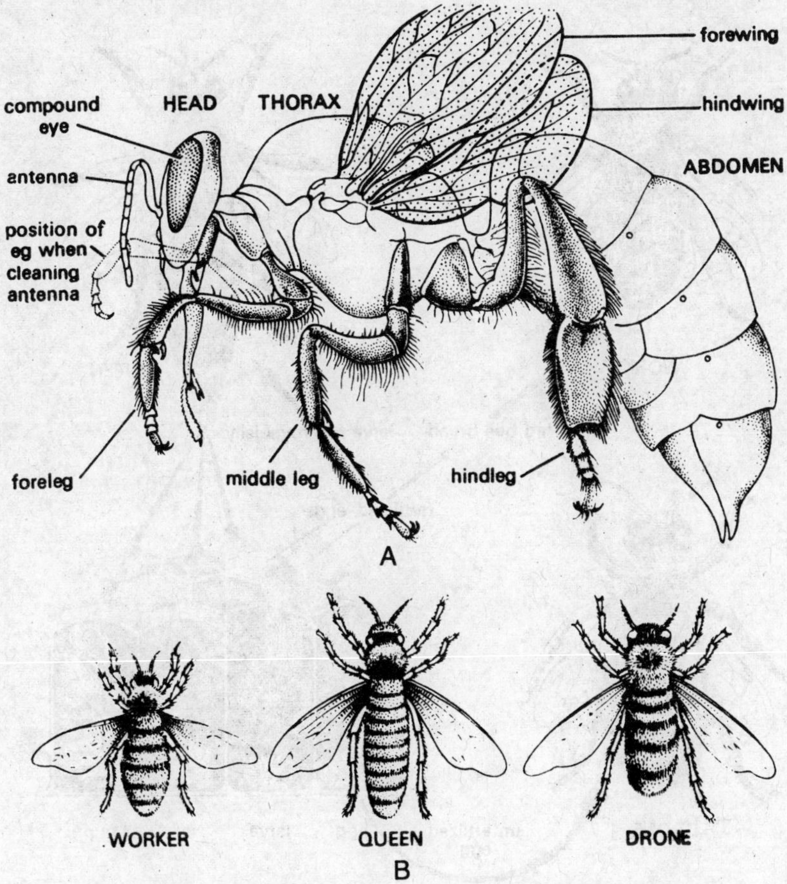

Fig. 11.55. A–Worker honey bee in lateral view; B–Castes of honey bee.

2. **Egg stage.** The queen takes a nuptial flight or mating flight with a number of drones. Copulation occurs in air, drones die after copulation. Queen returns back to old hive and starts laying egg at the rate of 100 to 200 eggs per day for about five years. The eggs are pinkish, bean-shaped and both fertilized as well as unfertilized. It usually lays one egg in one brood cell or chamber of the hive and later seals it.

3. **Larva.** After 3 days eggs hatch into larvae. A larva has head and segmented body containing spiracles. The larvae develop into queen or other castes of the colony depending upon the food they take; larvae fed on royal jelly develop into queen and those fed on honey and pollen develop into workers. The drones develop from the larvae which are derived from unfertilized eggs.

4. **Pupa.** Larva transforms into pupa. The pupa has mouth parts, eyes, rudiments of wings and walking legs. Its body is divisible into head, thorax and abdomen. After complete metamorphosis the pupa changes either into a queen in 13 days or worker in 18 days or drone in 21 days.

5. **Worker.** Smallest in size than the other castes. Body divisible into head, thorax and abdomen. Blackish or brownish in colour. It does all the duties in the colony, say collects nector, secretes wax, keeps the colony clean, defence, etc.

6. **Drone.** Intermediate in body size having large compound eyes. They copulate with the queen. Die after copulation.

7. **Queen.** Largest in body size. It is fertile female. Its job is to lay eggs only.

56. Legs of honey bee

Comments:

1. In honey bee all the three segments of thorax contain one pair of legs.

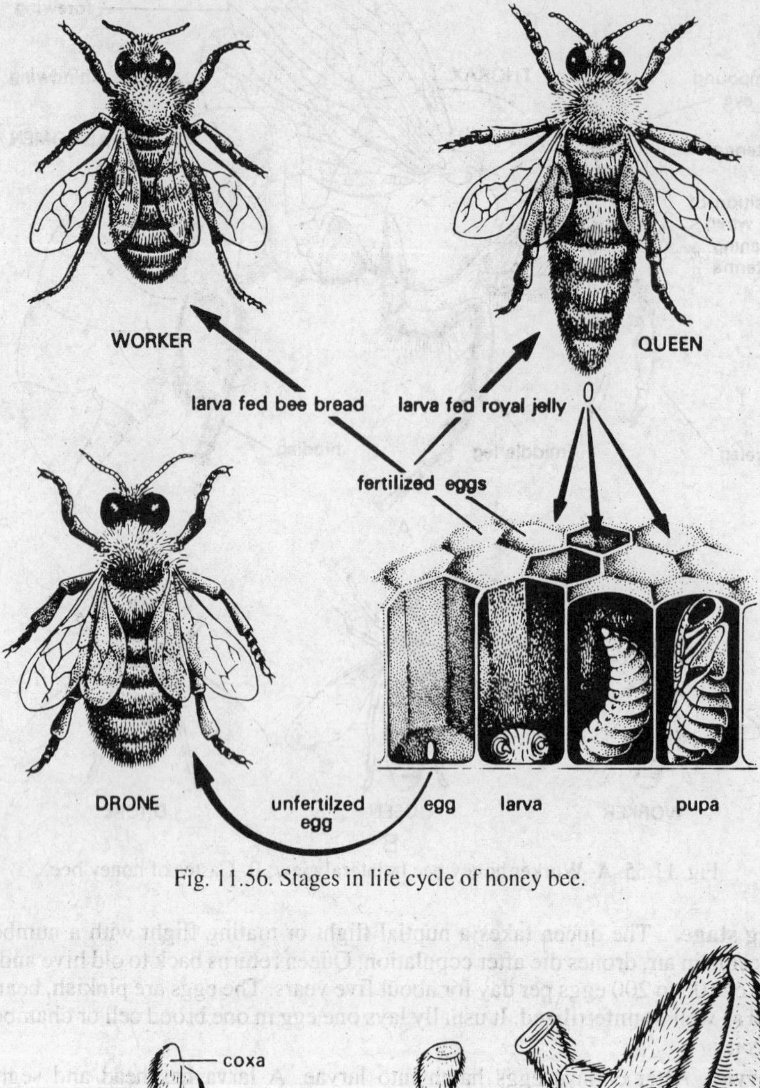

Fig. 11.56. Stages in life cycle of honey bee.

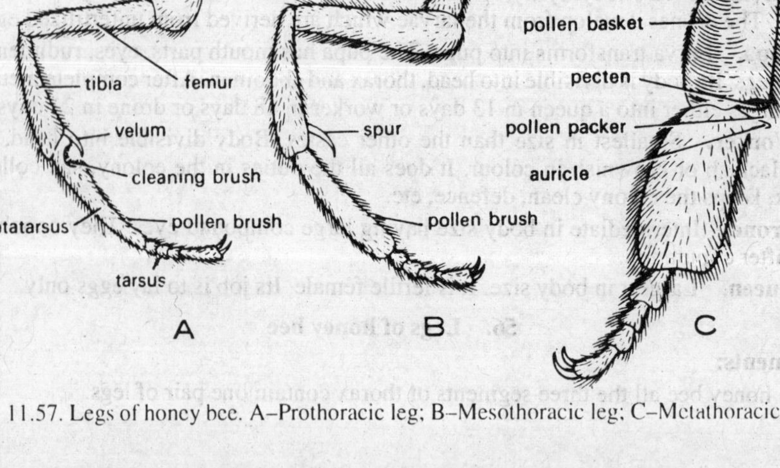

Fig. 11.57. Legs of honey bee. A–Prothoracic leg; B–Mesothoracic leg; C–Metathoracic leg.

2. Each leg is made of five parts: **coxa**, **trochanter**, **femur**, **tibia** and 5-jointed **tarsus** terminating into a pair of **claws** and **pulvillus**.

3. The **prothoracic leg** has a row of stiff bristles on tibia forming **eye brush** for cleaning the compound eyes, at the distal end of tibia is a movable spine called **velum** which can close over a notch on the tarsus to form an **antenna comb** through which the antenna is drawn for cleaning. Long bristles on the tarsus form a **pollen brush** for removing from the front part of the body.

4. The **mesothoracic leg** has a **pollen brush** on the tarsus, the end of tibia has a **spur-like spine** for removing pollen from the pollen basket and wax from abdomen.

5. The **metathoracic leg** has a large tibia with a cavity called **pollen basket** or **corbicula** used for storing pollen. At the distal end of tibia a row of stiff bristles called **pecten** below which a flat plate called **auricle** are found. The pecten and auricle form a **wax pincher** for removing wax from the abdomen. The outer surface has pollen brush and inner surface has pollen comb having rows of stiff spines; the pollen comb removes pollen from the body and fills it in the pollen basket.

Note: *The legs described above relate to the legs of worker bee only. In others, the legs are made of usual five parts with no specializations.*

57. Mouth parts of honey-bee

Comments:

1. Mouth parts of honey-bee are modified for **chewing** and **lapping** purposes.

2. Mouth parts consist of labrum, mandibles, maxillae and labium.

3. **Labrum** is small lies beneath the clypeus and fleshy epipharynx projects below the labrum.

4. **Mandibles** are smooth and spatulate situated on either side of the labrum.

5. Each maxilla consists of small cardo and stipes, vestigial maxillary palp and long blade-like galea. Lacinia is absent.

6. **Labium** consists of an elongated prementum and small mentum.

7. **Prementum** bears four segmented labial palpi, paraglossae and glossae.

8. **Paraglossae** are greatly reduced, while the **glossae** are hairy, greatly elongated and united forming the ligular tongue which terminates in a small lobe called **flabellum**.

9. Mouth parts of honey bee are modified for collecting nectar and pollens of flowers and moulding the wax.

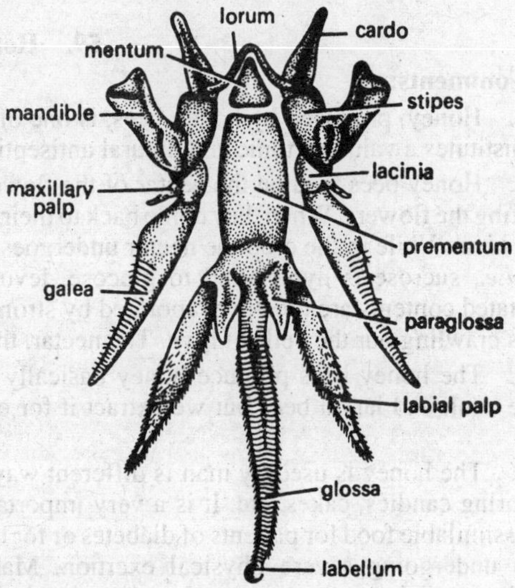

Fig. 11.58. Mouth parts of honey bee.

58. Sting of honey bee

Comments:

1. Sting of honey bee is a complicated structure.

2. It consists of two stylets articulated along their length to the hollow stylet sheath by a groove and rail arrangement.

3. The stylets are held in place by this arrangement and they can move only up and down.

4. The stylets and their sheaths bear barbs at the tips to make a wound.

5. Proximally the stylet sheaths have a dilated bulb.

6. Attached to the stylets proximally is median **poison sac** into which opens two **acid glands** and one **alkaline gland.**

7. The sting has a set of three chitinous plates on either side to serve as levers. The plates are (i) **oblong plate**, (ii) **triangular plate** and (iii) **quadrangular plate**..

8. In stinging, the muscles of plates drive the stylets and stylet sheath into a victim, the secretions of two types of glands mix and pass down the poison canal into the wound.

9. The sheath serves to guide the darts to open up the wound and to aid in conducting the poison.

10. The poison is secreted in acid and alkaline glands and stored in a **poison sac.**

11. The sting is a modified ovipositor, hence, found in queen and female worker bees only; absent in drones.

12. Queen bee uses it many times for stinging rival queens, while worker bees use their sting as a last resort because this always results in their death.

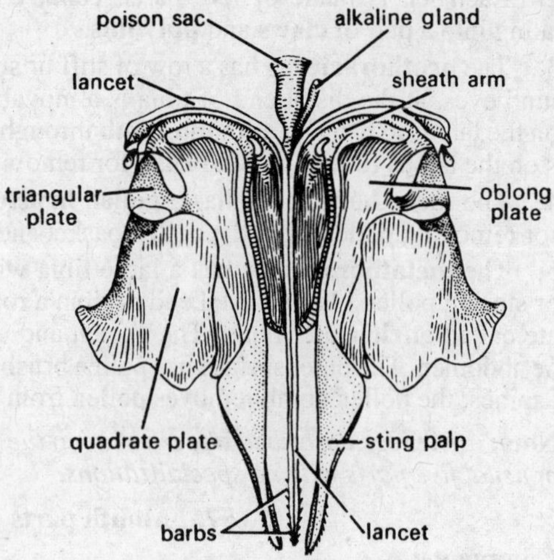

Fig. 11.59. Sting of honey bee.

59. Honey

Comments:

1. Honey, produced by honey bees, is one of the most important gifts to mankind by nature as it constitutes a valuable tonic and natural antiseptic for human body.

2. Honey bees feed on the nectar of the flowers of plants and store in their crops while they are visiting the flowers. When they come back to their hive, they regurgitate the nectar in the chambers of their hive. While in the crop the nectar undergoes some chemical changes due to its mixing with saliva, *i.e.,* sucrose is hyprolysed to glucose, levulose and fructose. The water contents of the regurgitated contents are mostly evaporated by strong current of air created by fanning of wings by the bees crawling our the cells of hive. The nectar, thus, ripens and becomes **honey.**

3. The honey bees produce honey basically to feed the adult and larval bees but we extract it for our use.

4. The honey is used by man is different ways; in preparing candies, cakes, etc. It is a very important easily assimilable food for patients of diabetes or for the persons undergoing severe physical exertion. Many ayurvedic medicines are diagnosed to be taken with honey.

5. The food value of honey can be estimated by the fact that 1 kg of honey is equal to 7 kg potatoes or 9 kg grapes or 6 kg bananas or 15kg pear or 10kg apples or 14kg of peaches.

6. The composition of honey is: levulose 38.19%, dextrose 21.28%, maltose and other sugars 8.81%, enzymes and pigments 2.21%, ash 1.0% and water 17.20%.

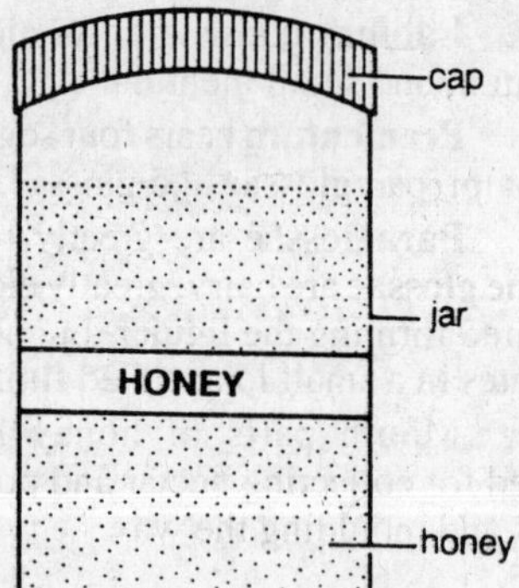

Fig. 11.60. Honey.

60. Bees-wax

Comments:

1. Bees-wax is a by-product of apiculture.
2. It is yellowish or greyish brown in colour.

3. It is insoluble in water but completely soluble in organic solvents like ether.

4. It is a secretion secreted by the wax glands in the abdominal segments of worker bees only.

5. It is chemically a mixtine of alcohols, fatty acids and paraffins.

6. It is used as important bases for industries concerned with manufacture of toilet goods and cosmetics.

7. It is also used in making candles, shaving creams, polishes, castings of models, carbon paper, crayons, etc.

61. *Corcyra cephalonica*

Classification:

Phylum	...	**Arthropoda**	→	Triploblastic; metamerically segmented; jointed appendages; body cavity haemocoel.
Class	...	**Insecta**	→	Air-breathing, terrestrial or aquatic; body divisible into head, thorax and abdomen; three pairs of legs; two pairs of wings.
Subclass	...	**Pterygota**	→	Wings usually present; abdomen devoid of appendages except genitalia and cerci; metamorphosis simple or complex.
Division	...	**Endopterygota**	→	Metamorphosis complex (complete); young stages known as larvae; wings develop internally.
Order	...	**Lepidoptera**	→	Body and wings completely covered by flat overlapping scales and hairs; antennae often clavate or serrate; hooked or knobbed or plumose; mouth parts siphoning type with long coiled proboscis; larvae caterpillars.
Family	...	**Galleriidae**	→	Labial palpi short in male, long in female; proboscis absent.
Genus	...	*Corcyra*		
Species	...	*cephalonica*		

Comments:

1. *Corcyra cephalonica* (Fig. 11.61) is commonly called **rice moth**.

2. It is the major pest of rice but it also feeds on maize, gram, sorghum, ground nut and cotton seeds.

3. It is a stored grain pest causing severe damage to rice and wheat.

4. The fertilized female moth lays nearly 150 eggs which after 2 to 15 days of incubation period hatch into small caterpillars which attain full grown stage within 30 to 35 days and then pupate in the form of cocoons. After 4 to 35 days of pupal period adult moths emerge.

Nature of damage: The small caterpillars are responsible for damage in rice grain and maize. In case of heavy infestation the stored stock of grains is converted into silken tubes to which grains adhere and give foul unpleasant smell. The grains become unfit for human consumption.

Control: Grains should be sterilized before storage. Godowns and store houses should be clean, dry and moistureless. Fumigation of store houses and godowns by carbon tetrachloride and carbon bisulphate before storage is done.

Distribution: *Corcyra cephalonica* is widely distributed throughout Asia including India and Pakistan, Europe and North America.

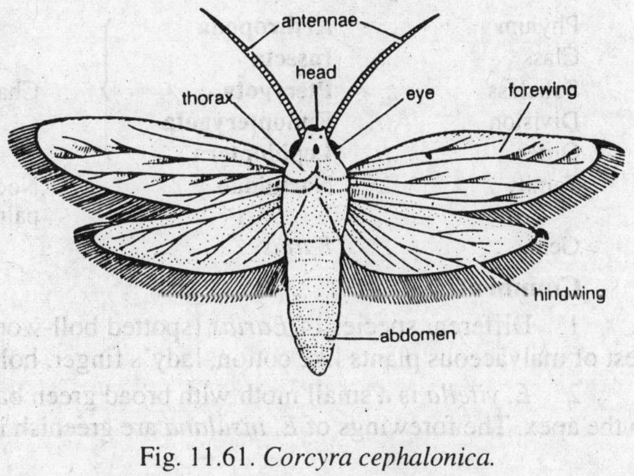

Fig. 11.61. *Corcyra cephalonica.*

62. *Chilo*

Classification:

Phylum	...	**Arthropoda**	Characters same as those of *Corcyra cephalonica*.
Class	...	**Insecta**	
Subclass	...	**Pterygota**	
Division	...	**Endopterygota**	
Order	...	**Lepidoptera**	
Family	...	**Crambidae**	→ Labial palpi long in both the sexes, moliniform; hind wing with the cell open.
Genus	...	*Chilo*	

Comments:

1. Two species of *Chilo* are very common; *C. infuscatellus* (sugarcane shoot borer) and *C. partellus* or *zonellus* (jowar and maize stem borer).

2. The mouth parts of the adult moths are **siphoning type**, while those of their caterpillars are **biting** and **chewing** type.

3. *Chilo infuscatellus* is a major pest of sugercane but it also feeds on bajra, maiza, etc.

4. The female moth lays eggs up to 300–400 in number in clusters of 10–35 eggs on the lower surface of sugarcane leaves. The eggs hatch into small caterpillars within 4–5 days, feed voraciously and become full grown within 3–4 weeks and then pupate. The adults emerge from pupae after 5–7 days of pupation. Adult moths live only for 2–4 days.

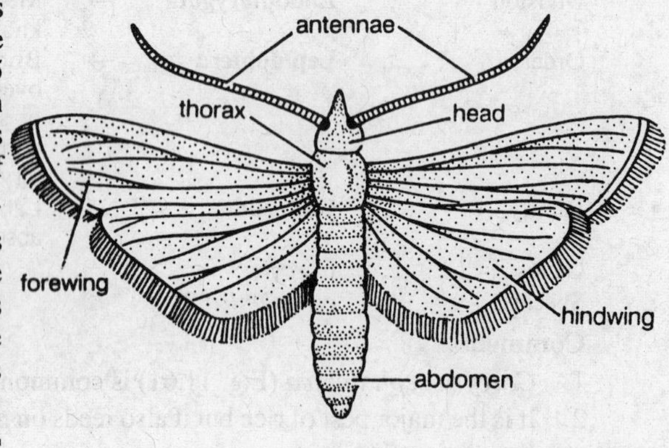

Fig. 11.62. *Chilo*.

Nature of damage: The caterpillars bore inside the shoot of sugarcane with the help of their biting and chewing mouth parts and cause **dead-heart**. They attack sugarcane when the plant is in young stage. Adult moths are non-damaging.

Control: Spraying of 10% BHC mixed with soil twice during April to June can control the pest. Attacked shoots should be removed from the field and destroyed.

Distribution: The pest is widely distributed in sugarcane growing areas of India, Pakistan, Burma, Afghanistan, Taiwan and Philippines.

63. *Earias*

Classification:

Phylum	...	**Arthropoda**	Characters same as those of *Corcyra cephalonica*.
Class	...	**Insecta**	
Subclass	...	**Pterygota**	
Division	...	**Endopterygota**	
Order	...	**Lepidoptera**	
Family	...	**Noctuidae**	→ Nocturnal moths; proboscis present rarely atrophied; palpi well developed; frenulum present.
Genus	...	*Earias*	

Comments:

1. Different species of *Earias* (spotted boll-worm), *e.g.*, *E. insulana*, *E. vitella*, etc., are the pest of malvaceous plants like cotton, lady's finger, hollyhock, shoe flower, etc.

2. *E. vitella* is a small moth with broad green band on the forewing extending from the base to the apex. The forewings of *E. insulana* are greenish in colour.

3. The female moth lays 200–400 eggs on the tender parts of the plant like flower, flower buds, tender leaves, leaf buds, etc., which hatch out within 3–7 days into small caterpillars and become fully grown within 7–15 days during which it moults five times. The full grown caterpillars pupate either on the plant or on the ground in fallen leaves. Adult moths emerge out from pupae within 21 to 84 days.

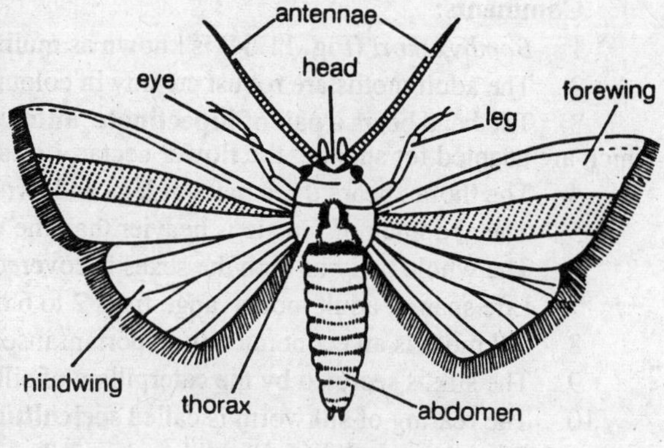

Fig. 11.63. *Earias.*

Nature of damage: The caterpillars, cut and chew the tender shoot by their biting and chewing mouth parts and bore through it and later on into the flowers and bolls. After attaining full growth they come out of the boll to pupate. They may damage 8–60% crop. In bolls a number of holes are formed pluged with excreta. Thus, their attack on cotton yield poor lint and those on vegetables like lady's finger damage the fruits.

Control: Spraying of 0.1% carbaryl after 15 days interval and dusting of 5% fenitrothion are effective in checking the pest population.

64. *Bombyx mori*

Classification:

Phylum	...	**Arthropoda**	Characters same as those of *Corcyra cephalonica.*
Class	...	**Insecta**	
Subclass	...	**Pterygota**	
Division	...	**Endopterygota**	
Order	...	**Lepidoptera**	
Family	...	**Bombycidae**	→ Antennae bipectinate in both the sexes.
Genus	...	*Bombyx*	
Species	...	*mori*	

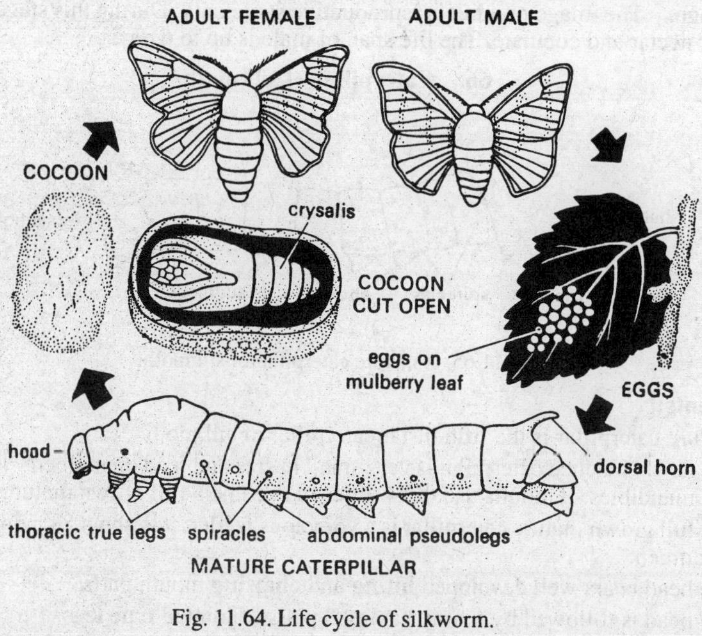

Fig. 11.64. Life cycle of silkworm.

Comments:

1. *Bombyx mori* (Fig. 11.64) is known as mulberry silkworm or silkmoth.
2. The adult moths are robust creamy in colour.
3. The head bears a pair of **bipectinate antennae**, a pair of compound eyes and mouth parts which are adapted for sucking the flower nectar, *i.e.*, **siphoning type.**
4. The thorax bears three pairs of legs and two pairs of heavy wings.
5. The abdomen of female is heavier than the male.
6. The whole body of both the sexes is covered with minute scales.
7. Life span of adult moths range from 2 to 6 days; females live longer than the males.
8. Silkmoth is an economically important insect as it is the source of **silk** for us.
9. The silk is secreted by the caterpillars of silkmoth as a protective house for them.
10. The rearing of silkworm is called **sericulture.**
11. The other species of silkmoths commonly used in sericulture in our country are *Antheraea assamensis* (Muga silkworm), *Antheraea mylitta* (Tassar silkworm) and *Attacus ricinii* (Eri silkworm) but belonging to the family Saturnidae.
12. The multivoltine hybrid varieties of silkmoth (*e.g.*, pure Mysore race, Nistari, Chotopole, etc.) are used extensively in sericulture in our country.

65. Silkmoth–Stages of life cycle

Comments:

1. The various stages in the life cycle of silkmoth (Fig. 11.64.) are egg, larva, pupa and adults.
2. **Egg.** A female moth after copulation lays nearly 300–400 eggs in clusters upon the leaves of mulberry and secretes a sticky secretion on the eggs which glues them on the leaves so that they cannot fall down. The eggs are small, rounded and whitish. Female moth dies after laying eggs.
3. **Larva**. Like those of other lepidopterans, the larva of silkmoth is also called **caterpillar**. The eggs hatch into small caterpillars within 10 days which feed on the leaves of mulberry by their **biting** and **chewing** mouth parts. They moult usually 4 times to become full grown caterpillars which are voracious feeders. Then they secrete a secretion around themselves called **cocoon.** Inside cocoon they undergo final moulting and transform into pupae.
4. **Pupa**. During pupation active metamorphosis takes place involving the processes of histolysis and histogenesis, and become **imago** or **adult** which emerge out.
5. **Imago.** The imago or adult is reproductive stage only. During this stage, the adult moths feed on flower nectar and copulate. The life span of male is up to 6 days.

66. Caterpillar of silkmoth

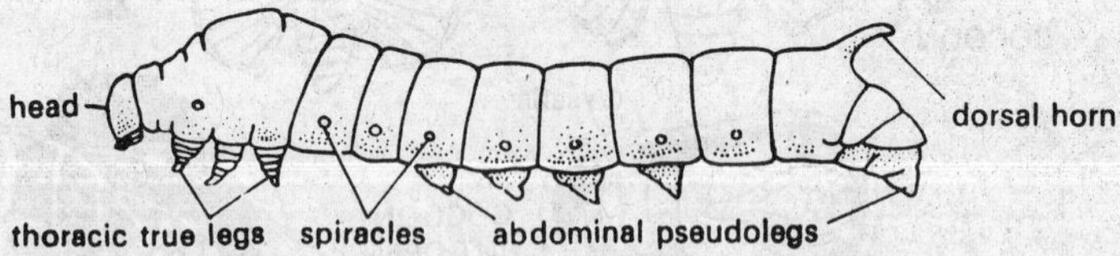

Fig. 11.65. A mature caterpillar of silkmoth.

Comments:

1. Mature caterpillar is the fifth instar caterpillar of silkmoth.
2. The newly hatched caterpillar is very small in size and feed on mulberry leaves by nibbling with their soft mandibles. It moults usually 4 times to become a full grown mature caterpillar.
3. The full grown mature caterpillar is a voracious feeder; it is about 8 cm long having a head, thorax and abdomen.
4. The head bears well developed **biting** and **chewing** mouth parts.
5. The head is followed by thorax having 3 pairs of jointed **true legs**.

6. The abdomen 10 segmented; its 3rd, 4th, 5th, 6th and last abdominal segments bear a pair of unjointed **pseudolegs**. The 8th segment has a **dorsal horn**. There are spiracles in thoracic and abdominal segments.

7. It has a well developed **silk gland** (modified salivary gland) which opens out at the base of hypopharynx by spinneret.

8. This caterpillar, after leading an active feeding life for 6–7 days, becomes inactive, stops feeding and starts secreting a sticky fluid from its silk gland which is fibrous and become hard in contact with air.

9. Thus, the inactive caterpillar gets enclosed in these fine filaments which form a case around its body called **cocoon**.

10. Inside the cocoon it moults finally to transform into **pupa** or **crysalis**.

11. Silk is, in fact, obtained from cocoon. A caterpillar secretes nearly 1000–1500 meters long silk fibre within 3 or 4 days.

12. The silk obtained from a cocoon is called **raw silk**.

67. Silk thread

Comments.

1. The silk thread is obtained from cocoons.
2. To start with silk is a pasty secretion of silk gland of the caterpillar.
3. This pasty secretion, after coming in contact with air, hardens to form strong pliable silk thread.
4. Chemically it consists of protein.
5. The silk thread is cemented by a gumy substance called sericin secreted by the Fillip's gland of silk gland.
6. The cocoons having silk thread are treated with hot water or steam is passed over it. During this process gumy substance sericin dissolves and the silk thread is then reeled off on a wheel. Such a silk is called **raw silk**.
7. This raw silk is then spinned and the spinned silken threads are used in weaving cloth.
8. According a data, India produces nearly 3700 tonnes of silk per year amounting to 5% of the total world produce.

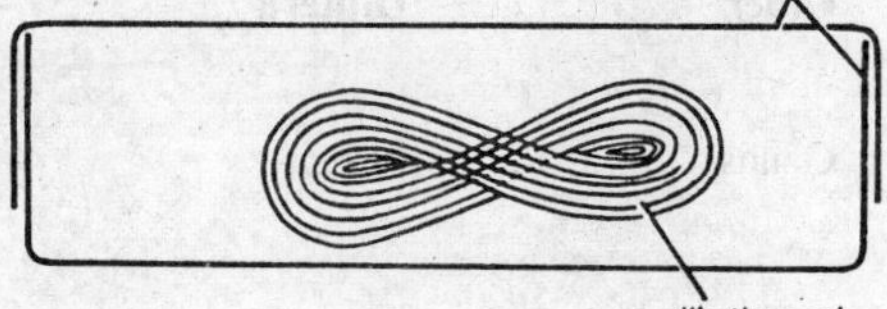

Fig. 11.66. Silk thread.

68. *Xenopsylla*

Classification:

Phylum	...	**Arthropoda**	Characters same as those of *Aulacophora*.
Class	...	**Insecta**	
Subclass	...	**Pterygota**	
Division	...	**Endopterygota**	
Order	...	**Siphonoptera**	→ Ectoparasitic on birds and mammals; body laterally compressed; mouth parts piercing and sucking; cerci absent.
Genus	...	*Xenopsylla*	
Species	...	*cheopis*	

Comments:

1. *Xenopsylla cheopis* (Fig. 11.67) is commonly called **rat flea**.
2. Body is laterally compressed without wings.
3. Integument is heavily sclerotized, brown in colour and armed with spines and bristles.
4. Head is set closely into thorax with front at tubercle and often with genal ctenidia.

5. Antennae are clavate, short and concealed in grooves.

6. Compound eyes and ocelli are absent.

7. Thorax is compact with the segmentation distinct often with pronotal ctenidia on the posterior margin; prosternum is large.

8. Walking legs are elongated, clawed and modified for jumping.

9. Mouth parts are **piercing** and **sucking**.

10. The adult flea is active and slips through hair with great ease.

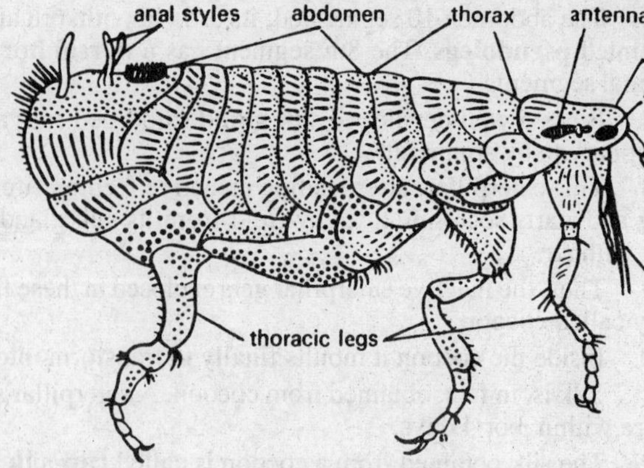

Fig. 11.67. *Xenopsylla.*

Habit and habitat: *Xenopsylla cheopis* is an ectoparasite of mammals specially rats. It transmits the **plague germs** from rats to man.

69. *Musca*

Classification:

Phylum	...	**Arthropoda**	Characters same as those of *Corcyra cephalonica.*
Class	...	**Insecta**	
Subclass	...	**Pterygota**	
Division	...	**Endopterygota**	
Order	...	**Diptera**	→ Soft bodied two-winged insects; hindwings reduced to club-shaped halteres; includes flies, mosquitoes, midges, etc.
Genus	...	*Musca*	

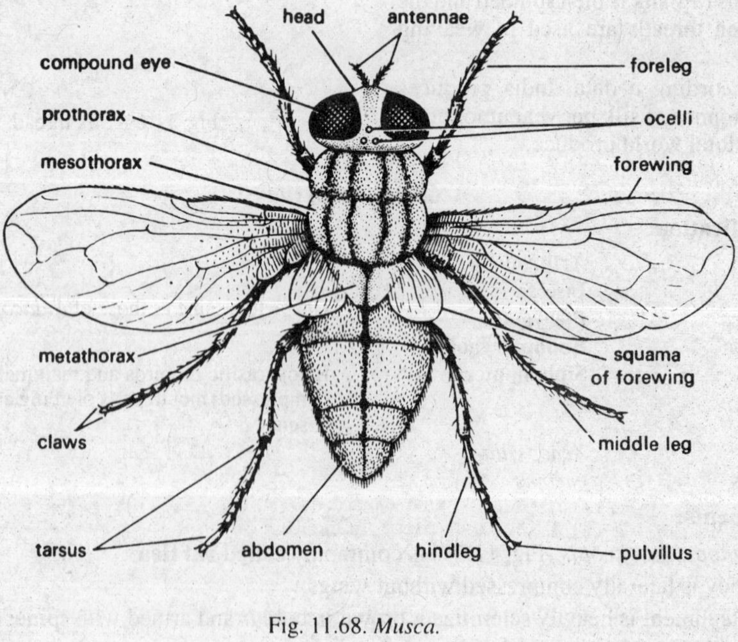

Fig. 11.68. *Musca.*

Comments:

1. *Musca* (Fig. 11.68) is commonly called **housefly**.
2. Body has distinct 3 regions–head, thorax and abdomen.
3. Head bears a pair of large compound eyes, ocelli, a pair of 3-jointed antennae and mouth parts which are of **sponging type**.
4. Thorax bears three segments–prothorax, mesothorax and metathorax; mesothorax bears a pair of transparent wings, metathoracic wings reduced into halteres; all thoracic segments bear a pair of legs terminating into claws having pulvilli.
5. Abdomen ten segmented.
6. Apart from its annoying habit the housefly carries and transmits the organisms of various diseases in man; it mostly transmits the germs causing intestinal disorders to produce diseases like typhoid, paratyphoid, diarrhoea, cholera, bacillary and amoebic dysentery, etc., It also transmits the germs of tuberculosis and poliomyelitis.

Habit and habitat: The houseflies are abundant and very active during summer and rainy seasons. They are **diurnal** and rest at night on ceilings, walls, elactric wires, ropes, curtains, clothes, etc. They sit upon dirty objects.

Distribution: World-wide.

70. Head and mouth parts of housefly

Comments:

1. Head of housefly (*Musca domestica*) bears a pair of large compound eyes, three ocelli, a pair of antennae and **sponging type** of mouth parts.
2. Antennae are aristate.
3. Mouth parts comprise labrum epipharynx, hypopharynx, short maxillary palp and a fleshy retractile proboscis.
4. Labrum is fused with the epipharynx and forms a narrow slender tube opening ventrally.
5. Hypopharynx is narrow blade-like containing salivary duct and fits into the tube formed by labrum-epipharynx.
6. Maxillary palps are small and unjoined.
7. Mandibles and maxillae are absent.
8. **Proboscis** consists of three parts, a basal large **rostrum**, middle **haustellum** and a distal **labella** or oral lobes.
9. **Labella** or oral lobes form an oral disc crossed by the pseudotracheal channels.
10. The pseudotracheae open externally by a double raw of tiny holes through which liquid food is taken in.

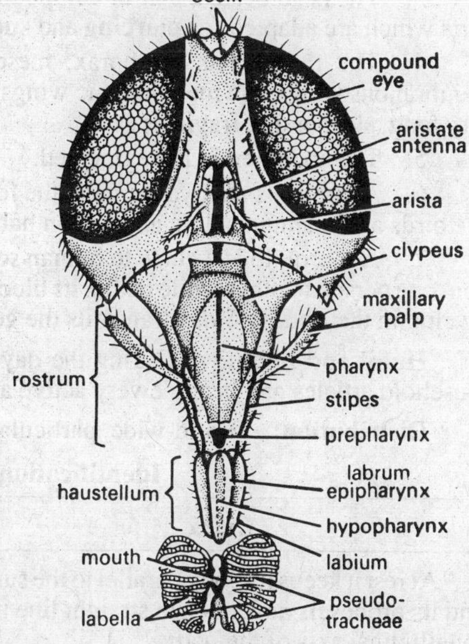

Fig. 11.69. Head and mouth parts of housefly.

71. *Culex*

Classification:

Phylum	...	**Arthropoda**	}	Characters same as those of *Corcyra*.
Class	...	**Insecta**		
Subclass	...	**Pterygota**		
Division	...	**Endopterygota**		
Order	...	**Diptera**	→	Characters same as those of *Musca*.
Genus	...	*Culex*		

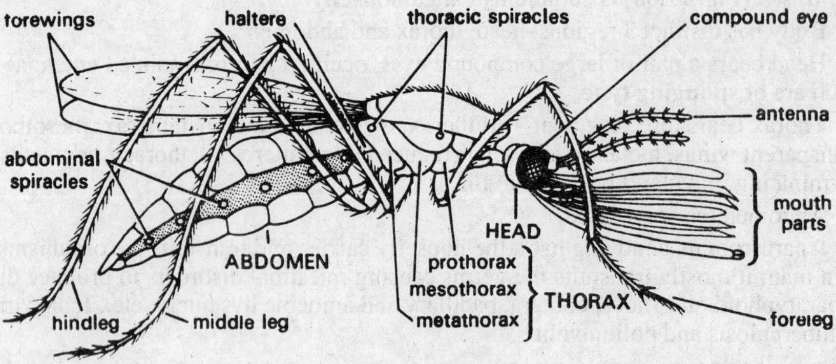

Fig. 11.70. *Culex*.

Comments:

1. *Culex* (Fig. 11.70) is a very common **mosquito**.
2. Its body has three distinct regions–head, thorax and abdomen.
3. The head bears a pair of compound eyes, no ocelli, a pair of filiform antennae and mouth parts which are adapted for **piercing** and **sucking**.
4. The thorax has prothorax, mesothorax and metathorax; mesothorax bears a pair of membranous wings, the metathoracic wings reduced as **halteres**. Each thoracic segment has pair of very long, slender and fragile legs.
5. The abdomen is 10 segmented.
6. Apart of its annoyance habit the female mosquito suck blood from warm-blooded animals like birds and mammals. It is noctural in habit and most abundant during rainy season.
7. A mosquito is a curse to human society because it transmits the germs of various diseases from one person to other while taking its blood meal and also transmit disease producing germs which develop in their body. *Culex* transmits the germs of **dengue fever**, **filariasis**, **encephalitis**.

Habit and habitat: During the day they take shelter in dark places and remain hidden in household articles and become very active at night. They lay eggs in stagnant water.

Distribution: World-wide, particularly common in tropics and subtropics.

Identification of *Culex* and *Anopheles*

Culex	*Anopheles*
1. At rest it keeps its body parallel to the surface and its proboscis never lies in straight line to the longitudinal axis of the body.	1. At rest it keeps its body at an angle of 45° to the surface and its proboscis remains in straight line to the longitudinal axis of the body.
2. Its body is longer and broader having stout legs.	2. Its body is smaller and slender having delicate legs.
3. Body covered by bristles which are sparsely distributed.	3. Body covered by bristles which are densely placed.
4. Wings with uniform colouraton.	4. Wings are darkly spotted.
5. The abdominal sterna covered with scales.	5. Scales on abdominal sterna absent.
6. Maxillary palps either longer than proboscis or very short.	6. Maxillary palps are equal in length to proboscis.

Anopheles female transmits disease germs of **malaria**, and **filariasis**.

72. Head and mouth parts of female *Culex* and *Anopheles*

Comments:

1. Head of female *Culex* and *Anopheles* bears a pair of antennae, a pair of compound eyes and **piercing** and **sucking** type of mouth parts.

2. **Antennae** in both *Culex* and *Anopheles* are many jointed with few small hairs on each joint.

3. **Maxillary palps** are extremely short and three jointed in *Culex*, while large and more or less equal to proboscis in *Anopheles*.

4. Food channel is formed by dorsal labrum-epipharynx and ventral hypopharynx.

5. Lateral needle - like mandibles and maxillae are contained within the grooved labium or proboscis which forms a long gutter.

6. **Mandibles** and **maxillae** are sharp and pointed stylets which pierce the tissues of organism and labrum-epipharynx sucks the blood. Labium or proboscis and maxillary palp do not enter the tissues.

7. Female mosquitoes feed on the blood of vertebrates exclusively.

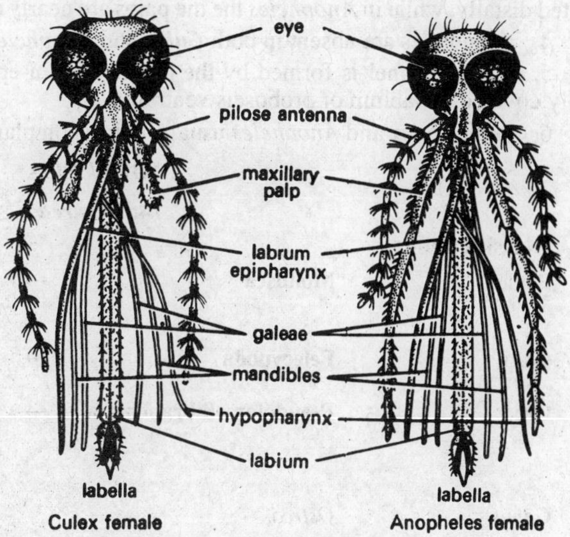

Fig. 11.71. Head and mouth parts of female *Culex* and *Anopheles*.

73. Head and mouth parts of male *Culex* and *Anopheles*

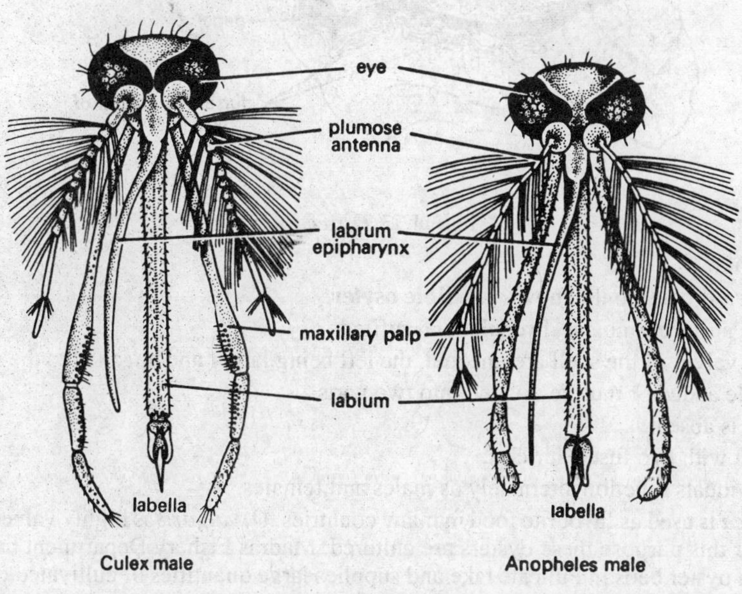

Fig. 11.72. Head and mouth parts of male *Culex* and *Anopheles*.

Comments:

1. Head of male *Culex* and *Anopheles* (Fig. 11.72) bears a pair of antennae, a pair of compound eyes and **piercing** and **sucking** type of mouth parts.
2. **Antennae** in both *Culex* and *Anopheles* are long and many jointed bearing tufts of hairs at the joints.
3. **Maxillary palps** are long and hairy. In *Culex* the palps are longer than proboscis and pointed distally, while in *Anopheles* the the palps are nearly equal to proboscis with club-shaped tips.
4. Mandibles are absent in both *Culex* and *Anopheles*.
5. Food channel is formed by the dorsal labrum-epipharynx and ventral hypopharynx and finally covered by labium or proboscis ventrally.
6. Male *Culex* and *Anopheles* usually subsist on plant juice.

MOLLUSCA

74. *Ostrea*

Classification:

Phylum	...	**Mollusca**	→	Body unsegmented, bilaterally symmetrical and consists of head, foot, mantle and visceral mass.
Class	...	**Pelecypoda**	→	Bivalved shell; body laterally compressed, head not distinct.
Order	...	**Pseudolamellibranchiata**	→	Shell valves frequently unequal, gills paired to form vertical folds; single large posterior adductor muscle present; foot rudimentary.
Genus	...	*Ostrea*		

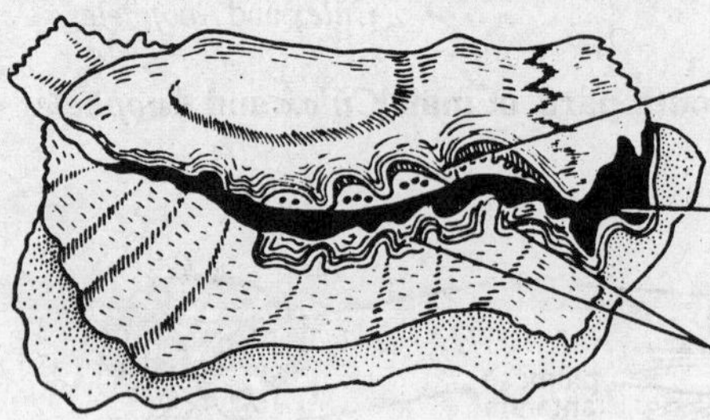

Fig. 11.73. *Ostrea.*

Comments:

1. *Ostrea* is commonly known as **edible osyter**.
2. Shell surface is coarse, irregular and ruffled.
3. Two valves of the shell are unequal, the left being larger and attached to the substratum.
4. Single adductor muscle divided into two parts.
5. Foot is absent.
6. Heart with two fused auricles.
7. Individuals function alternately as males and females.
8. *Ostrea* is used as favourite food in many countries. *O. vulgaris* is highly valued food in our country. So, for this purpose these oysters are cultured. Madras Fishery Department has developed and maintained oyster beds in Pulicate lake and supplies large quantities of cultivated oysters.

Habit and habitat: *Ostrea* is marine, sedentary animal found attached to rocks and stones, etc.

75. *Pinctada*

Classification:

Phylum	...	**Mollusca**	Characters same as those of *Ostrea*.
Class	...	**Pelecypoda**	
Order	...	**Pseudolamellibranchiata**	
Genus	...	*Pinctada*	

Comments:

1. *Pinctada* is commonly called **Indian pearl oyster**.
2. The left shell valve is large convex and permanently attached to the rocks.
3. The right shell valve is smaller and thinner, and covers the viscera.
4. The surface of the shell is coarse, irregular and ruffled.
5. There is a single adductor muscle.
6. The right and left mantle lobes are quite free.
7. Oyster's shell secretes **pearls** whenever any organism enters the shell and causes irritation. Because of its this characteristic, the pearl oysters are cultivated and artificially pearls are obtained from them.

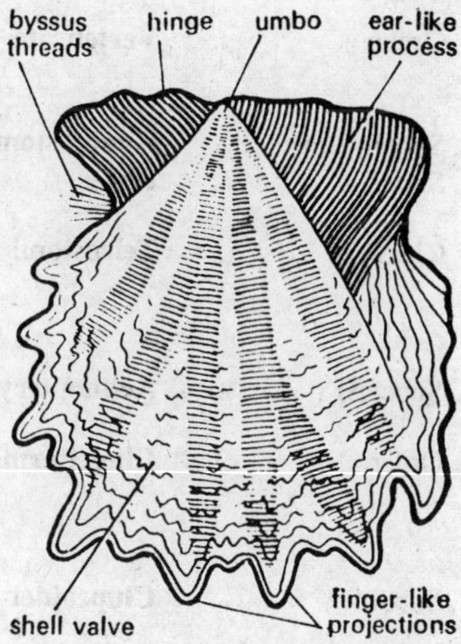

Fig. 11.74. *Pinctada*.

Habit and habitat: *Pinctada vulgaris* is widely distributed except in cold zones. It is very common in the Gulf of Kutch.

Distribution: It is widely distributed except in cold zones of the sea. It is commonly found at the Pacific coast of America, Northern Australia, East Indies and in the Persian Gulf.

76. T. S. of the shell and mantle of *Unio*

Comments:

1. Shell of *Unio* consists of three layers, *viz.*, **periostracum**, **prismatic** and **nacreous**.
2. **Periostracum** is the outermost, greenish-brown, thin, horny layer composed of **conchiolin** a chitin like substance.
3. **Prismatic** is the middle layer, composed of minute prisms or plates of calcium carbonate separated by thin layers of conchiolin.
4. **Nacre** or mother of pearl is the innermost layer, consists of alternate layers of calcium carbonate and conchiolin arranged parallel to the surface.
5. Mantle also comprises three layers, *i.e.*, outer, middle and inner.
6. Outer layer consists of nacre-secreting epithelial cells and lies close to the nacre layer.
7. Middle layer comprises thick connective tissue.

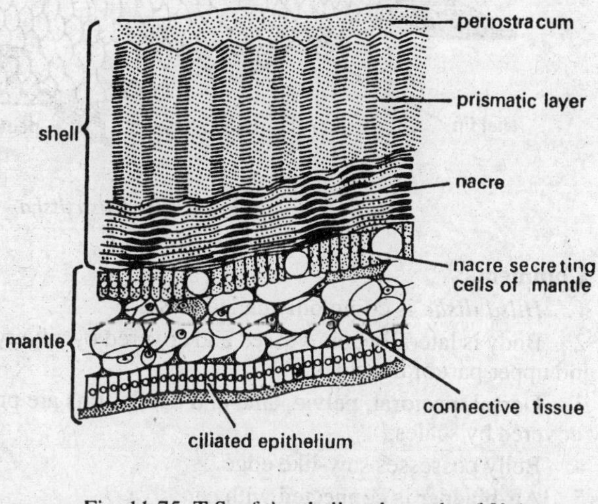

Fig. 11.75. T.S. of the shell and mantle of *Unio*.

8. Innermost layer composed of single layer of ciliated epithelium.

9. Mantle deposits nacreous layers over any foreign material that enters in between mantle and shell that finally becomes the **pearl**–a valuable gem.

PISCES

77. *Hilsa*

Classification:

Phylum	...	**Chordata**	→	Dorsal tubular nerve cord, notochord, and paired gill-slits are present.
Group	...	**Vertebrata**	→	Notochord is replaced by vertebral column; two pairs of appendages; circulatory system closed; hepatic portal system present; blood red containing R. B. Cs..
Subphylum	...	**Gnathostomata**	→	Jaws and paired appendages are present.
Series	...	**Pisces**	→	Presence of paired or unpaired fins supported by soft or spiny rays and gill-slits.
Class	...	**Teleostomi**	→	Endoskeleton bony; skull usually hyostylic, single external gill-slit on each side of head; mouth terminal; operculum and air-bladder present; cloaca and claspers absent.
Subclass	...	**Actinopterygii**	→	Radials of paired fins not arranged biserially; internal nares and squamosal absent.
Order	...	**Clupeiformes**	→	Scales are well-developed and cycloid; caudal fin homocercal; dorsal and anal fins are without spines; vertebral centra-completely ossified; Weberian apparatus absent.
Suborder	...	**Clupeoidei**	→	Largest otolith in the sacculus; mucous canals on the head; supra-orbital canal traversing the periotic; no adipose fin; parapophyses not ossified with centra; pectoral radials in one row; oviducts present.
Genus	...	*Hilsa*		
Species	...	*ilisha*		

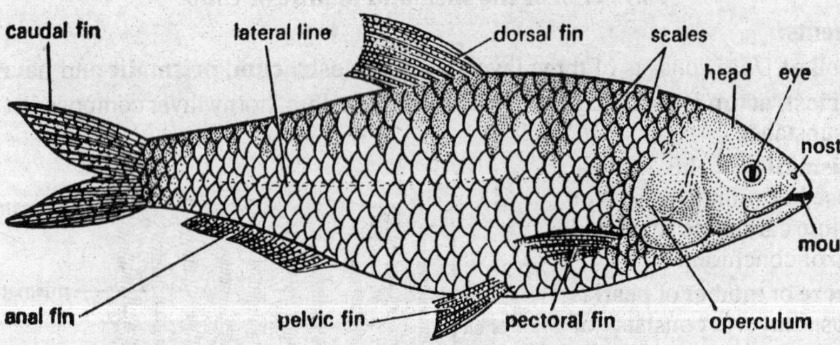

Fig. 11.76. *Hilsa ilisha.*

Comments:

1. *Hilsa ilisha* is commonly called **hilsa.**

2. Body is laterally compressed and covered by silvery scales. Vertical darker bars across the back and upper part of the sides.

3. Dorsal, pectoral, pelvic, anal and caudal fins are present. Caudal fin is deeply forked and partly covered by scales.

4. Belly possesses saw-like edge.

5. Air-bladder is connected with ear

6. This fish is used for food because it forms the delicious food for those who eat fish.

7. *Hilsa ilisha* swarm up in large rivers of India to spawn at the time of local monsoon. Mature fishes are available from May to October in the Ganga when they start leaving for sea.

Geographical distribution: *Hilsa* is found in temperate and tropical seas.

78. *Notopterus*

Classification:

Phylum	...	**Chordata**	Characters same as those of *Hilsa.*
Group	...	**Vertebrata**	
Subphylum	...	**Gnathostomata**	
Series	...	**Pisces**	
Class	...	**Teleostomi**	
Subclass	...	**Actinopterygii**	
Order	...	**Clupeiformes**	
Suborder	...	**Notopteroidei**	Cavum cranii reaching the ethmoid region; orbitosphenoid, opisthotic and basisphenoid present; air-bladder connected with ear, no supra-maxillary; symplectic and endopterygoid present; no oviduct; freshwater fishes.
Genus	...	*Notopterus*	
Species	...	*chitala*	

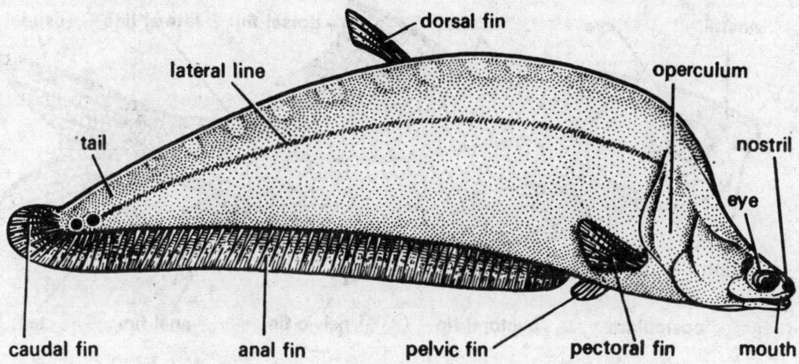

Fig. 11.77. *Notopterus chitala.*

Comments:

1. *Notopterus chitala* is commonly known as **chital**.
2. Body is strongly compressed and covered with **minute scales.**
3. It measures up to a meter in length.
4. The colour is coppery brown or greyish along the back with 15 or 16 silvery transverse bars.
5. Head is small and mouth is large.
6. Snout is obtuse and convex.
7. Dorsal fin is small. Pectoral and pelvic fins are very small.
8. Anal fin is much elongated and confluent with the caudal fin.
9. Musciferous channels on the head are well-developed.
10. The air-bladder is very large with several divisions.
11. Carnivorous. Food chiefly comprises worms and insects.
12. It is commercially important as food fish, its flesh is said to be uncommonly rich and well flavoured.

Geographical distribution: *Notopterus* lives in marshes and lakes of freshwater and brakish water of West Africa, India, Burma and Malaya. *N. chitala* is found exclusively in freshwaters of India.

79. *Labeo*

Classification:

Phylum	...	**Chordata**	}	Characters same as those of *Hilsa*.
Group	...	**Vertebrata**		
Subphylum	...	**Gnathostomata**		
Series	...	**Pisces**		
Class	...	**Teleostomi**		
Subclass	...	**Actinopterygii**		
Order	...	**Clupeiformes**	→	A peculiar Weberian apparatus connecting the ear with the air-bladder; air-bladder is connected with the alimentary canal by a duct; fins either without spines or dorsal, anal or pectoral have a spine each; ventral fins are abdominal.
Division	...	**Cyprini**	→	Body covered with scales or naked, never covered with bony plates; third and fourth vertebrae are not fused with each other.
Genus	...	*Labeo*		
Species	...	*rohita*		

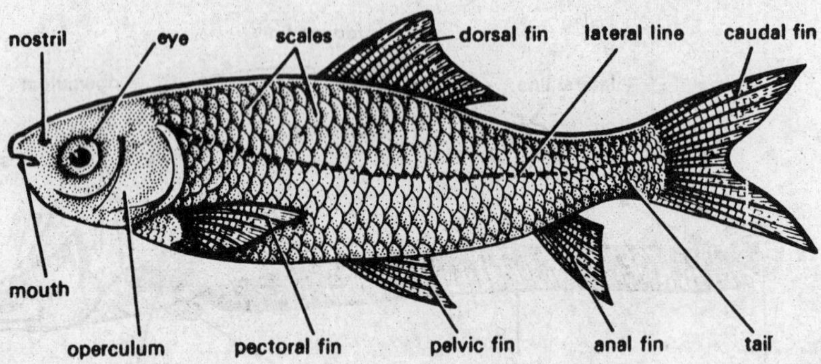

Fig. 11.78. *Labeo rohita.*

Comments:

1. *Labeo rohita* is the most famous carp and commonly known as **rohu** in Hindi.
2. Body is elongated with moderately rounded abdomen, measuring up to 1 meter in length.
3. Colour brownish grey to black above.
4. Scales are large and orange to reddish in colour in the centre.
5. Head is prominent with blunt snout.
6. Mouth is transverse and semi-oval.
7. Lips thick covering the jaws, one or both having an inferior transverse fold.
8. A soft and movable horny covering with a sharp margin on the inner side of the one or both lips.
9. Barbles are absent.
10. Air-bladder is usually large and divided into an anterior and posterior part.
11. It is economically important due to its food value. It is relished very much in food.

Geographical distribution: *Labeo rohita* is found all over Northern and Central India. It is now found in Godawari and Krishna rivers.

80. *Catla*

Classification:

Phylum	...	**Chordata**	Characters same as those of *Hilsa.*
Group	...	**Vertebrata**	
Subphylum	...	**Gnathostomata**.	
Series	...	**Pisces**	
Class	..	**Teleostomi**	
Subclass	...	**Actinopterygii**	
Order	...	**Cypriniformes**	Characters same as those of *Labeo.*
Division	...	**Cyprini**	
Genus	...	*Catla*	
Species	...	*catla*	

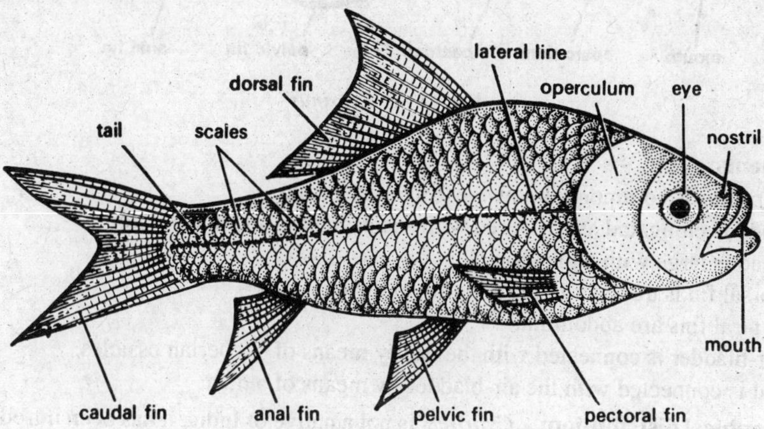

Fig. 11.79. *Catla catla.*

Comments:

1. *Catla catla* is the largest Indian carp and commonly known as **katla** in Hindi.
2. Body is elongated broad and stout, measuring more than a metre in length.
3. Colour blackish grey above and silvery on the sides.
4. Scales are pink or coppery in the centre on the dorsal side and whitish below.
5. Dorsal profile of the body is more convex.
6. Head is large having large rounded eyes.
7. Mouth is wide with prominent lips.
8. Dorsal fin is quite large. Caudal fin is bilobed.
9. Air-bladder is usually large and divided into an anterior and posterior part.
10. It is also important as a food-fish.

Geographical distribution: *Catla catla* is found throughout India, although rather scarce to south of the Krishna river.

81. *Cyprinus*

Classification:

Phylum	...	**Chrodata**	Characters same as those of *Hilsa.*
Group	...	**Vertebrata**	
Subphylum	...	**Gnathostomata**	
Series	...	**Pisces**	
Class	...	**Teleostomi**	
Subclass	...	**Actinopterygii**	

Order	...	**Cypriniformes**	Characters same as those of *Labeo*.
Division	...	**Cyprini**	
Genus	...	*Cyprinus*	

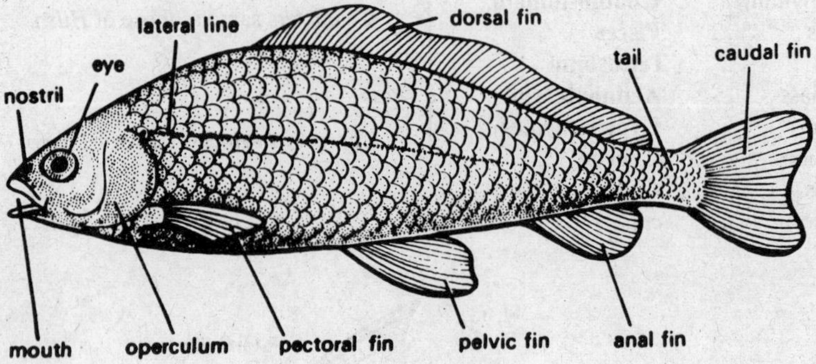

Fig. 11.80. *Cyprinus.*

Comments:

1. *Cyprinus* is commonly known as **European carp**.
2. Body is elongated and covered with scales, measuring about 75 cm.
3. Head is large with wide mouth.
4. Dorsal fin is quite large reaching up to the tail.
5. Ventral fins are abdominal.
6. Air-bladder is connected with the ear by means of Weberian ossicles.
7. Gut is connected with the air-bladder by means of a duct.

Geographical distribution: *Cyprinus* is not a native of India, it has been introduced into the lakes of Nilgiri mountain ranges from China and South East Asia. It was first introduced into Sri Lanka and later in South India.

82. *Cirrhina*

Classification:

Phylum	...	**Chordata**	Characters same as those of *Hilsa*.
Group	...	**Vertebrata**	
Subphylum	...	**Gnathostomata**	
Series	...	**Pisces**	
Class	...	**Teleostomi**	
Subclass	...	**Actinopterygii**	Characters same as those of *Labeo*.
Order	...	**Cypriniformes**	
Division	...	**Cyprini**	
Genus	...	*Cirrhina*	
Species	...	*mrigala*	

Comments:

1. *Cirrhina mrigala* (Fig. 11.81) commonly called **mrigal** is a freshwater carp of India.
2. The length of the body reaches up to 66 cm and weight up to 1.4 to 2.8 kg.
3. The body is covered by large cycloid scales. The scales are absent on the head.
4. The colour of the body is silvery, dark-grey along the back, sometimes with coppery tinge.
5. Mouth is wider with thinner lips.
6. Barbles small in fold of lip.
7. Pectoral, pelvic, and anal fins are orange in colour tipped with black.
8. Caudal fin is sharply forked.
9. It gives best sport on rod and line and is used as food.

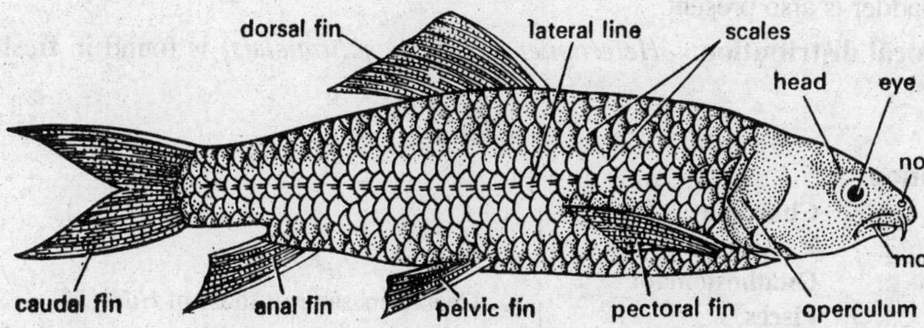

Fig. 11.81. *Cirrhina mrigala.*

Geographical distribution: *Cirrhina mrigala* is found throughout India.

83. *Heteropneustes (=Saccobranchus)*

Classification:

Phylum	...	**Chordata**	Characters same as those of *Hilsa.*
Group	...	**Vertebrata**	
Subphylum	...	**Gnathostomata**	
Series	...	**Pisces**	
Class	...	**Teleostomi**	
Subclass	...	**Actinopterygii**	
Order	...	**Cypriniformes**	→ Characters same as those of *Labeo.*
Division	...	**Siluri**	→ Body naked, scaleless; maxillary bone reduced supporting the barbules.
Genus	...	*Heteropneustes*	
Species	...	*foosilis.*	

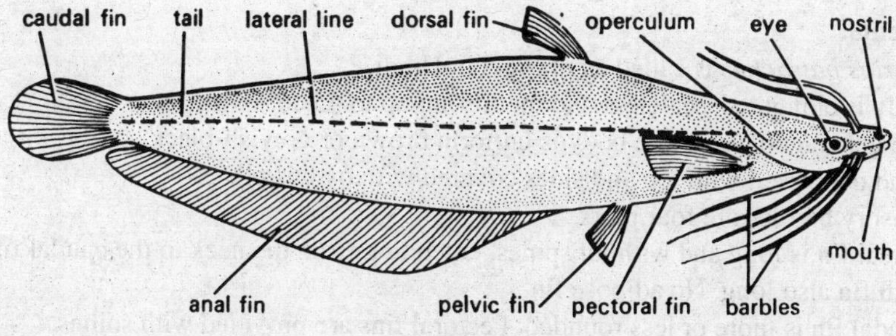

Fig. 11.82. *Heteropneustes foosilis.*

Comments:

1. *Heteropneustes foosilis* is commonly called **singhi** in Hindi.
2. Body is elongated and laterally compressed, measuring about 30 cm in length.
3. Skin without scales.
4. Head is flattened. Eyes with free circular margins.
5. Barbles are long and four pairs.
6. Dorsal fin is short without spine, ventral fin situated at the level of the dorsal fin.
7. Pectoral fins are strong with poison spine.
8. Anal fin is elongated, reaches up to the caudal fin separated from it by a notch.
9. Gill-opening is wide, the membranes not being confluent with the skin of isthmus and separated by deep notch.

10. Accessory breathing organs are present.
11. Air-bladder is also present.

Geographical distribution: *Heteropneustes* (=*Saccobranchus*) is found in freshwaters of India and Burma.

84. *Clarias*

Classification:

Phylum	...	**Chordata**	Characters same as those of *Hilsa*.
Group	...	**Vertebrata**	
Subphylum	...	**Gnathostomata**	
Series	...	**Pisces**	
Class	...	**Teleostomi**	
Subclass	...	**Actinopterygii**	
Order	...	**Cypriniformes**	→ Characters same as those of *Labeo*.
Division	...	**Siluri**	→ Characters same as those of *Heteropneustes*.
Genus	...	*Clarias*	
Species	...	*batrachus*	

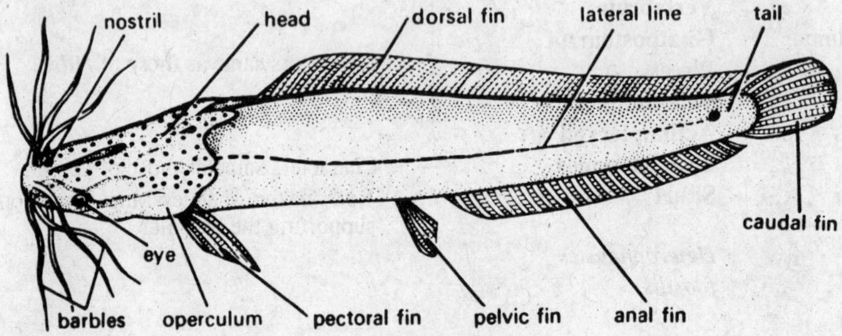

Fig. 11.83. *Clarias batrachus.*

Comments:

1. *Clarias batrachus* is called as **mangri** in Hindi.
2. Body is elongated, scaleless and measuring up to 45 cm in length.
3. The general colour of the body is uniform brown or greyish-black.
4. Head depressed with top and sides covered with osseus plates.
5. Sensory barbles are four pairs.
6. Dorsal fin is long and without spines, extending from the neck to the caudal fin.
7. Anal fin also long. No adipose fin.
8. Caudal fin is more or less rounded. Pectoral fins are provided with spines.
9. Accessory respiratory organs are branched tree-like, especially designed to take in oxygen from the air.
10. The air-bladder is connected with internal ear by Weberian ossicles.
11. It is highly nourishing and esteemed as food.
12. It is also used in laboratories for experimental purposes.

Geographical distribution: *Clarias* is found in Africa and South and West Asia. *Clarias batrachus* occurs in fresh and brakish waters throughout India.

85. *Wallago*

Classification:

Phylum	...	**Chordata**	Characters same as those of *Hilsa*.
Group	...	**Vertebrata**	
Subphylum	...	**Gnathostomata**	

Series	...	**Pisces**	Characters same as those of *Hilsa*.
Class	...	**Teleostomi**	
Subclass	...	**Actinopterygii**	
Order	...	**Cypriniformes**	→ Characters same as those of *Labeo*.
Division	...	**Siluri**	→ Characters same as those of *Heteropneustes*.
Genus	...	*Wallago*	
Species	...	*attu*	

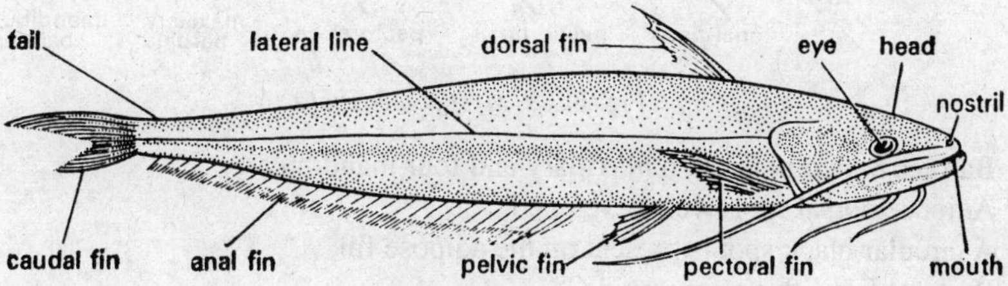

Fig. 11.84. *Wallago attu.*

Comments:

1. Body is elongated and laterally compressed with a straight back.
2. It may reach in length up to 2 metres but the specimens generally available are 60–90 cm in length.
3. Head is very large, trunk small and tail long and tapering.
4. Mouth is wide extending below and even behind the eyes.
5. Jaws are provided with teeth.
6. Barbles 4, two maxillary and two mandibular.
7. Eyes above the level of the mouth and not covered with skin.
8. Dorsal fin is short and spineless.
9. Anal fin is very long. Caudal fin bilobed.
10. Predatory in habit. It predates on fishes particularly young carps.

Geographical distribution: *Wallago attu* is found in rivers and lakes throughout India.

86. *Mystus*

Classification:

Phylum	...	**Chordata**	Characters same as those of *Hilsa*.
Group	...	**Vertebrata**	
Subphylum	...	**Gnathostomata**	
Series	...	**Pisces**	
Class	...	**Teleostomi**	
Subclass	...	**Actinopterygii**	
Order	...	**Cypriniformes**	→ Characters same as those of *Labeo*.
Division	...	**Siluri**	→ Characters same as those of *Heteropneustes*.
Genus	...	*Mystus*	
Species	...	*seenghala*	

Comments:

1. Body is elongated, measuring up to one meter in length.
2. The colour is brownish above with silvery sides.
3. The snout is distinctly long.
4. Mouth is transverse and terminal.
5. Upper jaw is generally longer.

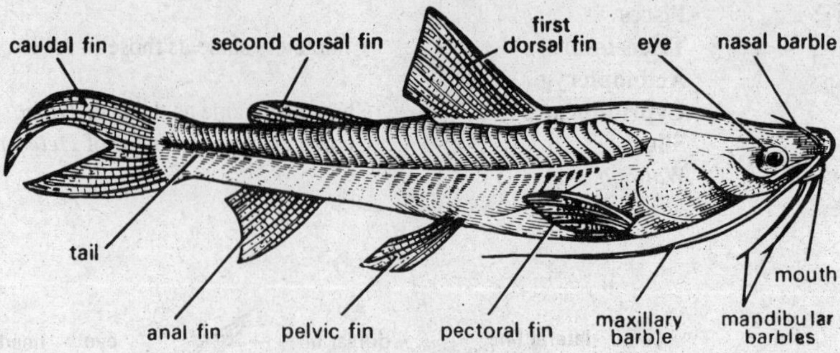

Fig. 11.85. *Mystus seenghala.*

6. Barbles 8, two nasal, two maxillary and four mandibular.
7. Adipose dorsal fin is well developed.
8. A circular black spot is present on the adipose fin.
9. Pectoral fin with a strong spine.
10. Predatory in habit, attacks small scarps, teleosts and prawns.
11. Employed as food by the poor natives.

Geographical distribution: *Mystus seenghala* is widely distributed in India being found in Jamuna, Ganga and the Deccan rivers as well as Assam rivers.

87. *Rita*

Classification:

Phylum	...	**Chordata**		Characters same as those of *Hilsa*.
Group	...	**Vertebrata**		
Subphylum	...	**Gnathostomata**		
Series	...	**Pisces**		
Class	...	**Teleostomi**		
Subclass	...	**Actinopterygii**		
Order	...	**Cypriniformes**	→	Characters same as those of *Labeo*.
Division	...	**Siluri**	→	Characters same as those of *Heteropneustes*.
Genus	...	*Rita*		
Species	...	*rita*		

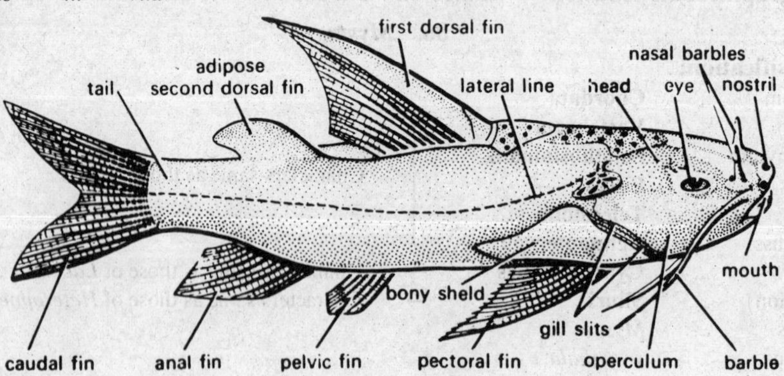

Fig. 11.86. *Rita rita.*

Comments:

1. Body is elongated, measuring from 45 cm to 1 meter in length.
2. There is no scale on the body.
3. Colour brownish olive green and lighter below.

4. Four barbles present on the head, two nasal barbles and two maxillary barbles.
5. Gill-slits are wide openings.
6. A bony shield present behind the gill-slits.
7. First dorsal fin with a large spine.
8. Pectoral fin is also spinous.
9. Pelvic and anal fins are small.
10. Caudal fin is forked.

Geographical distribution: *Rita rita* is found in freshwaters throughout India except in Madras and Mysore states.

88. *Anabas*

Classification:

Phylum	...	**Chordata**	Characters same as those of *Hilsa.*
Group	...	**Vertebrata**	
Subphylum	...	**Gnathostomata**	
Series	...	**Pisces**	
Class	...	**Teleostomi**	
Subclass	...	**Actinopterygii**	
Order	...	**Perciformes**	→ Fins usually with spines; two dorsal fins; ventral fins thoracic with not more than 6 rays; Weberian apparatus absent; physoclistic.
Genus	...	*Anabas*	
Species	...	*scandens*	

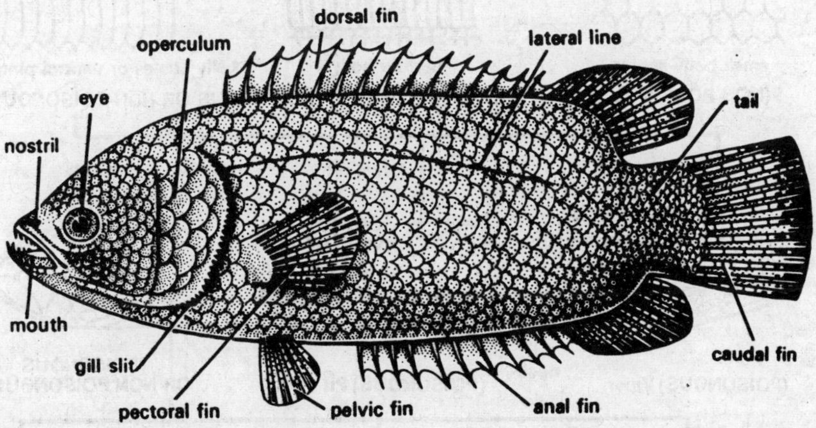

Fig. 11.87. *Anabas scandens.*

Comments:

1. *Anabas* is commonly known as **climbing perch**.
2. The body is laterally compressed and covered with ctenoid scales.
3. It attains the length of 20 cm.
4. Dorsal and anal fins are long and spinous.
5. Pectoral and pelvic fins are smaller. The caudal fin is well developed.
6. The operculum bears backwardly directed spines.
7. The accessory respiratory organs are well-developed.

8. The fish can live for a long time out of water. When it remains out of water it respires by the accessory respiratory organs.
9. It is predator, feeds on shrimps, gastropod shells and young fishes.
10. It is highly esteemed food fish at several places.
11. This fish can walk on land with the help of fin spines and opercular spines.

Geographical distribution: *Anabas scandens* is found in estuaries and freshwaters of India, Sri Lanka, Burma, Malaya Penninsula and Archipelago. It is also commonly found in South India.

REPTILIA

POISONOUS AND NON-POISONOUS SNAKES

The majority of snakes are non-poisonous and harmless animals. Some of the common poisonous snakes of India are cobra, king-cobra, krait, pit-viper, Indian viper, rattle-snake, russel's snake, coral-snake, and sea-snakes. The identification of poisonous and non-poisonous snakes is very simple. The following key will serve to distinguish **poisonous snakes** from **non-poisonous snakes**:

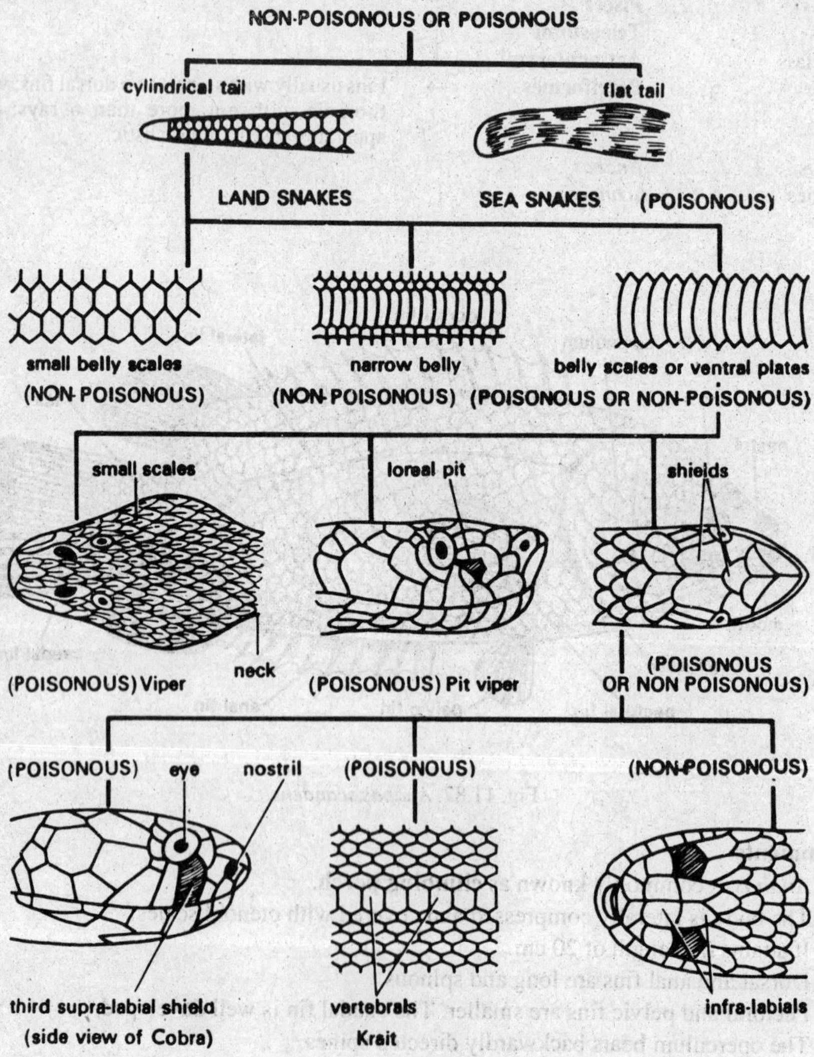

Fig. 11.88. Key to identification of poisonous and non-poisonous snakes.

Key to Identification of Poisonous and Non-poisonous Snakes.

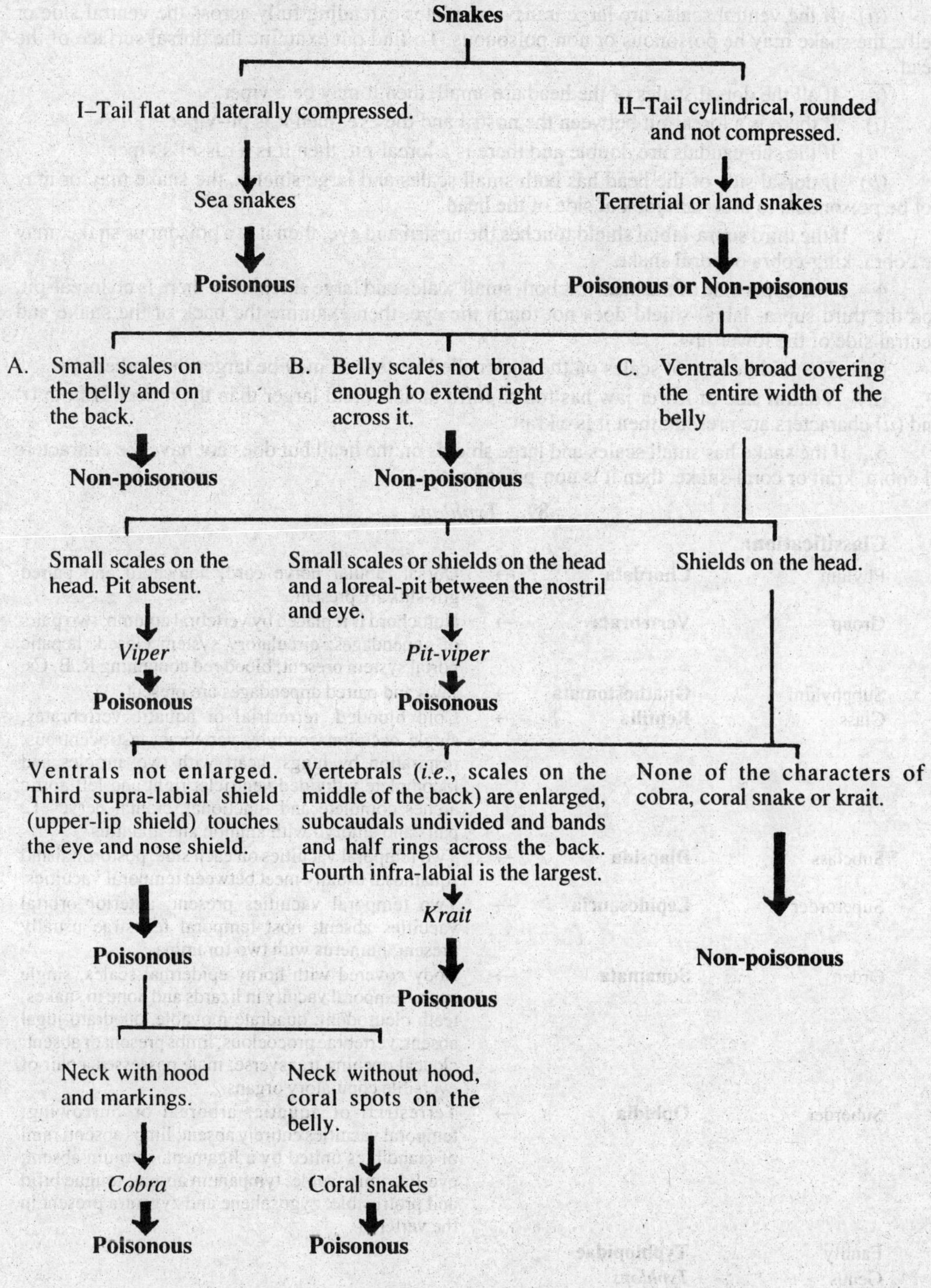

1. If it is a marine snake with the tail laterally compressed, it is poisonous.
2. In a terrestrial snake the tail is rounded or cylindrical and not compressed. Examine its ventral scales.

(*i*) If all the ventral scales are small or the ventral scales are somewhat broad, then it is a non-poisonous snake.

(*ii*) If the ventral scales are large transverse plates extending fully across the ventral side or belly, the snake may be poisonous or non-poisonous. To find out examine the dorsal surface of the head.

(*a*) If all the dorsal scales of the head are small, then it may be a viper.

(*i*) If there is a loreal-pit between the nostril and the eye, then it is pit-viper.

(*ii*) If the sub-caudals are double and there is a loreal-pit, then it is a russel's viper.

(*b*) If dorsal side of the head has both small scales and large shields, the snake may or may not be poisonous, to ascertain look at side of the head.

3. If the third supra-labial shield touches the nostril and eye, then it is a poisonous snake, may be cobra, king-cobra or coral snake.

4. If the upper side of the head has both small scales and large shields but there is no loreal-pit, and the third supra- labial shield does not touch the eye, then examine the back of the snake and ventral side of the lower jaw.

(*i*) The middle row of scales on the back called vertebrals may be larger then others.

(*ii*) Ventral side of lower jaw has fourth infra-labial shield larger than the others. If both (*i*) and (*ii*) characters are present, then it is a krait.

5. If the snake has small scales and large shields on the head but does not have the characters of cobra, krait or coral-snake, then it is non-poisonous.

89. *Typhlops*

Classification:

Phylum	...	**Chordata**	→	Dorsal tubular nerve cord, notochord and paired gill-slits are present.
Group	...	**Vertebrata**	→	Notochord is replaced by vertebral column; two pairs of appendages; circulatory system closed; hepatic portal system present; blood red containing R. B. Cs.
Subphylum	...	**Gnathostomata**	→	Jaws and paired appendages are present.
Class	...	**Reptilia**	→	Cold blooded, terrestrial or aquatic vertebrates; single occipital condyle; vertebrae gastrocentrous; respiration by lungs; heart with two auricles and incompletely divided ventricle; right and left aortic arches complete and functional, cranial nerves 12 pairs and embryo with amnion and allantois.
Subclass	...	**Diapsida**	→	Two temporal vacuities on each side; postorbital and squamosal usually meet between temporal vacuities.
Superorder	...	**Lepidosauria**	→	Two temporal vacuities present; anterior orbital vacuities absent; post temporal fenestrae usually present; humerus with two foramina.
Order	...	**Squamata**	→	Body covered with horny epidermal scales; single supra-temporal vacuity in lizards and none in snakes; teeth pleurodont; quadrate movable, quadrato-jugal absent; vertebrae procoelous; limbs present or absent; cloacal opening transverse; male possesses a pair of eversible copulatory organs.
Suborder	...	**Ophidia**	→	Terrestrial or aquatic, arboreal or burrowing; temporal vacuities entirely absent; limbs absent; rami of mandibles united by a ligament, sternum absent; eye-lids immovable; tympanum absent; tongue bifid and protrusible; zygosphene and zygantra present in the vertebra.
Family	...	**Typhlopidae**		
Genus	...	***Typhlops***		
Species	...	***braminus***		

Comments:

1. *Typhlops braminus* is commonly called **blind-snake** and is found in rotting vegetation in semi-tropical and tropical regions.
2. It is about 170–180 mm in length and brown with a shining chocolate hue above and lighter below in colour.
3. The body is cylindrical and covered with uniform imbricate scales all over.
4. There is no distinction between dorsal and ventral scales.
5. The head is not distinct from the body.
6. The tail is blunt with small point.
7. The eyes are small and more or less covered by scales.
8. Few teeth are present in the upper jaw only.
9. Parietal bone is paired.
10. Presence of a large rostral, nasal, ocular and preocular shields on the head.
11. Nasal shields are completely divided into an anterior lower and a posterior upper position.

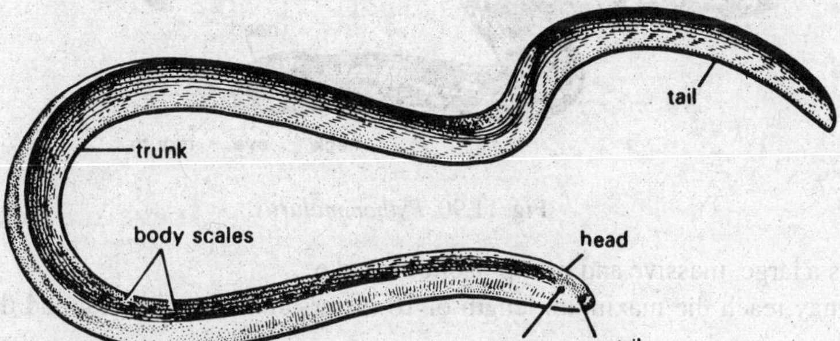

Fig. 11.89. *Typhlops braminus.*

12. Sebaceous glands are present on the body.
13. Some vestiges of pelvic bones are present inside the body.
14. Oviparous.
15. Carnivorous, feeding on worms, soft-bodied insects and their larvae.
16. Burrowing in habit.

Geographical distribution: *Typhlops braminus* is widely distributed in Asia, Malay islands, the islands in the Indina Ocean and in Southern Africa. It is found throughout India.

90. *Python*

Classification:

Phylum	...	**Chordata**	Characters same as those of *Typhlops.*
Group	...	**Vertebrata**	
Subphylum	...	**Gnathostomata**	
Class	...	**Reptilia**	
Subclass	...	**Diapsida**	
Superorder	...	**Lepidosauria**	
Order	...	**Squamata**	
Suborder	...	**Ophidia**	
Family	...	**Boidae**	
Genus	...	*Python*	
Species	...	*molurus*	

Comments:

1. *Python molurus* (Fig. 11.90) is commonly called **Indian python** or **ajgar**.

Fig. 11.90. *Python molurus.*

2. It is a large, massive and **non-poisonous** snake.
3. It may reach the maximum length of 10 meters and is the biggest and thickest snake of India.
4. It weighs to a maximum of 125 kg.
5. The colour is brown above with rhomboid dark-grey edged spots on the body and ventral side is greyishh with yellow-brown spots.
6. The scales are in 60 to 75 smooth rows, the ventrals are distinctly smaller.
7. Head is distinct from the neck and is covered with symmetrical shields or small scales.
8. Presence of a lancet-shaped brown mark on the head.
9. The eyes are without eyelids and the pupil is vertical.
10. The tail is short and prehensile.
11. Supra-labials are 11–13. The first two are full of pits and the sixth and seventh touching the eye.
12. Infra-labials are 16–18. Anterior ones are long and narrow and posterior ones have pits.
13. Parietal, loreal and temporal regions are covered with irregular scales.
14. The rostral scale on the head bears a deep pit on either side, which is probably thermoreceptive in function.
15. Rudiments of appendages in the form of two claw spurs are present near the cloaca.
16. Carnivorous, feeding on reptiles, birds and mammals.
17. Oviparous.

Geographical distribution: *Python molurus* is distributed in India and also in Indo-China.

91. *Eryx*

Classification:

Phylum	...	**Chordata**	Characters same as those of *Typhlops*.
Class	...	**Vertebrata**	
Subphylum	...	**Gnathostomata**	

Class	...	**Reptilia**	Characters same as those of *Typhlops*.
Subclass	...	**Diapsida**	
Superorder	...	**Lepidosauria**	
Order	...	**Squamata**	
Suborder	...	**Ophidia**	
Family	...	**Boidae**	
Genus	...	*Eryx*	
Species	...	*johnii*	

Comments:

1. *Eryx johnii* is commonly called **Indian** or **John sand-boa** or **dumuhi** and found burried in the soil.
2. Body is short and flat and chocolate in colour.
3. Head is not distinguished from the neck.
4. Eyes are very small with a vertical pupil.
5. Nostril is like a slit.
6. Tympanum is absent.
7. Tail is blunt and looks like the head end.
8. Only three scales on the head are enlarged and remaining are small.
9. Body scales are 40–55 in rows often keeled particularly on the tail.
10. The ventral scales do not run across the body.
11. Carnivorous, feeding on frogs, lizards and mice, etc.
12. It is subterranean and **harmless** snake.
13. It sometimes crawls back towards the tail end and its blunt tail end resembles head, hence, called **double-headed** snake.
14. It kills its prey by constricting the prey in its body coils.

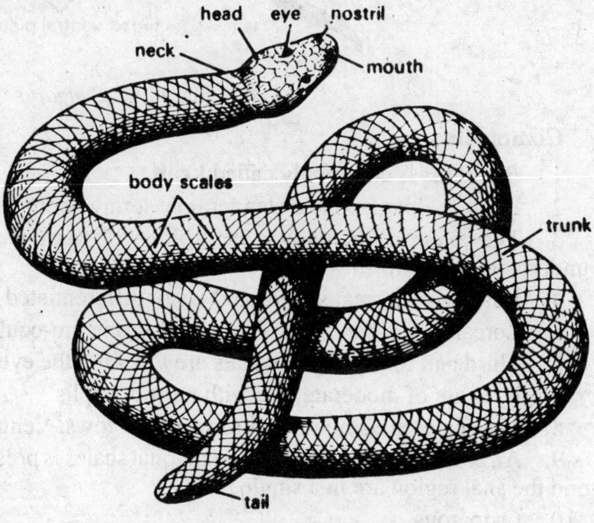

Fig. 11.91. *Eryx johnii* (Dumuhi).

Geographical distribution: *Eryx johnii* is distributed in sandy parts of India, Sri Lanka and Africa.

92. *Bungarus*

Classification:

Phylum	...	**Chordata**	Characters same as those of *Typhlops*.
Group	...	**Vertebrata**	
Subphylum	...	**Gnathostomata**	
Class	...	**Reptilia**	
Subclass	...	**Diapsida**	
Superorder	...	**Lepidosauria**	
Order	...	**Squamata**	
Suborder	...	**Ophidia**	
Family	...	**Elapidae**	
Genus	...	*Bungarus*	

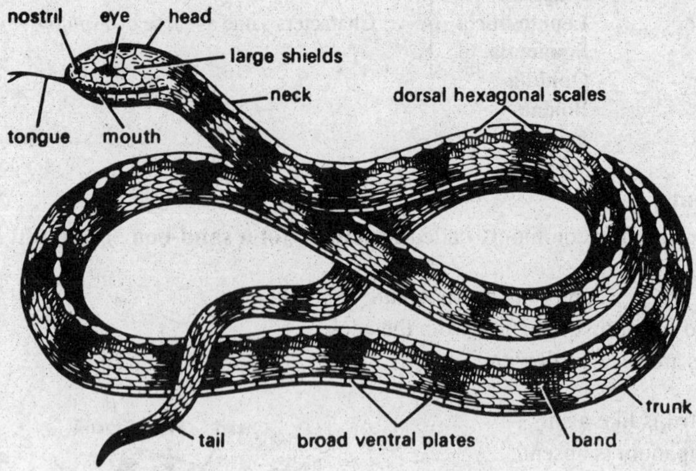

Fig. 11.92. *Bungarus.*

Comments:

1. *Bungarus* is commonly called **krait**.
2. Body is elongated and slender, measuring about 1 meter in length.
3. The colour of the body is steel-blue with narrow cross bars or white specks dorsally and the underparts are uniform white.
4. Head with normal shields and is not differentiated from neck.
5. Loreals are absent. Two post-oculars, one pre-ocular and seven supra-labials are present.
6. Third and fourth supra-labials are touching the eye.
7. Eyes are of moderate size with narrow puplis.
8. The scales are smooth and form 13–17 rows. Ventrals are 194–234 and 42–52 caudals.
9. An enlarged chain of dorsal hexagonal scales is present on the dorsal side and ventral scales beyond the anal region are in a single row.
10. Oviparous.
11. Carnivorous, feeding on rats, lizards and snakes.
12. It is nocturnal in habit and lives in cracks in walls or in heaps of bricks, etc.
13. *Bungarus* is **poisonous** and its venom is **neurotoxic**.
14. The krait "*Bungarus*" seems to cause more deaths in India than any other snake, since it is very common and often creeps into the houses.

Geographical distribution: *Bungarus* is found in South Eastern Asia, India and Malaya.

93. *Naja*

Classification:

Phylum	...	**Chordata**	Characters same as those of *Typhlops*.
Group	...	**Vertebrata**	
Subphylum	...	**Gnathostomata**	
Class	...	**Reptilia**	
Subclass	...	**Diapsida**	
Superorder	...	**Lepidosauria**	
Order	...	**Squamata**	
Suborder	...	**Ophidia**	
Family	...	**Elapidae**	
Genus	...	*Naja*	
Species	...	*naja*	

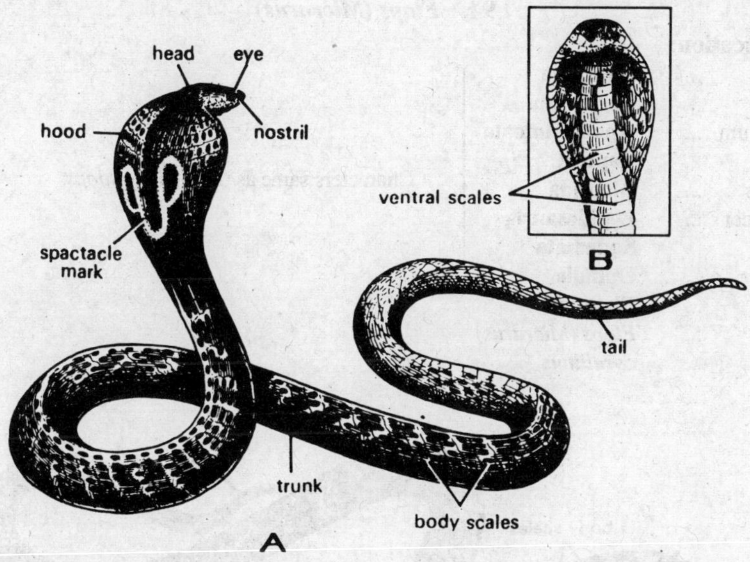

Fig. 11.93. *Naja naja.* A–Hood expanded; B–Ventral veiw of hood.

Comments:

1. *Naja naja* (Fig. 11.93) is commonly called **Indian cobra** or **nag**.
2. Body is elongated measuring one and a half to two meters in length.
3. The colour of the body is brown or blackish.
4. Body is covered with smooth oblique scales without pits and are arranged in 15–25 rows. The subcaudals form only two rows.
5. Head is not differentiated from the neck. Neck is dilatable and the cervical ribs are elongated. The expansion of the neck and cervical ribs form the **hood**.
6. The upper surface of the hood bears a **binocoellate mark** to which people call a mark of **spectacle** or the figure of ten.
7. The lower surface of the hood bears two dark round spots running to four scales surrounded by white lateral borders.
8. Eyes are very small and with round pupils.
9. Each nostril lies between two nasals and the inter-nasal.
10. Loreal is absent. Frontal shield is truncated. Three postocular scales are present.
11. Third supra-labial is large and touches the eye and the nasal.
12. Poison fangs are followed by 1–3 small teeth.
13. Tail is cylindrical and tapering posteriorly.
14. Oviparous.
15. Carnivorous, feeding on frogs, lizards, rats and other snakes.
16. *Cobra* is **deadly poisonous** and its venom is neurotoxic and fatal. When it bites there is light pain and swelling, irritaion and death due to respiratory failure.
17. *Cobra* raises its hood when alarmed and the hood sways back and forth for striking the object. During this period, it produces hissing sound.
18. *Cobra* is diurnal, shy, living in holes, under stones, mud walls and in thick vegetation.

Geographical distribution: *Naja* is widely distributed from Transcaspia of China and Malay islands. *Naja naja* is found all over India.

94. *Elaps (Micrurus)*

Classification:

Phylum	...	**Chordata**	Characters same as those of *Typhlops*.
Group	...	**Vertebrata**	
Subphylum	...	**Gnathostomata**	
Class	...	**Reptilia**	
Subclass	...	**Diapsida**	
Superorder	...	**Lepidosauria**	
Order	...	**Squamata**	
Suborder	...	**Ophidia**	
Family	...	**Elapidae**	
Genus	...	*Elaps (Micrurus)*	
Species	...	*corallinus*	

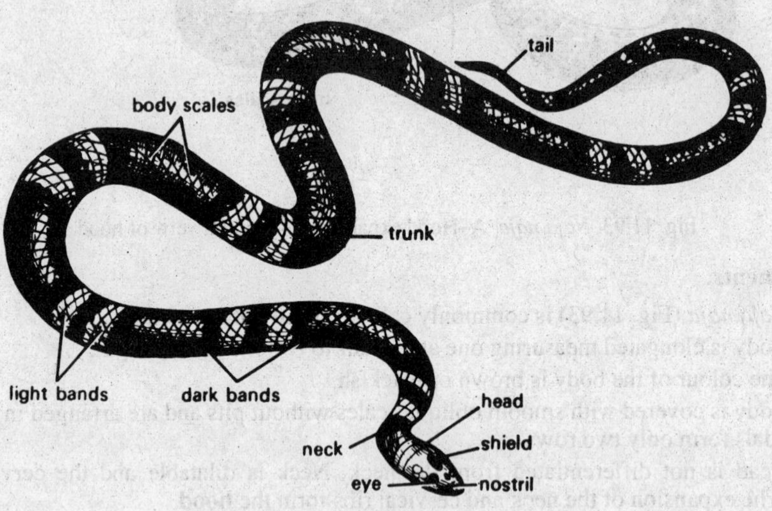

Fig. 11.94. *Elaps*.

Comments:

1. *Elaps* or *Micrurus* is commonly called **coral-snake.**
2. Body is elongated, cylindrical and measuring about one to one and a half meter in length.
3. The general colour of the body is extremely pretty black and red rings alternate with one another.
4. The scales of the body are smooth and form 15 rows.
5. The head is small and not distinct from the tail.
6. Eyes are small with vertical pupil.
7. Maxillaries carry no teeth behind the poison fangs.
8. The gape of the mouth is so limited that these snakes although possessing strong poison fangs are practically harmless to man.
9. The tail is short and round.
10. **Poisonous**, poison is **neurotoxic**.
11. Carnivorous, feeding upon lizards and other snakes.

Geographical distribution: *Elaps* is an entirely American genus. It is distributed in Tropical South America and the Lesser Antilles.

95. *Viper*

Classification:

Phylum	...	**Chordata**	Characters same as those of *Typhlops.*
Group	...	**Vertebrata**	
Subphylum	...	**Gnathostomata**	
Class	...	**Reptilia**	
Subclass	...	**Diapsida**	
Superorder	...	**Lepidosauria**	
Order	...	**Squamata**	
Suborder	...	**Ophidia**	
Family	...	**Viperidae**	
Genus	...	*Viper*	
Species	...	*russelli*	

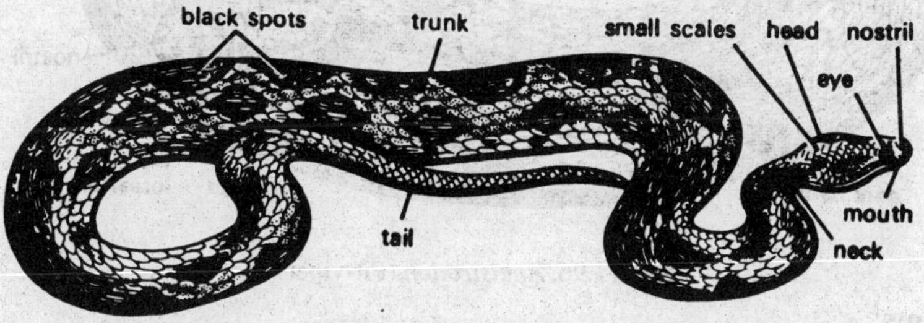

Fig. 11.95. *Viper russelli.*

Comments:

1. *Viper russelli* is commonly called **Russell's viper** or **daboia.**
2. Body is elongated and measures up to one and a half meter in length.
3. The general colour is pale brown above with three longitudinal series of black spots and underparts are yellowish white.
4. The scales form about 30 rows on the body.
5. Head is triangular and covered with very small imbricate scales on the upper surface.
6. The snout is angulate and the nasal opening is quite prominent.
7. The eyes are with golden iris and elliptical pupil.
8. Supra-labials are 10–12. Fourth supra-labial is the largest and it does not touch the eye.
9. Viviparous.
10. Carnivorous, feeding mainly on small mammals, *e.g.*, mice and rats and also on lizards and birds.
11. Very **poisonous.** Venom is **haemotoxic.**
12. *Viper russelli* is found in rocky bushy regions where the colouration on the skin is in keeping with the surroundings.

Geographical distribution: *Viper russelli* is distributed in India, Sri Lanka, Burma and Thailand.

96. *Ancistrodon (Agkistrodon)*

Classification:

Phylum	...	**Chordata**	Characters same as those of *Typhlops.*
Group	...	**Vertebrata**	
Subphylum	...	**Gnathostomata**	
Class	...	**Reptilia**	
Subclass	...	**Diapsida**	
Superorder	...	**Lepidosauria**	
Order	...	**Squamata**	

Suborder	...	**Ophidia**	Characters same as those of *Typhlops*.
Family	...	**Viperidae**	
Genus	...	*Ancistrodon*	
Species	...	*himalayans*	

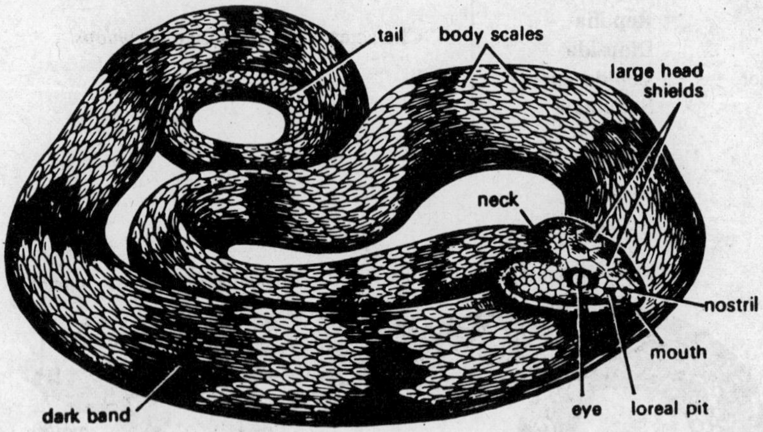

Fig. 11.96. *Ancistrodon* (Pit-viper).

Comments:

1. *Ancistrodon himalayans* is commonly called **Himalayan pit-viper.**
2. The general colour is bluish-brown above with dark-brown or black spots appearing like cross bars and the ventral side is brown with black and white spots.
3. Body is not much elongated measuring about 70 cm in length.
4. The scales covering the body have apical sensory pits.
5. Mouth is grey with black spots.
6. The upper lip is slightly raised in the rostral region.
7. Supra-labials are 5–7, 3rd and 4th are enlarged but not touching the eye.
8. Post-oculars are two, the last greatly enlarged and pre-oculars are three.
9. A **loreal-pit** is present in between the eye and the nostril on each side.
10. Eyes are big with golden iris and vertical pupils.
11. Tail is cylindrical without rattles.
12. Viviparous.
13. **Poisonous** and its poison is **haemotoxic.**
14. *Ancistrodon* is quite a timid snake, remains hidden under stones, etc.

Geographical distribution: *Ancistrodon himalayans* is distributed in hilly regions of Eastern parts of India, Sri Lanka and Java. It is very common in Kashmir and is found even at an altitude of 4000 meters.

97. *Crotalus*

Classification:

Phylum	...	**Chordata**	Characters same as those of *Typhlops*.
Group	...	**Vertebrata**	
Subphylum	...	**Gnathostomata**	
Class	...	**Reptilia**	
Subclass	...	**Diapsida**	
Superorder	...	**Lepidosauria**	
Order	...	**Squamata**	
Suborder	...	**Ophidia**	

Family	...	**Crotalinae**
Genus	...	*Crotalus*

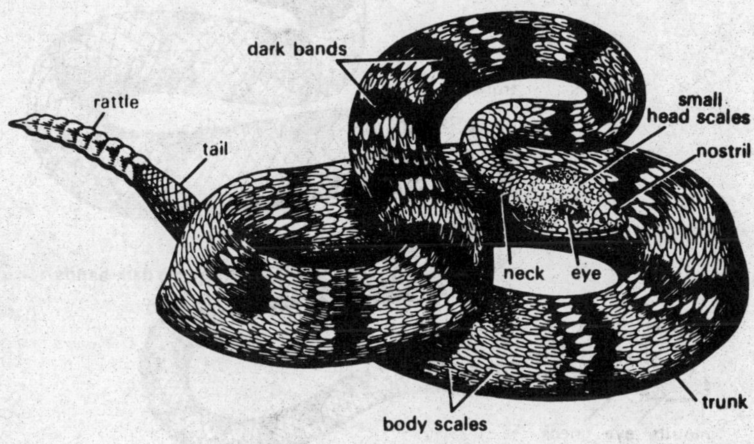

Fig. 11.97. *Crotalus* (Rattle-snake).

Comments:

1. *Crotalus durissus* is commonly called **rattle-snake.**
2. The general colour is grey-brown with dark bands on the body forming a handsome pattern.
3. Body is very elongated reaching a length of about two and a half meters.
4. Head is distinct from neck.
5. The upper surface of the head is covered with small scales.
6. Presence of two pairs of inter-nasals.
7. Presence of a **rattle** at the end of the tail.
8. Rattle is formed by the union of 10–12 hollow segments of dry horny skin.
9. Rattle causes a noise which is useful to the snake as an instrument of warning off any approaching enemy.
10. Viviparous.
11. **Deadly poisonous.**
12. Carnivorous and nocturnal. They chiefly feed upon small mammals.
13. During the cold months they hibernate underground, often in considerable numbers.

Geographical distribution: *Crotalus* is distributed in U. S. A.; *C. horridus* is the common rattle snake of U. S. A., *C. confluentus* is found in Western America and *C. durissus* the common species in South Eastern North America.

98. *Hydrophis*

Classification:

Phylum	...	**Chordata**	Characters same as those of *Typhlops.*
Group	...	**Vertebrata**	
Subphylum	...	**Gnathostomata**	
Class	...	**Reptilia**	
Subclass	...	**Diapsida**	
Superorder	...	**Lepidosauria**	
Order	...	**Squamata**	
Suborder	...	**Ophidia**	
Family	...	**Hydrophiidae**	
Genus	...	*Hydrophis*	

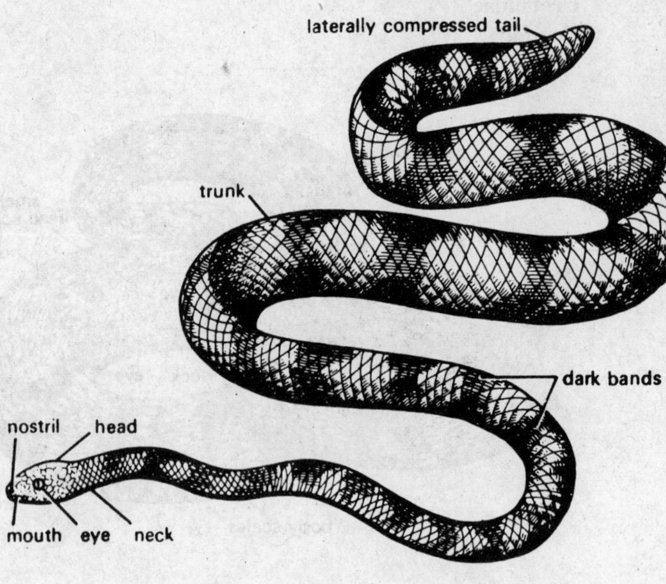

Fig. 11.98. *Hydrophis.*

Comments:

1. *Hydrophis* (Fig. 11.98) is commonly called **sea-snake.**
2. The body is long and laterally compressed posteriorly.
3. Head and neck are very slender.
4. The general colour is dark olive-green above with yellowish cross bars and white below.
5. The ventral scales are small.
6. Tail is **laterally campressed** and acts like paddle in swimming.
7. Eyes are small with rounded pupil.
8. Loreal-shield is absent. One pre-ocular, two post-oculars and 7–8 supra-labials of which 3rd and 4th are touching the eye.
9. Presence of 14–18 maxillary teeth behind the poison fangs.
10. Viviparous.
11. Carnivorous, feeding on fishes, etc.
12. *Hydrophis* is **deadly poisonous** and its venom is **neurotoxic.**

Geographical distribution: *Hydrophis* occurs in the Bay of Bengal and Malay Archipelago.

99. *Natrix*

Classification:

Phylum	...	**Chordata**	Characters same as those of *Typhlops.*
Group	...	**Vertebrata**	
Subphylum	...	**Gnathostomata**	
Class	...	**Reptilia**	
Subclass	...	**Diapsida**	
Superorder	...	**Lepidosauria**	
Order	...	**Squamata**	
Suborder	...	**Ophidia**	
Family	...	**Colubridae**	
Series	...	**Aglypha**	
Genus	...	*Natrix*	
Species	...	*piscator*	

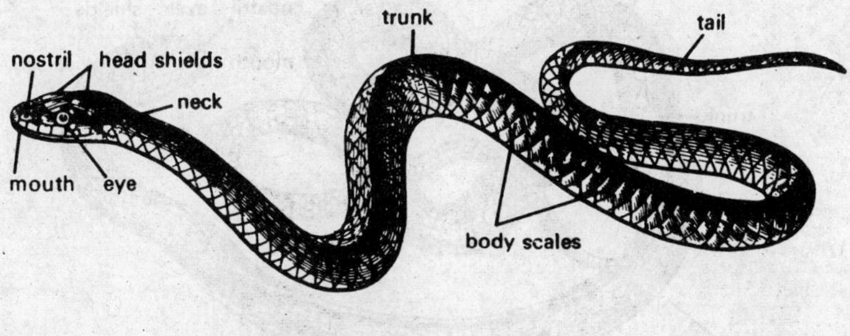

Fig. 11.99. *Natrix piscator.*

Comments:

1. *Natrix piscator* is commonly called **water-snake** or **panisamp** and found in freshwater ponds and in paddy-fields during rainy season.
2. Body is stout having a flat head.
3. The colour is olive with black spots arranged in the chess-board manner.
4. Scales are keeled and form 19 rows. Ventrals are rounded and subcaudals are paired.
5. Eyes are round and seen surrounded by a white circle.
6. Tympanum is absent.
7. Maxillary teeth 22–28 are enlarged posteriorly.
8. Supra-labials are 9. 4th and 5th are touching the eye.
9. **Non-poisonous** (fangs are absent).
10. Oviparous.
11. Carnivorous, feeds on fishes, frogs and tadpoles, etc.

Geographical distribution: *Natrix* is distributed widely in United States, Mexico and Central America, North Africa and India. *N. piscator* is found all over India.

100. *Zamenis (Ptyas)*

Classification:

Phylum	...	**Chordata**	Characters same as those of *Typhlops.*
Group	...	**Vertebrata**	
Subphylum	...	**Gnathostomata**	
Class	...	**Reptilia**	
Subclass	...	**Diapsida**	
Superorder	...	**Lepidosauria**	
Order	...	**Squamata**	
Suborder	...	**Ophidia**	
Family	...	**Colubridae**	
Series	...	**Aglypha**	
Genus	...	*Zamenis (Ptyas)*	
Species	...	*mucosus*	

Comments:

1. *Zamenis mucosus* (Fig. 11.100) is commonly called **dhaman** or **rat-snake of India.**
2. It grows to a length of more than two meters.
3. It is brown above often with black cross bands on the posterior part of the body and tail and underparts are yellowish.
4. The scales on the body form 16–17 rows.
5. Presence of a prominent **dorsal ridge** of the back bone along the mid-dorsal line.
6. Head is distinct from the neck.
7. The eyes are large with round pupil.

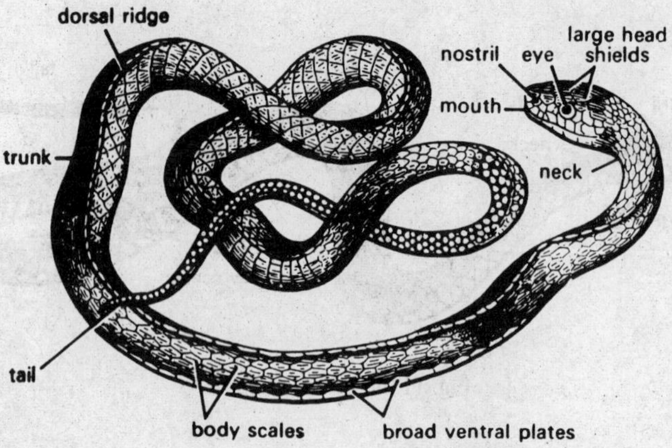

Fig. 11.100. *Zamenis macosus*..

8. Fourth and fifth supra-labials are touching the eye.
9. Teeth are present but fangs are absent.
10. Tail is long and prehensile.
11. It can climb a tree and lives mainly in the plains feeding on frogs and toads, lizards, mammals and even the snakes.
12. **Non-poisonous** (fangs absent)
13. Oviparous.

Geographical distribution: *Zamenis* is very widely distributed. It is found in France, Italy, North-West Africa, America and India. *Z. mucosus* is found all over India in plains and also at some altitudes.

101. *Dendrophis*

Classification:

Phylum	...	**Chordata**	Characters same as those of *Typhlops*.
Group	...	**Vertebrata**	
Subphylum	...	**Gnathostomata**	
Class	...	**Reptilia**	
Subclass	...	**Diapsida**	
Superorder	...	**Lepidosauria**	
Order	...	**Squamata**	
Suborder	...	**Ophidia**	
Family	...	**Colubridae**	
Series	...	**Aglypha**	
Genus	...	*Dendrophis*	

Comments:

1. *Dendrophis* (Fig.11.101) is commonly called **tree-snake** and arboreal in habit.
2. Body is elongated and slender, may reach the length of two meters.
3. The scales are keeled and form only 13 or 15 rows.
4. The scales of the vertebral row are enlarged.
5. The ventrals have a pair of suture-like lateral keels and notch on each side which are helpful in climbing.
6. The eyes are covered by a transparent membrane which is fused with it.
7. Tympanic membrane is absent.
8. Limbs and limb girdles are absent.
9. Oviparous.

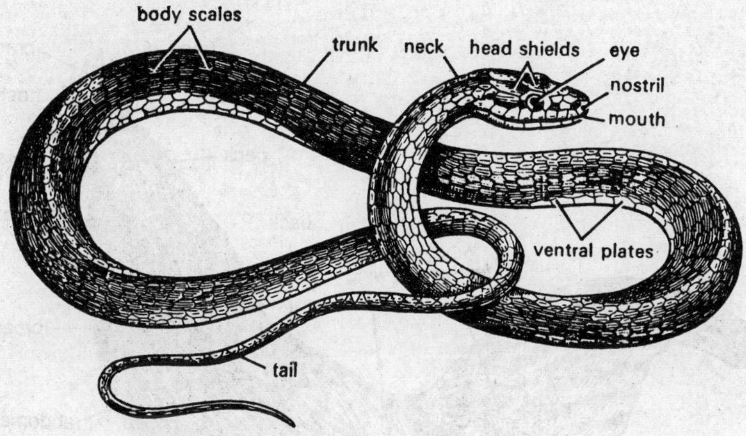

Fig. 11.101. *Dendrophis.*

10. Harmless (fangs are absent).
11. Carnivorous.

Geographical distribution: *Dendrophis* is found in South-East Asia and Australia.

AVES

102. *Gallus*

Classification:

Phylum	...	**Chordata**	→	Dorsal tubular nerve cord, notochord, and paired gill-slits are present.
Group	...	**Vertebrata**	→	Notochord is replaced by vertebral coloumn; two pairs of appendages; circulatory system closed; hepatic portial system present; blood red containing R. B. Cs.
Subphylum	...	**Gnathostomata**	→	Jaws and paired appendages are present.
Class	...	**Aves**	→	Warm-blooded with an exoskeleton of feathers; fore-limbs modified into wings; hind-limbs for walking, perching or swimming; skull monocondylic, modern birds without teeth, a horny beak present; heart four chambered; oviparous; embryo with amnion, allantois and yolk-sac.
Subclass	...	**Neornithes**	→	Tail short ends in a pygostyle; rectrices are arranged in a semicircle around the pygostyle, teeth absent; metacarpals fused with distal carpals to form carpometacarpus; claws absent in fore-limbs; sternum well developed with a keel; vertebrae heterocoelous.
Order	...	**Galliformes**	→	Beak short; graminivorous terrestrial birds with short and powerful flight; feet usually adapted for scratching and running.
Genus	...	*Gallus*		

Comments:

1. *Gallus* (Fig. 11.102) is commonly called **fowl.**
2. *G. gallus*, the **red jungle fowl** is the ancestor of all domestic breeds of fowls.
3. It is a moderate sized graminivorous terrestrial bird.
4. Its wings are strong, powerful but wing feathers are not much developed. Therefore, it can fly only to a short distance.
5. Its tail feathers are long and well developed.

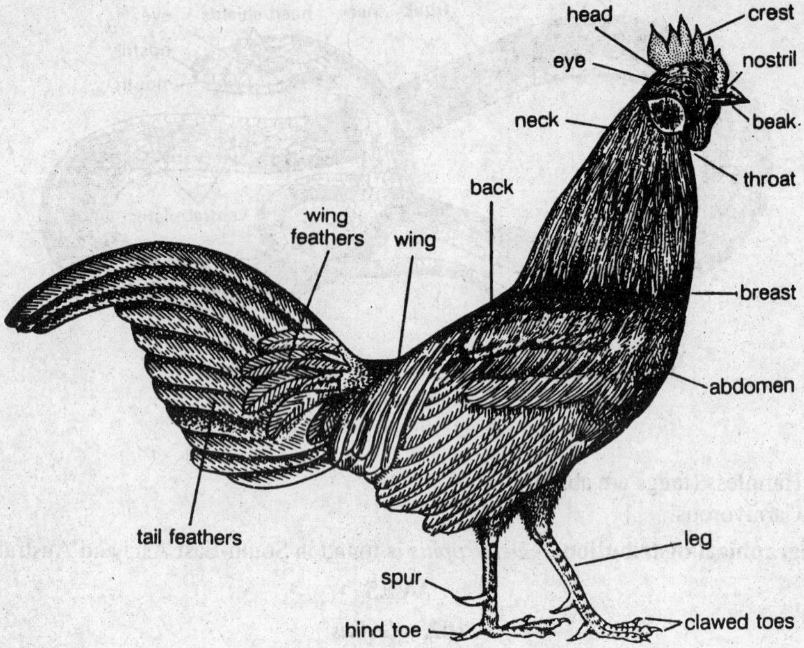

Fig. 11.102. *Gallus gallus*.

6. Its beak is adapted for graminivorous habit.
7. Its legs are adapted for scratching the ground.
8. It is domesticated for egg production and for meat.
9. It is a poor layer and the eggs are small-sized.
10. The present day breeds of poultry are all believed to be descended from the wild jungle fowl of the genus *Gallus*.

Distribution: *Gallus gallus* is found chiefly in the Himalayan terrain and foot hills extending to south.

MAMMALIA

103. *Lepus*

Classification:

Pnylum	...	**Chordata**	→	Dorsal tubular nerve cord, notochord, and paired gill-slits are present.
Group	...	**Vertebrata**	→	Notochord is replaced by vertebral column; two pairs of appendages; circulatory system closed; hepatic portal system present; blood red containing R. B. Cs.
Subphylum	...	**Gnathostomata**	→	Jaws and paired appendages are present.
Class	...	**Mammalia**	→	Warm blooded; skin covered with hairs; mammary glands in females; pinnae present; teeth heterodont, thecodont, and diphyodont; skull with two occipital condyles; vertebrae gastrocentrous; limbs either plantigrade, digitigrade or unguligrade; presence of muscular diaphragm; heart four chambered; R. B. Cs. non-nucleated; kidneys metanephros; penis always present; viviparous.

Subclass	...	**Theria**	→	Placental mammals; ear usually with external pinna; teeth present both in youngs and adults; cloaca absent; mammary glands with nipples; testes in scrotal sacs; females viviparous.
Infraclass	...	**Eutheria**	→	Marsupial pouch absent; mammary glands well developed with nipples; cloaca absent; corpus callosum absent; vagina single; viviparous; young always nourished for a considerable time in the uterus by allantoic placenta and born in a relatively advanced state.
Order	...	**Lagomorpha**	→	Herbivorous; incisors evergrowing chisel-shaped 4 in upper and 2 in lower jaws; canines absent.
Genus	...	*Lepus*		

Comments:

1. *Lepus* (hare) is an allied genera of *Oryctolagus* (rabbit).
2. Its body colour is brown above and white below.
3. Its pinnae are longer having black tips and fore- and hind-limbs of almost equal size.
4. Its upper front incisor teeth without groove.
5. Its upper lip is divided into two by a median vertical cleft so that its upper incisor teeth can be seen from outside.
6. It is nocturnal and remains active during night for feeding.
7. It cannot be domesticated and leads a vagabond life.
8. It damages our crops by eating them and cutting them voraciously.

Habit and habitat: It is solitary in habit and nocturnal. It is nomadic living in temporary shelters in bushes or in grass and cannot be domesticated. It is a fast runner at about 60 km per hour.

Distribution: Found throughout India; its North Indian species is *Lepus ruficaudatus*, while South Indian species is *Lepus nigricollis*.

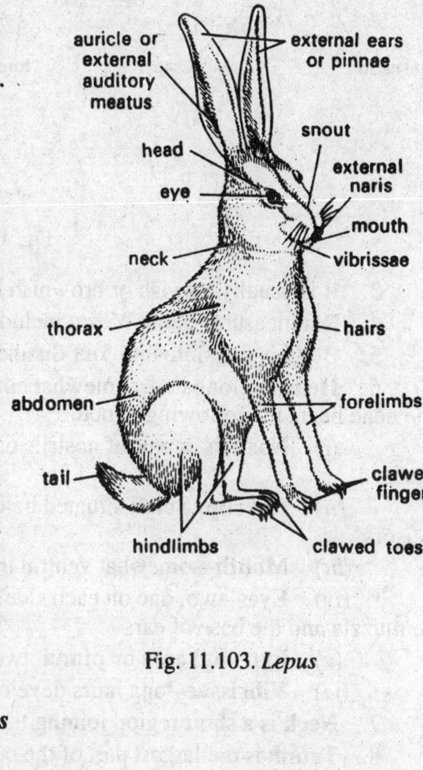

Fig. 11.103. *Lepus*

104. *Rattus*

Classification:

Phylum	...	**Chordata**		Characters same as those of *Lepus*.
Group	...	**Vertebrata**		
Subphylum	...	**Gnathostomata**		
Class	...	**Mammalia**		
Subclass	...	**Theria**		
Infraclass	...	**Eutheria**		
Order	...	**Rodentia**	→	Limbs with five toes and claws; incisors long, chisel-shaped, rootless; canines absent, diastema present.
Genus	...	*Rattus*		

Comments:

1. *Rattus rattus* (Fig. 11.104) is commonly called **common rat.**
2. The body of rat is slender with thick-set fine fur mixed with stiff hairs.

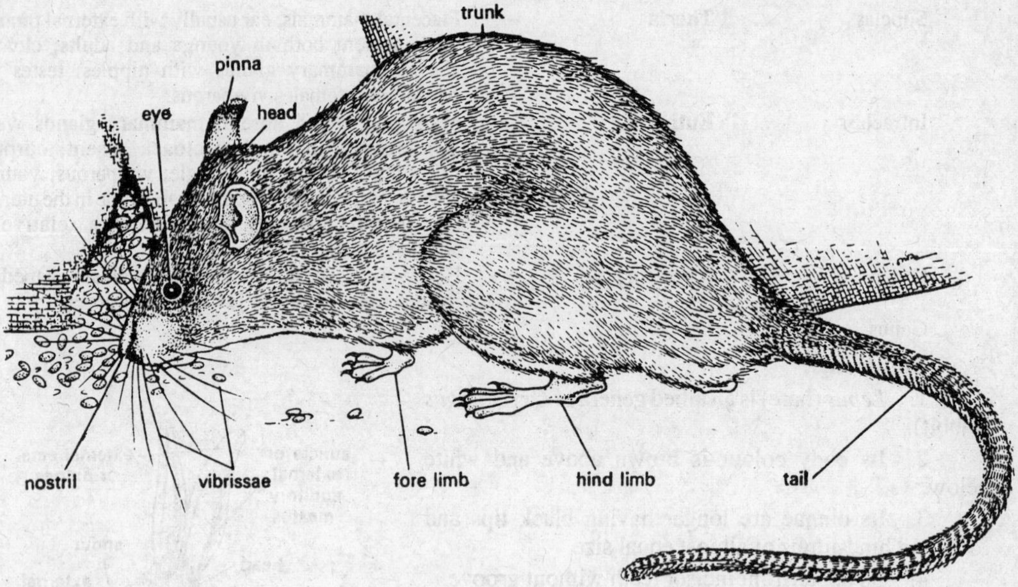

Fig. 11.104. *Rattus rattus.*

3. It is usually grayish or brownish in colour.
4. Rat measures about 25 cm including tail.
5. Body is divisible into four distinct regions; **head**, **neck**, **trunk** and **tail.**
6. **Head** is elongated, somewhat conical in shape and tapering anteriorly to a terminal nose. The head bears the following structures:

(*i*) **Nostrils**–a pair of nostrils or nares lying at the tip of the nose which is slightly elevated.

(*ii*) **Harelip**–a cleft situated below the nostrils in the upper tip exposing the two upper incisors.

(*iii*) **Mouth**–somewhat ventral in position, bounded by an **upper** and a **lower lip.**

(*iv*) **Eyes**–two, one on each side. These are dorso-lateral situated about mid-way between the muzzle and the base of ears.

(*v*) **External ears or pinna**–two lying at the base of head in the postero-lateral area.

(*vi*) **Vibrissae**–long hairs developed on the face at the sides of the snout.

7. **Neck** is a short region joining the head to the trunk.
8. **Trunk** is the largest part of the body. It is divided into **thorax** and **abdomen.** Trunk bears two pairs of limbs, *viz.*, **fore-limbs** and **hind-limbs.** It also bears certain apertures.

(*i*) **Fore-limbs** include the **arm** and **hand.** The hand comprises the palmer region and the digits four in number. The first digit or thumb is greatly reduced.

(*ii*) **Hind-limbs** consists of **thigh**, **shank** and **foot.** The foot bears five digits.

(*iii*) **Nipples**–In females one pair of thoracic nipples, two pairs of axillary nipples and one pair of abdominal nipples and two pairs of inguinal nipples are present on the ventral surface.

(*iv*) In **female** urethral opening and vulval opening and anus lie close together in the pelvic region.

(*v*) In **male** penis is cylindrical which can be retracted into a loose sheath of skin, the prepuce. Two large testes are situated in the scrotal sacs.

(*vi*) **Anus** is a small median aperture below the base of the tail.

9. **Tail** is elongated, cylindrical and tapering and often slightly longer than body. It is nearly devoid of hairs and covered with overlapping vestigial scales. It acts as a balancing organ.

10. Viviparous.

11. The rat is chiefly nocturnal in habit and thrives best in warm climate. It inhabits holes and burrows in the houses and cultivated fields.

12. The original food of rat was perhaps various kinds of grains and seeds but at present time the rat consumes practically anything that contains nourishment and comes within the reach of its sharp teeth.

13. Rat also acts as a carrier of diseases like plague and typhus fever.

Geographical distribution: *Rattus rattus* is found throughout the world. It prefers warm and dry climate.

Economic importance: *Rattus rattus* is domestic rat and causes serious damage to our household goods including clothes and stored grains. Other rodents of allied species cause severe damage to our agricultural crops, transmit various germs of diseases like bubonic plague, rabies, rat bite fever, etc., to man and domestic animals. Total annual economic losses due to rats are believed to over 20 million tones in the field and over 33 million tons in storage of world's cereal production.

Cultural, biological, chemical and integrated techniques are used in controlling the rat nuisence.

CHAPTER 12

EXPERIMENTAL PHYSIOLOGY

The experimental physiology includes tests of carbohydrates, proteins, lipids (fats), enzymes, estimation of haemoglobin, counting of R.B.C., W.B.C., differential count of W.B.C., study of coagulation of blood, bleeding time of blood, clotting time of blood, preparation of haemin crystals, effect of osmolarity of salt solutions and haemolytic agents on R.B.Cs., study of circulation of blood in the capillaries of foot of frog, recording of heart beat in situ; recording of muscular contraction, determination of rate of respiration, reflex action, wiping reflex, knee-jerk reflex, test of urine for urea, proteins, ketones and sugar, and demonstration of osmosis.

CARBOHYDRATES

The carbohydrates are the compounds of carbon, hydrogen and oxygen. Chemically the carbohydrates are polyhydroxy aldehydes or ketones and, therefore, they are also referred to as aldoses or ketoses. The carbohydrates are of common occurrence in plants and animals.

Carbohydrates are, however, classified into **monosaccharides,** *viz.*, simple sugars like *glucose* and *fructose,* **di-** and **tri-saccharides** or **oligosaccharides,** *viz.,* compound sugars like *sucrose, lactose, maltose,* etc., and **polysaccharides,** *viz.,* complex sugars like *glycogen, starch, cellulose,* etc. Monosaccharides and oligosaccharides are water soluble crystalline substances having a sweet taste and are collectively known as sugars. Polysaccharides are insoluble amorphous substances and are, thus, called non-sugars. The laboratory tests for carbohydrates done are as under : .

EXPERIMENT No. 1 A

General Test for Carbohydrates

Before proceeding with the specific tests, check the solubility of given sample in water if it is in crystalline or powdered form. For this, dissolve a little of the sample in water in a test tube. If it dissolves completely, the given sample is water soluble. Now proceed with the specific test given below :

Molisch's Test

Take two test tubes, mark them A and B. Now take 5 ml solution of the given sample in test tube A. Add 2 drops of Molisch's reagent in it and shake the test tube so that the two solutions mix up thoroughly. Then take conc. H_2SO_4 in test tube B and pour it gently in test tube A along its side so that the two solutions do not mix.

Observation. A **pink** or **purple** ring appears at the junction of the two solutions.

Chemistry involved. H_2SO_4 when reacts with carbohydrates, it forms **furfural derivatives,** which on reaction with Naphthol of Molisch's reagent gives pink or purple ring at the surface because acid is heavier and does not mix with solution until stirred.

Inference. The given sample is a carbohydrate.

EXPERIMENT No. 1 B

Colour Test for Fructose

1. Foulger's Test

Procedure. Take 0.5 ml **test solution*** in a test tube. Add 3 ml of Foulger's reagent in it, boil and shake the test tube gently.

Observation. Blue colour appears.

Inference. Test solution contains **fructose.**

2. Seliwanoff's Test

Procedure. Take 3 ml Seliwanoff's reagent in a test tube, add 1 ml of the test solution and boil for 30 seconds.

Observation. Red to orange colour appears.

Inference. Presence of **fructose** confirmed in test solution.

EXPERIMENT No. 1 C

Colour Test for Glucose

Moor's Test

Procedure. Take 4 ml of test solution in a test tube. Add an equal volume of 2% sodium hydroxide solution in it and boil.

Observation. Solution turns yellow in the beginning and reddish brown later. It is due to the formation of **caramel,** a condensation product of glucose.

Inference. The test solution contains **glucose.**

EXPERIMENT No. 1 D

Colour Tests for Reducing Monosaccharides

1. Barfoed's Test

Procedure. Take 5 ml of Barfoed's reagent in a test tube. Add 4 ml test solution in the test tube, stirr and boil for one to two minutes. Let it stand for at least 15 minutes.

Observation. Precipitate of orange to red coloured cuprous oxide is formed.

Inference. Test solution contains **monosaccharide** which acts as a reducing agent.

2. Picric Acid Test

Procedure. Take 3 ml of test solution in a test tube. Add 1 ml of saturated picric acid solution and 1 ml of 1% or 4% NaOH solution in the test tube and warm it.

Observation. Red colour develops.

Inference. The test solution contains **reducing monosaccharide** because picric acid is reduced to picrylic acid which gives red colour.

3. Benedict's Test

Procedure. Take 5 ml of Benedict's reagent in a test tube. Add 0.5 ml of test solution in the test tube and mix it well. Heat to boil for two minutes then cool under tap-water.

Observation. Green, yellow or red precipitate is formed.

Inference. The test solution contains **reducing monosaccharide** because the precipitate is due to the formation of cuprous oxide.

---

* Test solution is the sample provided for detection of substances or self prepared solution from given sample in crystalline or powdered form.

4. Fehling's Test

Procedure. Take 5 ml of Fehling's solution in a test tube, add 2 ml of test solution in the test tube and boil.

Observation. Yellow brick-red precipitate appears.

Inference. The test solution contains **reducing monosaccharide** because the red colour of precipitate is due to cuprous oxide.

5. Methylene Blue Test

Procedure. Take 3 ml of distilled water and 0.5 ml of 40% NaOH solution in a test tube. Add a drop of 1% methylene blue in the test tube. Boil and cool. The colour of solution remains blue. Then add 1 ml of test solution to it, boil and cool again.

Observation. Blue colour of solution disappears.

Inference. The test solution contains **reducing monosaccharide** because methylene blue is reduced to leuco-methylene blue which is colourless.

6. Tommer's Test

Procedure. Take 2 ml of 0.5% copper sulphate solution in a test tube and add 2 ml of test solution in the test tube and mix. Then add 2 ml of 40% NaOH solution in the test tube, boil and cool.

Observation. Yellow or red precipitate appears.

Inference. The test solution contains **reducing monosaccharide** and cuprous oxide is formed due to reduction of $CuSO_4$.

7. Nylander's Test

Procedure. Take 5 ml of test solution in a test tube and add 0.5 ml of Nylander's reagent in it and boil for 3 minutes and cool.

Observation. Black precipitate develops.

Inference. The test solution contains **reducing monosaccharide** because bismuth sub-nitrate is reduced to bismuth which precipitates black.

8. Potassium Ferricyanide Test

Procedure. Take 3 ml of 1% potassium ferricyanide solution in a test tube and add 1 ml of 40% NaOH solution in it. Boil the solution and then add nearly 10 drops (but drop by drop) of the test solution in the test tube and keep on boiling.

Observation. The yellow colour of ferricyanide begins to fade and finally decolourizes.

Inference. The test solution contains **reducing monosaccharide.**

EXPERIMENT No. 1 E

Colour Test for Sucrose (Disaccharide)*

Fearson's Test

Procedure. Take 4 ml of test solution in a test tube and add 4 drops of 10% methylamine hydrochloride solution in it. Boil it for 30 seconds and then add 5 drops of 20% NaOH solution in it.

Observation. Yellow colour appears which turns red.

Inference. The test solution contains **sucrose.**

---

* Lactose and maltose (disaccharides) are reducing sugars which give positive response to tests like Benedict's, Fehling's, Barfoed's and Tommer's.

EXPERIMENT No. 1 F

Colour Test for Starch (Polysaccharide)

Iodine Test

Procedure. Take 2 ml of test solution/suspension in a test tube, boil and cool at room temperature. Then add 1 or 2 drops of iodine solution in it.

Observation. Blue colour appears which disappears on heating and blue colour appears again on cooling.

Inference. The test solution/suspension contains **starch.**

EXPERIMENT No. 1 G

Colour Test for Glycogen (Polysaccharide)

Iodine Test

Procedure. Take 2 ml of test solution/suspension in a test tube, boil and cool at room temperature. Then add 1 or 2 drops of iodine solution in it.

Observation. Red or wine red colour appears.

Inference. The test solution/suspension contains **glycogen.**

Method of writing such experiments in your practical note book.

S.No.	*Name of Test*	*Experimental Procedure*	*Observations*	*Inference*
1.	Molisch's Test	Took two test tubes and marked them A and B. Then took 5 ml given sample in test tube A, added 2 drops of Molisch's reagent in it and shook the test tube well to mix the two solutions in the test tube. Then took conc. H_2SO_4 in test tube B and poured it gently in test tube A along its side.	A pink or purple ring appeared at the junction of solution and acid in test tube A.	Given sample contained **carbohydrate.**
2.	Foulger's Test	Took 0.5 ml test solution in a test tube, added 3 ml Foulger's reagent in it, boiled and shook the test tube gently.	Blue colour appeared.	Given sample contained **fructose.**
3.	Seliwanoff's Test	Took 3 ml Seliwanoff's reagent in a test tube, added 1 ml of the test solution and boiled for 30 seconds.	Red to orange colour appeared.	Presence of **fructose** confirmed in the given sample.
4.	Moor's Test	Took 4 ml test solution in a test tube, added equal volume of 2% NaOH solution in it and boiled.	Solution turned yellow and then reddish brown.	Test solution contained **glucose.**
5.	Barfoed's Test	Took 5 ml Barfoed's reagent in a test tube, added 4 ml test solution in it, stirred and boiled for 1 to 2 minutes. Allowed it to stand for 15 minutes.	Orange red precipitate formed.	Test solution contained reducing **monosaccharide.**

S.No.	Name of Test	Experimental Procedure	Observations	Inference
6.	Fearson's Test	Took 4 ml test solution in a test tube, added 4 drops of 10% methylamine hydrochloride solution in it, boiled for 30 seconds and then added 5 drops of 20% NaOH solution in it.	Yellow colour appeared which turned red.	Test solution contained **sucrose.**
7.	Iodine Test	Took 2 ml test solution/suspension in a test tube. Boiled and cooled at room temperature. Then added 1 or 2 drops of iodine solution in it.	Blue colour appeared which disappeared on heating and again blue colour appeared on cooling.	Test solution contained **starch.**
			Red or wine colour appeared	Test solution contained **glycogen.**

PROTEINS

Proteins are complex organic compounds found in animals and plants. They are composed of carbon, hydrogen, oxygen, nitrogen and sulphur in their molecules. Proteins are, in fact, the polymers of amino acids linked together by peptide bonds. They have high molecular weight and contain free amino ($-NH_2$) and carboxyl ($-COOH$) groups. On hydrolysis, they break up into peptones, proteoses, peptides and amino acids.

The commonest protein used in laboratory experiments are the egg white or egg albumen. For this, 5% solution of egg white is prepared. For preparing 5% solution of egg albumen, take 5 ml egg albumen, add 95 ml distilled water in it and filter overnight through cotton in a funnel. Proteins, however, show both colour and precipitation reactions.

EXPERIMENT No. 2 A

Colour Tests for Protein

1. Biuret Test

Procedure. Take 3 ml of test solution in a test tube. Add 1 ml of 40% NaOH solution in the test tube to make the test solution strongly alkaline and stirr it thoroughly. Then add 2 drops of 1% $CuSO_4$ solution.

Observation. Violet or pink colour appears which disappears if excess of $CuSO_4$ solution is added.

Inference. Presence of **protein** confirmed in the test solution.

2. Millon's Test

Procedure. Take 5 ml of test solution in a test tube. Add 3 or 4 drops of Millon's reagent in the test tube, mix and heat the mixture till it boils.

Observation. White precipitate appears which turns yellow on heating.

Inference. Presence of **protein** confirmed in the test solution.

3. Ninhydrin Test

Procedure. Take 5 ml of test solution in a test tube. Add 0.5 ml of Ninhydrin solution in the test tube, boil for 2 minutes and cool.

Observation. Blue colour appears.

Inference. The test solution contains **protein.**

4. Xanthoproteic Test

Procedure. Take 2 ml of test solution in a test tube. Add 0.5 ml of conc. HNO_3 to it and heat till it boils. Then cool the test tube slowly under tapwater and add 40% NaOH to make it alkaline.

Observation. After boiling solution turns yellow* and after adding NaOH solution yellow colour changes to orange.

Inference. Presence of **protein** confirmed in the test solution.

EXPERIMENT No. 2 B

Colour Tests for Individual Amino acids

1. Test for Arginine

Procedure. Take 3 ml of test solution in a test tube and add to it 2 drops of 1% α - Naphthol solution. Then add 1 ml of 40% NaOH solution and 2 drops of sodium hypobromide solution.

Observation. Brick red colour appears.

Inference. The test solution contains **arginine.**

2. Test for Tyrosine

Procedure. Take 3 ml of test solution. Add 2 ml of mercuric sulphate solution in the test tube by a pipette and boil carefully. Then add 2 drops of 1% sodium nitrite solution to it and heat gently.

Observation. First yellow precipitate develops which turns red after adding sodium nitrite and heating.

Inference. Presence of **tyrosine** confirmed in the test solution.

3. Test for Tryptophan

Procedure. Take 3 ml of test solution in a test tube and add 3 ml of glyoxalic reagent in it. Then take another test tube containing 5 ml of conc. H_2SO_4. Now add the solution of first test tube very carefully in the test tube containing H_2SO_4 in such a manner that the two solutions do not mix and rotate the test tube gently.

Observation. Purple or violet colour develops at the junction of two solutions.

Inference. Presence of **tryptophan** confirmed in the test solution.

EXPERIMENT No. 2 C

Precipitation Tests for Proteins

Test-1

Procedure. Take 3ml of test solution in a test tube and add to it 5% mercuric chloride solution drop by drop.

Observation. White turbidity appears in the beginning which becomes thick and granular white precipitate. (The precipitate would dissolve if excess of mercuric chloride is added).

Inference. The test solution contains **proteins.**

Test-2

Procedure. Take 3 ml of test solution in a test tube. Add 5% ferric chloride solution drop by drop in the test tube.

Observation. In the beginning turbidity appears which precipitates later on.

Inference. The test solution contains **proteins.**

Test-3

Procedure. Take 3 ml of test solution in a test tube and add to it 5 drops of freshly prepared tannic acid solution.

---

* While handling test tube containing HNO_3, the finger tips become yellow. This indicates and confirms the presence of protein in human skin.

Observation. Turbidity develops and brown non-granular precipitate forms.

Inference. The test solution contains **protein.**

Note: *Write the experiments performed in your practical note book in the same way as suggested for carbohydrates.*

LIPIDS (FATS AND OILS)

The lipids (fats and oils) are the organic compounds which are insoluble in water but abundantly present in animal and plant tissues. They are soluble in organic solvents like ether, chloroform, benzene, hot alcohol and petroleum ether. They contain long chains of aliphatic hydrocarbons or benzene ring in their molecules. The lipids are non-polar and hydrophobic. They have greasy feel and low melting point. The fats are hydrolysed by boiling acids and alkalies. They exist in nature as esters of glycerol and different fatty acids. .

The most suitable material for testing fat in laboratory is olive oil, coconut oil or castor oil.

EXPERIMENT No. 3

1. Colour Test with Sudan III

Procedure. Take a test tube containing 0.5 ml of ether or chloroform and add to it 0.5 ml of test sample drop by drop. When fully dissolved add one drop of Sudan III reagent in the test tube.

Observation. Red colour appears.

Inference. The test sample contains **fat.**

2. Emulsification Test

Procedure. Take 3 ml of test sample in a test tube and add 2 drops of oleic acid. Shake well, now add 2 drops of this liquid to another test tube having 3 ml of dilute caustic soda.

Observation. The acid dissolves in alkali forming soap which within short time becomes an emulsion.

Inference. The test sample contains **fat.**

3. Formation of Acrolein

Procedure. Take a clean and dry test tube and put in it 0.5 gm of powdered potassium bisulphate ($KHSO_4$) or sodium bisulphate ($NaHSO_4$). Now pour 3 or 4 drops of test sample in the test tube. Mix thoroughly with a glass rod and heat the test tube.

Observation. An irritating smell of acrolein comes.

Inference. The test sample contains **fat.**

4. Solubility Test

Procedure. Take five test tubes and mark them 1,2,3,4 and 5. Now, add 5 ml of water, absolute alcohol, ether, benzene and chloroform respectively in test tubes marked 1 to 5. Then to each test tube pour 2 or 3 drops of test sample. Shake them well, allow to stand and observe.

Observation. In test tube 1 oil drops are seen floating on the surface of water, in test tube 2 oil drops settle at the bottom of alcohol and becomes miscible after heating, while in test tubes 3,4 and 5 test sample is miscible.

Inference. The test sample contains **fat** because it is not soluble in water and soluble only in organic solvents.

Note: *Adopt the same method for writing these experiments in your practical note book as suggested for carbohydrates.*

ENZYMES

EXPERIMENT No. 4

Test for Salivary Amylase

The enzyme salivary amylase is present in the saliva. It is also known as **ptyalin.** It changes starch into glucose and maltose. Salivary amylase acts steadily at body temperature but it acts optimally between 40 to 50°C. The optimum pH for activity of amylase is between 6 to 7 and that of salivary amylase of man is 6.6.

For the demonstration of amylase activity of salivary glands, saliva of man and salivary glands of cockroach are the best materials.

Preparation of enzyme extract. For the preparation of extract of salivary glands, 5–10 salivary glands are taken out from the cockroaches. The salivary glands are ground in a mortar with little thymol. The solution is diluted with 50 % glycerol to make 10 % solution of salivary gland extract. The extract is stored in a bottle with toluene.

Procedure. Take 5 drops of extract in an incubation tube and add 5 drops of 1% starch solution, mix it and then cover the mixture with a layer of toluene. In another incubation tube take 5 drops of boiled extract, add 5 drops of starch solution. This will serve as the control experiment, Incubate both the tubes in an oven at 37°C for about one hour. After one hour take a small part of the incubated mixture and add iodine solution. No change in colour of iodine solution will indicate the presence of **amylase enzyme.** Take another part of the extract, add few drops of Benedict's solution and heat it till it boils. A brick-red precipitate will confirm the presence of amylase enzyme in the extract. Repeat the above two tests with the control mixture. In the first test violet colour will confirm the presence of starch as such which has not been digested.

Starch always gives bluish-violet colour with iodine solution. In the experimental solution the enzyme extract is mixed with iodine solution, the amylase present in the enzyme extract hydrolyses the starch into maltose and glucose, therefore, there is no change in the colour of iodine. In the control experiment amylase is destroyed by boiling the enzyme extract, the starch is not hydrolysed into maltose and glucose, therefore, it gives bluish-violet cvolour with the iodine solution. In the second test with Benedict's solution a red precipitate of cuprous oxide appears. In this glucose is formed. Since glucose is a reducing sugar, it reduces copper sulphate into cuprous oxide.

Inference. Saliva contains **amylase.**

EXPERIMENT No. 5

Test for Enzyme Invertase

The enzyme invertase is found in the small intestine of verteberates. it acts on sucrose and hydrolyses it into glucose and fructose. It works at a pH of about 7.

For the demonstration of invertase activity a living frog or rat are the best animals chosen for laboratory work.

Preparation of enzyme extract. For the preparation of enzyme extract, first pith a living frog or chloroform a living rat. Then open the abdomen and take out small intestine in a petridish containing distilled water. Cut open the intestine along its length, wash it with distilled water so that intestimal contents are removed. Now put the intestine in a mortar, add 2 ml of 5% glycerol and grind till a homogenous suspension is formed. Keep this suspension in a reagent bottle and add 1 to 2 ml of toluene (Toluene acts as preservative). Then keep the reagent bottle away from light for 5 to 6 days at room temperature. Filter the suspension through muslin cloth and now the suspension (enzyme extract) is ready for the desired test.

Procedure. Take three test tubes and mark them A, B and C. Take 5 ml sucrose solution in test tubes A and B both, then add 2 ml enzyme extract solution in test tube B only. Take 2 ml enzyme extract solution in test tube C, boil, cool and then add 5 ml sucrose solution in it. Leave all the test tubes for 24 hours at room temperature. Now divide the contents of these test tubes in two equal halves. Mark them A_1, A_2 for the contents of test tube A; B_1, B_2 for the contents of test tube B and C_1, C_2 for the contents of test tube C.

Then test the contents of these test tubes as mentioned below :

(i) Test the contents of test tube A_1 with **Fearson's** reagent, it would give yellow colour which turns red. This confirms **sucrose.**

(ii) Test the contents of test tube A_2 with **Benedict's** reagent for monosaccharides resulted due to hydrolysis of sucrose. It gives negative result.

(iii) Test the contents of B_1 and C_1 with **Fearson's** reagent. Contents of B_1 would give negative result, while contents of C_1 would give positive result for sucrose.

(iv) Test the contents of B_2 and C_2 with **Benedict's** reagent. The contents of B_2 would give positive result, while contents of C_2 would give negative result.

Inference. Because enzyme activity of invertase is destroyed by boiling the contents of test tube C_1 and C_2, the contents of test tubes A_1 and C_1 give positive results for sucrose, while contents of test tube B gave negative result for sucrose. In the same way, contents of test tubes A_2 and C_2 gave negative results, while contents of test tube B_2 gave positive result showing the presence of reducing monosaccharides (glucose and fructose) which have been formed due to hydrolysis of sucrose due to the enzymatic activity of invertase. Thus, the tests with the contents of test tubes B_1 and B_2 confirm the enzymatic acvtivity of **invertase.**

EXPERIMENT No. 6

Test for Pepsin Enzyme

Pepsin is the most important enzyme of the gastric juice. It hydrolyses proteins into proteoses and peptones. It is secreted by the gastric glands in the gut in inactive form which is converted into active pepsin by the hydrochloric acid present in the stomach. Pepsin requires a rather definite pH for its optimum activity, The optimum pH for the activity ranges between 1.5 to 2.2.

Preparation of substrate. To test the presence of pepsin, acid solution of casein is generally used as substrate. To prepare the substrate take 2.0 gm of vitamin free casein in a beaker, add 5.0 ml of NaOH of 0.1 normality, dilute it up to 50 ml with distilled water and heat it at 60°C. Stirr the mixture well until the casein is completely dissolved. Transfer this mixture into another beaker containing a mixture of 5 ml of HCI of 1 normality and 4.5 ml of distilled water. In this way a 2% solution of casein having 1.5 pH is prepared.

Preparation of enzyme extract. For preparing the enzyme extract, dissect three or four frogs and take out the stomach portion, keep the stomach portion in a petridish, wash it thoroughly with distilled water and add a little thymol to check the bacterial action. Ground the stomatch portion in a mortar and add 50% glycerine so as to prepare approximately 10% solution of stomach tissue. Centrifuge the extract and then filter. Store the filterate in the test-tube with toluene.

Procedure. Take 5 drops of extract in a test tube and add equal amount of substrate solution (acid solution of casein) and incubate it at 37°C for 8–12 hours. Perform control experiment by taking 5 drops of substrate with equal amount of distilled water and incubate it at 37°C. To confirm the presence of pepsin take 2 drops of incubated mixture, add 1 drop of 10% sodium acetate solution. Casein is precipitated by addition of sodium acetate but in the presence of pepsin casein is hydrolysed into proteoses and peptones which cannot be precipitated by sodium acetate. Repeat the confirmatory test in the control experimental tube and compare the results of the two tubes.

The presence of pepsin can also be demonstrated by dissolving the **fibres of fibrin** in an acidic medium. Take a few fibres of fibrin, two drops of enzyme extract and two drops of buffer of pH 2.0 on a microslide and cover it by the cover slip. Keep the microslide in the incubation bath at 37°C for 8–12 hours. After incubation for 8–12 hours observe the microslide under the microscope; dissolved fibrin will indicate the presence of pepsin.

Inference. Presence of enzyme **pepsin** in the stomach of frog is confirmed.

EXPERIMENT No. 7

Test for Trypsin Enzyme

Trypsin is a proteolytic enzyme found in the intestine, pancreas and the liver of animals. In pancreas it is in the inactive condition which is called **trypsinogen** but in intestine it is converted into the active trypsin by the alkaline action of bile. The optimum pH of trypsin is about 8.0 and it is more effective than pepsin. Trypsin hydrolyses the protetins which are not hydrolysed by the gastric juice.

Preparation of substrate solution. Prepare 20% alkaline solution of casein. For this take 1 gm vitamin free casein and dissolve it in 50 ml of buffer of 7.2 pH.

Preparation of enzyme extract. Take a part of pancreas or intestine and prepare the enzyme extract as described in the case of pepsin.

Procedure. Take 5 drops of extract, add 5 drops of substrate solution and to this also add 5 drops of buffer of 7.8 pH. Perform control experiment by taking 5 drops of substrate and 5 drops of distilled water in a test-tube. Incubate both the tubes in the incubation oven at 37°C for 4–6 hours.

To confirm the presence of trypsin add few drops of 1% acetic acid in the experimental test-tube. Absence of any precipitate will confirm the presence of trypsin. Acetic acid precipitates the casein. Since the casein has been hydrolysed, therefore, it does not form any precipitate and shows the presence of trypsin. In the control experimental tube there is no extract, therefore, it does not contain trypsin and the casein is not hydrolysed and the addition of acetic acid gives the precipitate.

Inference. Presence of **trypsin** in pancreas or intestine is confirmed.

ESTIMATION OF HAEMOGLOBIN BY HAEMOMETER

EXPERIMENT No. 8

Object. *To estimate the amount of haemoglobin in (i) frog's blood; (ii) human blood; and (iii) blood of any other vertebrate animal.*

Requirements. *(i)* Haemometer (Haldane's haemoglobinometer); *(ii)* decinormal (N/10) HCI (1.2 ml of conc. HCI dissolved in 100 ml of distilled water); *(iii)* distilled water; *(iv)* living frog or any other living vertebrate.

Apparatus. The haemometer consists of two sealed lateral comparison tubes containing a suspension of acid haematin. These are held in a black frame against a white ground glass. Besides, a graduated test-tube of the same diamter is also provided which can fit in the haemometer in between the two side tubes for comparison. A micropipette of 20 cmm is also provided. The other things

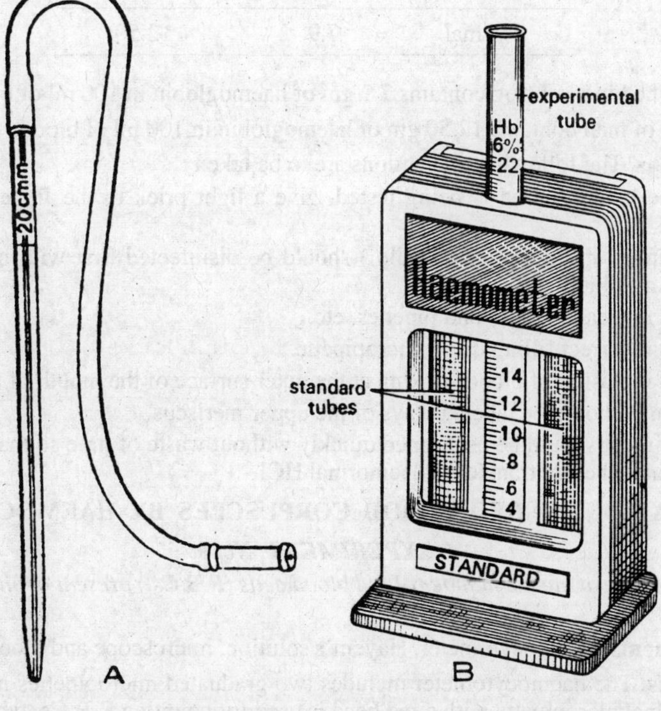

Fig. 12.1. A–20 cmm micropipette; B–Haldane's haemometer.

provided are a small glass rod stirrer, a small bottle brush and a dropper and also a small bottle to contain the decinormal acid solution.

Principle. The method to estimate the haemoglobin content of blood is based on the principle of making an acid haematin solution of blood under experimentation in the graduated tube and then comparing it with the sealed comparison tubes containing the standard acid haematin.

Procedure. The graduated tube is first cleaned with distilled water and then with methylated spirit or 90% alcohol. It is thoroughly dried up before being used. Now with the help of a dropper the N/10 HCI solution is filled in the graduated tube up to 2 gms mark. The micropipette is now filled up by sucking fresh blood of the vertebrate under experimentation up to the mark of 20 cmm. The small amount of blood adhering to the outerside of micropipette should be wiped off by sterilized cotton. The blood of micropipette is now added to the N/10 HCI solution in the graduated tube. The micropipette should be introduced carefully into the tube and its lower mouth should pass right up to the bottom into HCI solution. When blood has been expelled the micropipette is rinsed twice or thrice by distilled water. Every time the contents of micropipette should be expelled into the graduated tube. The acid haematin solution is now thoroughly stirred with the help of a glass rod and then allowed to stand at least for 10 minutes. Afterwards the acid haematin solution is gradually diluted by adding distilled water in a dropwise manner. With the addition of each drop of distilled water the solution should be stirred and its colour matched with that of the standard sealed tubes. This should be continued till the colour of the acid heamatin solution just fades away as compared to that of the standard comparison tubes. The reading before the colour just fades is taken as the correct and final reading.

Observations. The readings are tabulated in the following manner :

Readings	*Frog's blood*	*Human blood*
I	7.8	12.45
II	7.10	12.60
III	7.9	12.50
Final =	7.9	12.50

Results. The blood of frog contains 7.9 gm of haemoglobin in 100 ml of blood.

The blood of man contains 12.50 gm of haemoglobin in 100 ml of blood.

Precautions. The following precautions are to be taken :

1. In case human blood is being tested, give a light prick to the finger tip of the person providing blood.
2. The finger and the prick (needle) should be disinfected first with methylated spirit or alcohol (90% or absolute).
3. Avoid uncleaned tubes and pipettes, etc.
4. Avoid incorrect filling up of micropipette.
5. Avoid inclusion of blood sticking at the outer surface of the mouth of the pipette.
6. Readings should be taken always of the upper meriscus.
7. Experiment should be performed quickly without waste of time so that fresh blood is not allowed to coagulate before transfer to decinormal HCI.

ENUMERATION OF RED BLOOD CORPUSCLES BY HAEMOCYTOMETER

EXPERIMENT No. 9

Object. *To determine the number of red blood cells (R.B.Cs.) present in one cubic mm volume of blood.*

Requirements. Haemocytometer, Hayem's solution, microscope and blood.

Apparatus. The haemocytometer includes two graduated micropipettes in which dilution of blood is done. One micropipette with a red bead is used for counting R.B.Cs, while the other with a white bead for counting W.B.Cs. Each micropipette bears at least 3 graduations; those on the micropipette for R.B.Cs are 0.5, 1 and 101. The graduations on the micropipette for W.B.Cs are 0.5, 1 and 11. Besides these micropipettes the haemocytometer includes a glass slide; 7.5 cm long by 3.75 cm wide. It bears two counting chambers with a cover slip. One side of the slide bears two central platforms bordered by an edged-shaped groove. Beyond this groove on either side is a lateral platform further demarcated by a lateral groove. Each central platform bears one counting chamber, each counting chamber is formed by several straight perpendicular and horizontal lines enclosing squares of various measurements; the biggest square is 3 × 3 mm with an area of 9 sq.mm. It is divisible into 9 squares of 1 × 1 mm each, the central square is again divisible into 25 smaller squares. Each square measures 1/5 × 1/5 = 1/25 sq.mm. Each of these squares is further divisible into 16 smaller squares so that there are 400 smallest squares in all. Each with an area of 1/400 sq.mm.

Hayem's solution has the following composition :

(*i*)	Mercuric chloride ($HgCl_2$)	=	0.5 gm
(*ii*)	Sodium chloride (NaCl)	=	1.0 gm
(*iii*)	Sodium sulphate (Na_2SO_4)	=	5.0 gm
(*iv*)	Distilled water (H_2O)	=	200 ml

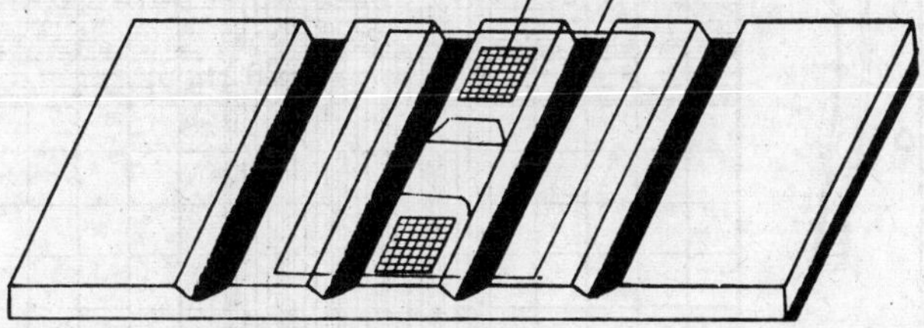

Fig. 12.2. Glass slide with counting chambers.

The mercuric chloride acts as corrosive sublimate and fixes the R.B.Cs present in the blood. The other ingredients act isotonically so that the R.B.Cs may not burst due to haemolysis. The Hayem's solution also serves to dilute the blood.

Procedure. Blood is obtained either directly from the heart of a frog or from some body part externally in case of man. Human blood is generally taken out from a finger (avoid thumb and first finger) which is thoroughly washed and cleaned by spirit or alcohol (absolute or 90%). It is now pricked by an ordinary injection needle which is also sterilized with spirit and dried before use. The finger is pricked quickly and effectively. It should not be pressed hard to let out blood so as to avoid other body juices also oozing out. For this reason also the first one or two drops of oozing blood should be avoided and wiped off.

Now take the micropipette meant for R.B.Cs which is already rinsed with alcohol or spirit or ether and thoroughly dried. Suck the blood in the micropipette up to 0.5 mark taking care that air bubbles are not included. The excess of blood, if any, may be run out by touching the mouth of micropipette to the palm. The blood which is sticking to the outer side of micropipette should also be carefully cleaned. The micropipette should now be transferred to the container of Hayem's solution which is carefully sucked up to 101 mark. The micropipette is now held horizontally between the fore-finger and thumb or palm surfaces of the hand and rotated several times so that blood thoroughly mixes with Hayem's fluid. The red bead in the micropipette also helps in mixing. In this way dilution of blood becomes 200 times.

Before starting to count R.B.Cs in this diluted blood, place the cover-slip on the counting chambers. The cover-slip is supported upon the side platform but remains separated from the central platforms by a distance of 0.1 mm. First reject 3 or 4 drops of mixture from the micropipette. Now apply the tip of the micropipette between the cover-slip and the platform and allow few drops of blood mixture to flow in the narrow space between the cover-slip and the counting chambers. If necessary both chambers may be filled in this manner. Blood mixture remains filled up between the cover-slip and the counting chambers because of capillary action. Air should not be taken into and also pouring excess blood mixture so that the H-shaped groove remains free from it.

When the counting chambers are properly flooded the slide may be kept aside for a few minutes so that the R.B.Cs settled down on the bottom floor of the two counting chambers. Now transfer the slide gently and carefully under the microscope without disturbing the settled R.B.Cs and start counting them.

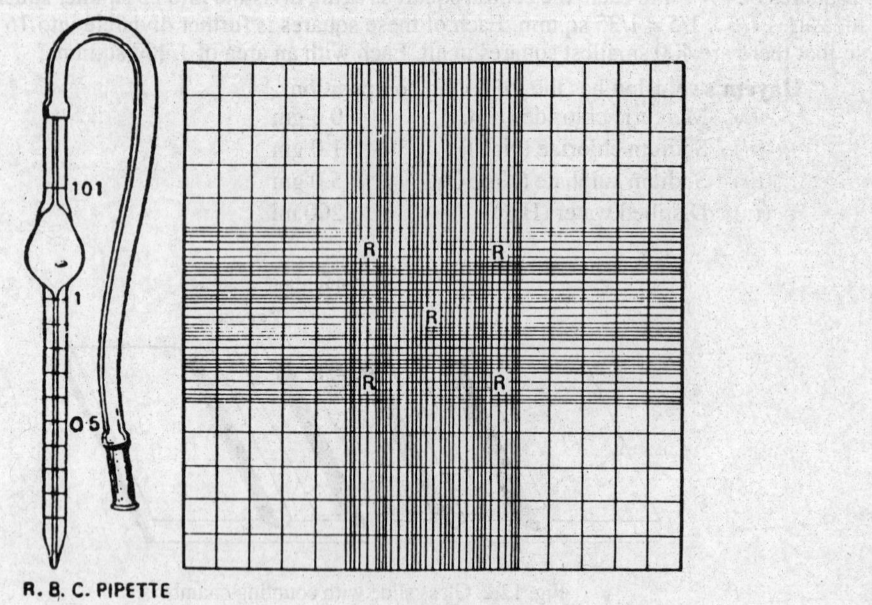

Fig.12.3. R.B.C. pipette and small squares in the counting chamber.
R denotes small squares for red cell count.

Counting. It is not necessary to count the R.B.Cs in all the 400 smallest squares or even in the 25 smaller squares, count them only in five smaller squares, *i.e.,* in the 1st, 5th, 13th, 21st and 25th. The R.B.Cs lying on the lower and right sides of a square are to be added in the total, while those lying on the upper and left sides are to be rejected.

Calculations. The calculations may be done by the following method :

$$\text{Number of R.B.Cs per cubic mm} = \frac{\text{Number of cells counted} \times \text{dilution} \times 4000}{\text{Number of small squares counted}}$$

Suppose total R.B.Cs for 5 smaller squares are equal to $A + B + C + D + E$. Therefore, total R.B.Cs in 80 smallest squares $= A + B + C + D + E$.

$$\therefore \text{ R.B.Cs in one smallest square} = \frac{A + B + C + D + E}{80}$$

$\therefore$ R.B.Cs in 400 smallest squares $= 5\,(A + B + C + D + E)$

But the height of blood film $=$ 0.1 mm

Dilution of blood $=$ 200 times

So that one cubic mm of blood will contain

$= 200 \times 10 \times 5\,(A + B + C + D + E)$ R.B.Cs.

$= 10{,}000\,(A + B + C + D + E)$ R.B.Cs.

Thus, in order to avoid the detailed calculations to get quick results four zeros are added to the total R.B.Cs of 5 small squares (80 smallest squares).

Tabulation. The readings are tabulated in the following table :

Frog's blood

Smaller squares	*R.B.Cs in I chamber*	*R.B.Cs in II chamber*
I A	25	15
II B	30	20
III C	15	18
IV D	20	25
V E	22	20

Human blood

Smaller squares	*R.B.Cs in I chamber*	*R.B.Cs in II chamber*
I A	83	105
II B	87	97
III C	74	79
IV D	96	86
V E	98	83

	Frog		*Human*	
Total R.B.Cs =	112	98	438	450
Average =	$\frac{112 + 98}{2} = 105$		$\frac{438 + 450}{2} = 444$	

Results. The result is given as follows :

(*i*) 1 cubic mm of frog's blood contains 10,50,000 R.B.Cs.

(*ii*) 1 cubic mm of human blood contains 44,40,000 R.B.Cs.

ENUMERATION OF WHITE BLOOD CORPUSCLES (W.B.Cs) BY HAEMOCYTOMETER

EXPERIMENT No. 10

Object. *To determine the number of white blood cells (W.B.Cs) per cubic mm in the human blood.*

Requirements. Haemocytometer, glacial acetic acid and solution of gentian violet in water or distilled water. The glacial acetic acid haemolyses the red cells, while the gentian violet slightly stains the nuclei of white blood corpuscles (W.B.Cs), the W.B.Cs may, therefore, be easily recognised.

Procedure. The procedure of counting of W.B.Cs is the same as that of the R.B.Cs. The W.B.Cs are counted in the four corners of 1 square millimetre in the central ruled area on both the sides of the counting chambers of the haemocytometer. The W.B.Cs are recognised by the refractile appearance and by the slight colour given to them by the stain contained in the diluting fluid. The cells touching the boundary lines are not counted.

Calculations. The calculations may be done by the following method :

$$\text{Number of W.B.Cs per cubic mm} = \frac{\text{Number of cells counted} \times \text{dilution} \times 10}{\text{Number of 1 sq. mm counted}}$$

Since the dilution is 20 times and the cubic capacity of the area counted is 1/10 cubic millimetre, the total volume is 1/200 cubic millimetre. Say for instance, the number of W.B.Cs in 4 outer squares is (10 + 12 + 13 + 10) = 45.

In other words number of W.B.Cs in 1/200 cubic millimetre = 45.

Therefore, the number of W.B.Cs in 1 cubic millimetre = 45 × 200 × 1/2

= 4500 W.B.Cs.

Normal healthy man has 4000 to 6000 W.B.Cs per cubic millimetre of blood.

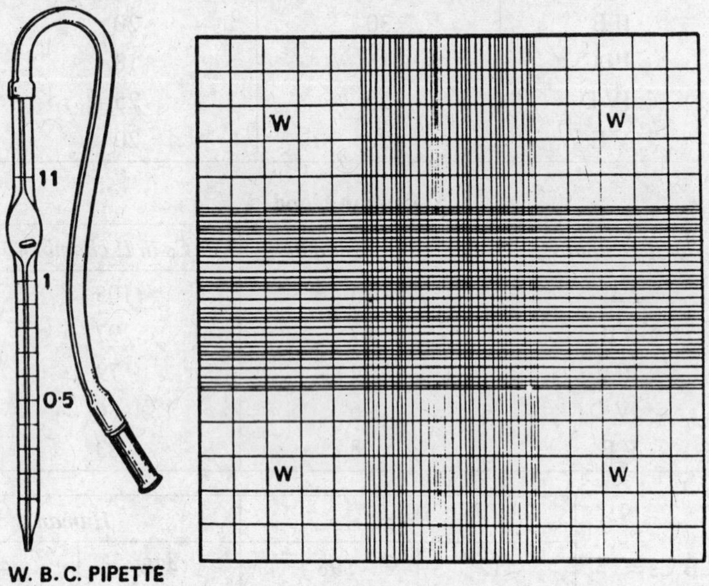

Fig.12.4. W.B.C. pipette. Big squares in the counting chamber.
W denotes big squares for white cell count.

DIFFERENTIAL COUNT OF WHITE BLOOD CORPUSCLES

EXPERIMENT No. 11

Object. *To determine the relative proportions (differential count) of white blood corpuscles in the human blood.*

Requirements. Sterilized needle; clean slides; Leishman's stain or Wright's stain and microscope.

Procedure. Prick the finger with a sterilized needle, wipe away the first drop of blood. Take the second drop on the end of a clean slide, place the slide on a smooth surface holding it steady with the left hand. Hold a second slide or drawing slide at 45° just in front of the drop of the blood, draw this slide in such a way that the blood spreads along its edge. Push the second or drawing slide forward at 45° on the first slide without exerting any pressure, a film of blood will be made on the first slide. Dry the blood film in air, but do not heat it.

Stain the blood film with Leishman's stain or Wright's stain. Wait and after one minute carefully add distilled water to the stain, the water should be twice the volume of the stain used. Mix the water and stain well by sucking the mixture in and out of a pipette. After a time a greenish metallic scum forms on the surface of the mixture. Allow the diluted stain to act from 7 to 10 minutes. Drain off the stain and wash film for 10 seconds with distilled water. The film should be rose pink in colour, if it is too purple, then wash again with distilled water. Dry the slide and examine it under the microscope for white blood corpuscles.

First examine the stained film under low power of microscope to get an idea of leucocytes. Then examine under high power or oil immersion lens after thinly smearing the stained film with oil. In a longitudinal strip of the film count the various kinds of white blood corpuscles from top to the bottom. Count 100 or 200 white corpuscles noting on a paper each type of white blood corpuscles encountered, and calculate the percentage.

Calculations. The following formula can be adopted to calculate the percentage of different types of white blood corpuscles :

$$\frac{\text{Number of type cells}}{\text{Total number of white blood corpuscles}} \times 100$$

The normal proportion of white blood corpuscles is as follows :

Neutrophils	...	50 – 70	per cent	or	3000 – 6000	per cu mm
Eosinophils	...	1.4	per cent	or	150 – 300	per cu mm
Basophils	...	0.5 – 1	per cent	or	0 – 100	per cu mm
Lymphocytes	...	20 – 0	per cent	or	1500 – 2700	per cu mm
Monocytes	...	2 – 8	per cent	or	300 – 600	per cu mm

Precautions. It is absolutely essential for differential count of white blood corpuscles that the film of blood is well spread with no clogging of corpuscles.

COAGULATION OF BLOOD

EXPERIMENT No. 12

Object. *To observe the coagulation of blood.*

Requirements. Test-tube and cotton wool.

Principle and procedure. Bleeding takes place if the skin is cut, but after a time blood coagulates to close the cut, thus, preventing much loss of blood. Take blood of a mammal in a large test-tube and close the tube with cotton wool. It will be seen that blood forms a soft jelly and a faint yellow liquid appears. In a few hours the liquid or **serum** increases in amount and the solid jelly or **clot** becomes smaller. The serum consists of blood plasma without its fibrin and almost no blood corpuscles. The clot is scarlet on the surface and bluish-black within. The clot is made of **fibrin**, an insoluble protein derived from the soluble fibrinogen of the blood, and entangled in the fibrin are almost all the blood corpuscles. The fibrin is a mass of fine interlacing fibres.

The blood contains an inactive enzyme **prothrombin** which is kept in check by **heparin** (antiprothrombin) found in blood. Hence, the blood flows freely in blood vessels and does not coagulate. If an injury occurs, blood begins to escape from it, but the injured tissue cells and blood platelets on contact with air release a substance called **thromboplastin**. Thromboplastin soon neutralizes heparin, and it combines with calcium ions in the blood to act on prothrombin changing it into **thrombin.** Thrombin is an active enzyme, it acts on the fibrinogen of blood converting it into **fibrin.** The fibrin forms a network of fibres which entangle the corpuscles to form a clot at the site of the injury.

BLEEDING TIME OF BLOOD

EXPERIMENT No. 13

Object. *To determine the bleeding time of blood.*

Requirements. Sterilized needle, stop-watch and blotting or filter paper.

Procedure. Prick the base of the finger nail with a sterilized needle and let a drop of blood appear. Every 30 seconds dab the blood drop with a piece of blotting or filter paper without touching the skin otherwise the blood may be wiped off. The time from the appearance of the drop of blood to the time when bleeding stops and the filter paper is no longer stained is taken as the bleeding time. Normal bleeding time is 2 to 5 minutes.

CLOTTING TIME OF BLOOD

EXPERIMENT No. 14

Object. *To determine the clotting time of blood.*

Requirements. Sterilized needle; small test-tubes, stop-watch, water bath, syringe and glass slides.

Principle and procedure. It is the time taken for blood to clot in vitro at a standard temperature, or it is the length of time from the moment blood is collected till the appearance of clotting.

Following three methods are used for counting the clotting time of blood.

A. Note the time, then take 1 ml venous blood in a small test-tube, put the test-tube in a water bath at 37°C. Every 30 seconds shake the test-tube and tilt it slowly. Note the time when blood stops flowing in the tube and its surface line no longer changes on tilting. The time taken is the clotting time, it is normally 5 to 10 minutes.

B. Take 5 ml venous blood in a syringe and start a stop-watch. Put 1 ml of blood in each of two small equal-sized test-tubes, put the test-tubes in a water bath at 37°C. After some time tilt the test-tubes every half minute, when the test-tube can be tilted through 90° or even inverted without spilling the blood then the blood has coagulated, note the time.

C. Put a drop of blood on a clean slide and note the time. Every half minute pull a clean needle slowly through the drop of blood. When a fine thread of fibrin can be pulled up by the point of the needle, then coagulation has started, note the time.

PREPARATION OF HAEMIN CRYSTALS

EXPERIMENT No. 15

Object. *To prepare haemin crystals.*

Requirements. Sterilized needle, glass slide, cover glass, acetic acid (glacial) and spirit lamp.

Principle. Haemoglobin imparts a red colour to blood, it is a conjugated protein made of a non-protein pigment **haematin** or haeme, and a colourless protein known as **globin.** The haematin is made of a pigment called porphyrin which is combined with iron. A hydrochloride of haematin forms **haemin.**

Procedure. Take some dried blood or a few drops of fresh blood, add a little glacial acetic acid and heat gently, on cooling dark brown haemin crystals (also known as Teichmann's crystals) are formed. Examine haemin crystals under a microscope, they appear as shining rhombic plates, prisms and star-shaped clusters. These crystals contain haematin and iron. Haemin crystals test is used in medico-legal work to differentiate fresh or dried blood marks from any other red colouring material.

Precaution. In the case of old dried blood it is necessary to boil with glacial acetic acid, and at boiling to add a crystal of sodium cloride.

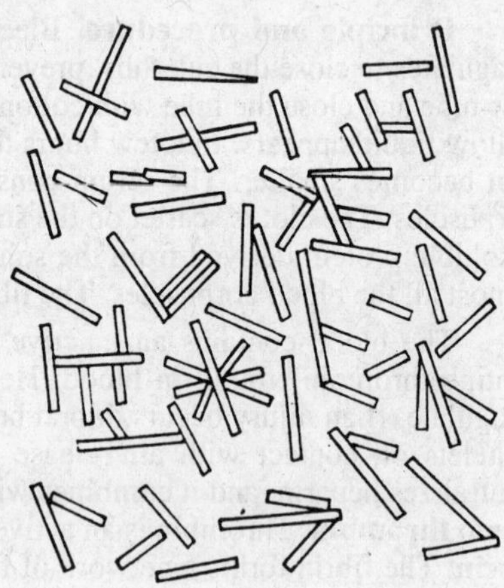

Fig. 12.5. Haemin crystals.

EFFECT OF OSMOLARITY OF SALT SOLUTIONS AND HAEMOLYTIC AGENTS ON RED BLOOD CORPUSCLES

EXPERIMENT No. 16

Object. *To study the effect of osmolarity of salt solutions and haemolytic agents on red blood corpuscles.*

Requirements. Test-tubes, sodium chloride solutions of different percentages, ether or benzene or chloroform and potassium hydroxide.

Principle and procedure. Erythrocytes are membranous envelopes filled with haemoglobin. The envelope or plasma membrane is semipermeable, hence, substances in solution can pass in and out of the corpuscles. If normal red blood corpuscles are placed in normal sodium chloride solution which has the same concentration as the contents of the corpuscles, then no exchange takes place because the osmotic pressure of corpuscles and of the saline solution is the same. The salt solution is said to be **isotonic** or of equal concentration with the corpuscles. If normal red blood corpuscles are put in a dilute solution of sodium chloride, they absorb water from the salt solution, and they swell up, become spherical, and finally rupture. The dilute salt solution is **hypotonic** because its osmotic pressure is lower than that of the interior of the corpuscles. If normal red corpuscles are placed in strong salt solution in which the concentration of sodium chloride is higher than in the red corpuscles, then the corpuscles shrink and become wrinkled. The strong salt solution is **hypertonic** because its osmotic pressure is greater than that of the interior of the corpuscles and the fluid passes from the corpuscles into the salt solution.

A. Put 1 ml of venous blood in each of three small test-tubes. To the first test-tube add 1 ml of 0.85% sodium chloride solution (isotonic). To the second test-tube add 1 ml of 0.5% sodium chloride solution (hypotonic). To the third test-tube add 1 ml of 1.5% sodium chloride solution (hypertonic). Shake the test-tubes and note the results.

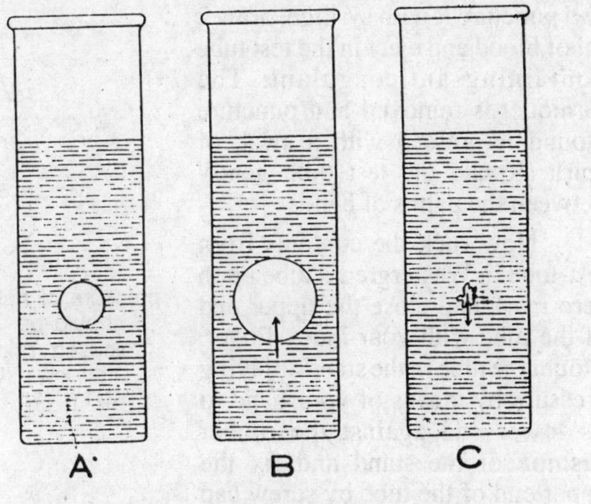

Fig. 12.6. Osmolarity of red corpuscles in different saline solutions. A – isotonic; B – hypotonic; C – hypertonic.

In a hypotonic solution the red corpuscles rupture, in rupturing the haemoglobin passes from the red corpuscles into the plasma, the plasma becomes red in colour. This is called **haemolysis or laking of blood**. The substances inducing haemolysis are called haemolytic agents. Some common haemolytic agents are hypotonic solutions, fat solvents, snake venoms, and some toxins.

B. Take a drop of blood in a test-tube, add a fat solvent (such as ether, benzene, chloroform or bile salts), shake the test-tube, the blood will be haemolysed because the fat component of the plasma membrane of red corpuscles is dissolved and haemoglogin is set free.

C. Take a drop of blood in a test-tube, add a few drops of 0.2% potassium hydroxide, the red corpuscles are dissolved.

EXPERIMENT No. 17

Object. *To determine erythrocyte sedimentation rate (E.S.R.) of blood.*

Requirements. Westergren's apparatus with stand, a test-tube containing 3.8% sodium citrate solution as anticoagulant, a dry test-tube, rectified spirit, sterilized 2 ml syringe with needle, cotton and torniquet.

Description of apparatus. Westergren's apparatus consists of Westergren's tube which is graduated and open at both the ends and Westergren's stand. The stand can accomodate six Westergren's tubes and it has a rubber cushion at the base on which the lower ends of Westergren's tubes rest. There is a screw cap or a screw clip at the upper end of stand which fits upon the tubes (Fig. 12.7).

Principle. "Erythrocyte Sedimentation Rate (E.S.R.) is the rate at which the red cells settle". It is, therefore, governed by the factors which control rouleaux formation. When blood is mixed with a suitable anticoagulant and allowed to stand in a narrow vertical tube, the red cells settle down as a result of rouleaux formation. The red cells are negatively charged and move randomly in plasma. When they strike each other, they tend to stick to each other to form **rouleaux,** *i.e.*, they cling together and resemble a pile of coins. When a rouleaux is formed, the density of its mass increases and it sinks down quickly. In fact, rouleaux formation is affected by the following factors :

(i) Fibrinogen, globulin, albumin and lipid content of plasma, (ii) Shape and size of red cells and number of red cells.

Procedure. Take 0.5 ml of 3.8% sodium citrate (anticoagulant) in a syringe and eject in a test tube. Apply torniquet to the upper arm so that the anticubital (veins opposite to elbow joint) become prominent. Clean the skin over veins with a cotton swab of rectified spirit. Now select a prominent vein and puncture it with syringe, draw 2 ml of blood and eject in the test tube containing anticoagulant. The torniquet is removed and puncture wound is pressed with a swab of spirit. Rotate the test tube gently between the palms of hands.

Now suck the contents from test-tube in Westergren's tube up to zero mark and close the upper end of the tube with your index finger. Mount the tube in the stand applying pressure by means of your hand to its lower end against the rubber cushion of the stand and fix the upper end of the tube by screw cap or screw clip. Now the tube will be held vertically in the stand. Allow the tube to stand for an hour and read the upper level of red column (red blood cells). Above the red column, there will be clear plasma. Again observe and read the red column after two hours. Note the readings for calculation.

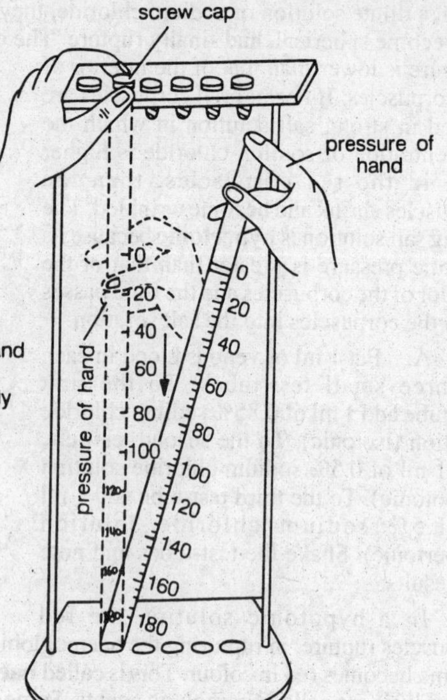

Fig. 12.7. Westergren's apparatus for the determination of E.S.R.

Result.

Suppose the reading at the end of 1st hour be X mm

Suppose the reading at the end of 2nd hour be Y mm.

Then, the mean average of E.S.R. will be $\frac{X + Y}{2} = Z$ mm per hour

Inference. The E.S.R.* is either normal, above normal or below normal.

Discussion. The normal range of E.S.R. determined by this method is as under :

At the end of 1st hour 3 to 5 mm in Males; 4 to 7 mm in Females.

At the end of 2nd hour 7 to 15 mm in Males; 12 to 17 mm in Females.

If in a person, the E.S.R. value changes in two consecutive tests carried out at definite intervals, it suggests the increased activity of the diseased process. Thus, indication of change in E.S.R. is more significant than the actual rate observed for a person as this change suggests the presence of a disease, though it does not suggests the nature of disease.

However, an increase in E.S.R. indicates quick rouleaux formation which is caused by a peculiar pattern of plasma molecules. The following physiological and pathological conditions cause a variation from normal E.S.R.

A decreased E.S.R. is very uncommon but increased E.S.R. suggests :

(i) Pathological conditions like rheumatic fever, tuberculosis, malignancy and similar other chronic conditions which are associated with tissue break down, and anaemia.

(ii) Physiological conditions like pregnancy, menstruration and parturation.

* *Students while working in the class may note E.S.R. after every 15 minutes and plot a graph against time. This will help them in reporting E.S.R. at the end of an hour or so.*

Precautions. While observing E.S.R. by the method mentioned above, following precautions should be taken:

(i) The tube should be kept perfectly vertical otherwise acceleration in E.S.R. will take place.

(ii) The blood sample taken should be used within two hours.

(iii) Air bubbles in the blood column of Westergren's tube should be avoided.

(iv) Blood showing any sign of coagulation should not be used.

(v) The test should be performed at a room temperature of 22°C to 27°C as temperature effects E.S.R.

(vi) The exposure of experiment to direct sunlight should be avoided.

Note: E.S.R. can also be determined by **Wintrob's method** which is based on the same principle as described above.

BLOOD PRESSURE AND ITS MEASUREMENT

Blood pressure. Arterial blood pressure is the force of pressure which the blood is exerting against the walls of the blood vessels in which it is contained. The blood pressure in the arteries varies during the **cardiac cycle.** During ventricular systole, when the left ventricle is forcing blood into the aorta the pressure rises to a peak, referred to as **systolic pressure.** During diastole the pressure falls, the lowest value it reaches is referred to as **diastolic pressure.**

The systolic blood pressure is produced by the heart muscles which drive the contents of the ventricle into the already stretched arteries. During diastole the arteries are kept partly distended because the peripheral resistance of the arterioles prevents all the blood running off into the tissues. Thus, the blood pressure depends partly on the force and volume of blood pumped by the heart and partly on the contraction of the muscles in the walls of the arterioles. This contraction is maintained by vasoconstrictor nerves which are controlled by the vasomotor centre in the medulla oblongata of the brain.

The vasomotor centre adjusts the peripheral resistance to maintain the blood pressure relatively constant. It, however, changes slightly in physiological variations of exertion as during exercise, with mental changes of anxiety and emotion, in sleep and when eating. For this reason, it is always suitable to measure blood pressure when a person is relaxed and resting. The measurement of blood pressure is very significant in correctly estimating the state of blood circulation and, thus, the working of heart.

Measurement of blood pressure. Blood pressure is measured by a blood pressure instrument called **sphygmomanometer** and a stethescope is also needed.

Description of sphygmomanometer (Fig.12.8). It consists of a long glass **manometer** capillary tube. The upper end of this tube is open while its lower end is slightly bend. The manometer is mounted in the middle of a **dark scale** having markings on its both sides. The lower bend end of manometer is connected to the ventral side of a large **bulb** filled with mercury. The upper open end of the manometer is exposed to atmospheric pressure but is guarded by a valve to check the spilling of mercury from the open end. The upper end of the bulb is connected with a **nozel** and the nozel is connected with a rubber tube having narrow lumen. The free end of rubber tube is connected to a flexible rubber bag enclosed in a **cuff**. The cuff is further connected to a **rubber pumping bulb** with a rubber tube. The neck of the bulb is provided with an **adjusting screw;** the tightening or untightening of this screw ensures the flow of air into the cuff (inflation) or out of the cuff (deflation).

Method of using sphygmomanometer. Open the adjusting screw at the proximal end of the pumping bulb so that air is removed from the cuff and tubes. Then wrap the cuff around the upper arm just above the elbow joint and fix it with the help of attached hook in a manner that neither it is too light nor too loose. While wrapping the cuff care should be taken that the tubes attached to it must face toward the lower end. Then tighten the adjusting screw and pump the air by pumping rubber bulb rapidly so that mercury is raised up to 110–120 mm Hg mark in the manometer. Observe the sound of arterial blood in brachial artery with the help of a stethescope. Then again pump the air slowly and keep on observing the sound of arterial blood in the brachial artery with stethescope to disappear. As and when sound is not felt, note the mercury level in the manometer and mark the reading. This is the **diastolic pressure.** Then start loosing the adjusting screw slowly and keep on observing carefully the return of normal arterial sounds through the stethescope. As and when it is felt, close the adjusting screw and mark the reading of mercury in manometer. Pump a little more air and again note the proper arterial sound to reappear. This would be the **systolic pressure.** Now, loose the screw, remove all air and open the cuff and keep the apparatus in a safe place.

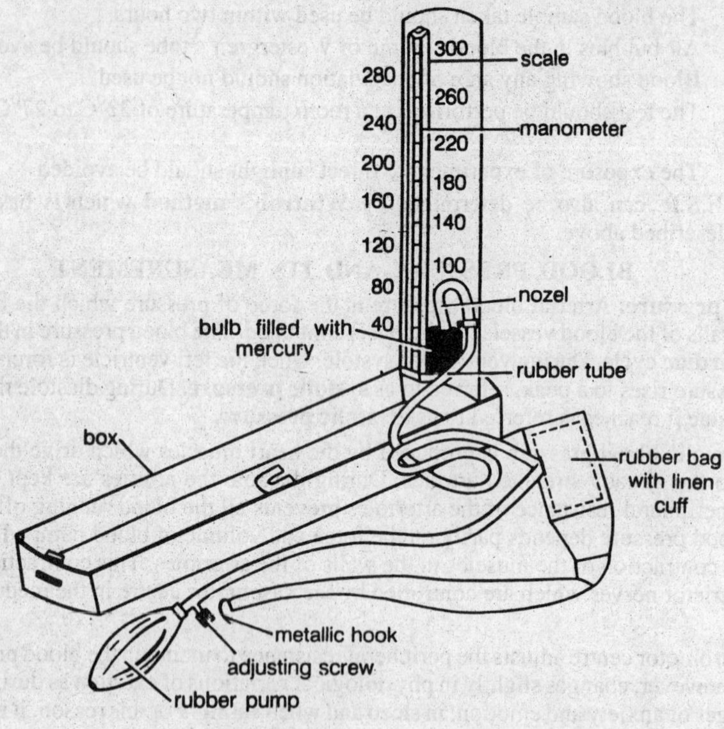

Fig. 12.8. Sphygmomanometer.

The difference in pressure between systole and diastole is called the **pulse pressure** which is normally 30–50 mm Hg. The lower limit of systolic pressure in a normal adult is estimated about 105 mm Hg and the upper limit about 150 mm Hg. In women the blood pressure is from 5 to 10 mm Hg less than in men.

Normal Blood Pressure Range (In mm Hg)

		Diastolic	Systolic
In infants	...	50	70 – 90
In childhood	...	60	80 – 100
During adolescent stage	...	60	90 – 110
In young adult	...	60 – 70	110 – 125
As age advances it is increased	...	80 – 90	130 – 150

CIRCULATION OF BLOOD

EXPERIMENT No. 18

Object. *To study the circulation of blood in the capillaries of the web of foot of frog.*

Requirements. Soft wooden board, moist cloth, microscope and live frog.

Procedure. Take a soft wooden board and make an aperture of about 2 cm, about 4 cm away from one end. Cover a live frog with moist cloth bag and tie it up to the wooden board placing the web between the toes of the foot on the aperture, the foot may be pinned down with web well stretched. Then place the wooden board with the frog on the stage of the microscope so that the aperture in the wood lies on the hole of the microscope stage or platform. Examine the web under the microscope. Blood corpuscles will be seen moving in capillaries, and arterial vessels can be distinguished from veins by the jerky movement of blood corpuscles in the former. Large chromatophores will also be seen in the web.

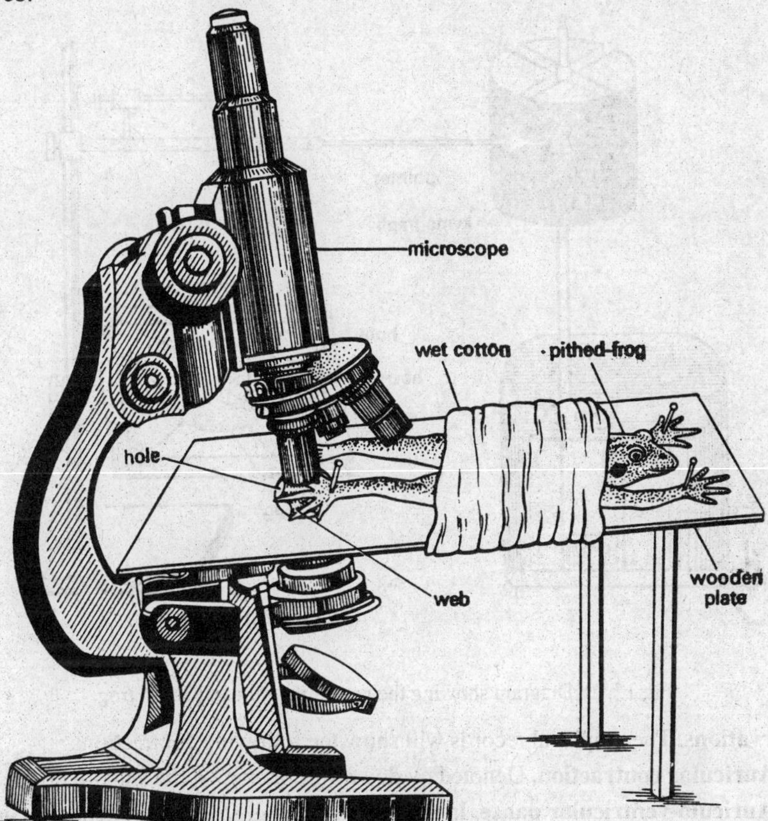

Fig.12.9. To show blood circulation in the web of foot of frog.

Alternatively, the frog may first be pithed, in pithing the spinal cord is cut by a sharp pin inserted just behind the head and the brain is crushed. Then place the web of the foot in position on the microscope.

Observations. Blood corpuscles are seen moving in capillaries. The arterial capillaries can be distinguished from veins by the jerky movement of blood corpuscles in the former.

RECORDING OF HEART BEAT IN SITU

EXPERIMENT No. 19

Object. *To record frog's heart beat in situ.*

Requirements. Sherrington's revolving drum, Sheerington's gaskell, heart lever, stand thread, small bent pin, board with cork sheet and live frog.

Procedure. First pith the medulla oblogata connected with the spinal cord (rotating a needle at this point), then expose and clear the heart but the heart is to be kept in situ, By destroying the nervous system there will be no effect of nervous system on heart and beat will be autonomic. Add drop by drop 0.6% NaCI solution, then on the muscle at the tip of the ventricle a hook made up of from a small pin is passed through the apex without injuring the ventricular cavity. The heart is gently lifted by a thread attached to the hook. The free end of the thread is tied to the heart lever with the help of the screw. The pointer of the lever is adjusted in a horizontal manner and pointer on the long arm of the lever is adjusted just to touch the smooth surface of the drum in order to reduce the friction of different speeds of the drum. The normal cardiographic records are registered on the smooth surface.

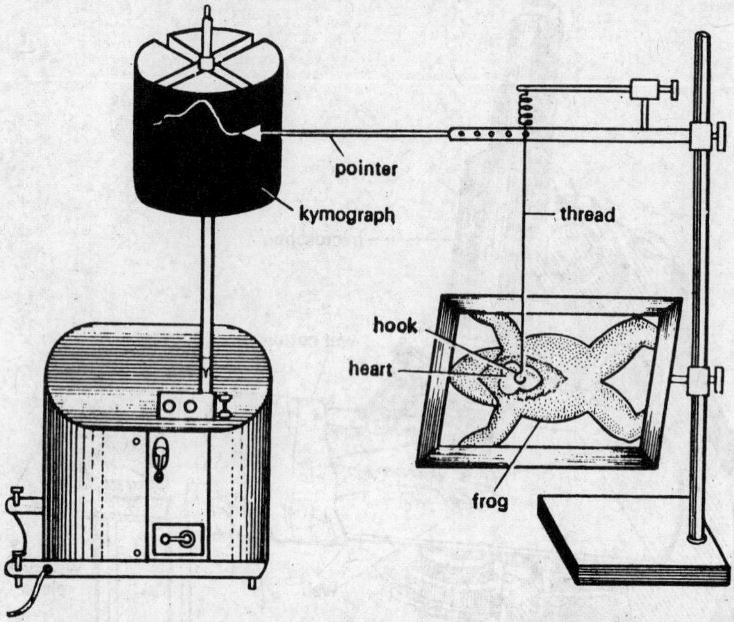

Fig.12.10. Diagram showing the recording of heart beat of frog.

Observations. The graphical records will show the following contractions :

(a) **Auricular contraction.** Denoted by downward facing on the smooth drum.

(b) **Auriculo-ventricular pause.** In the form of a small straight or slightly inclined line.

(c) **Ventricular contraction.** Just like an auricular contraction but the amplitude and height of contraction downward record is much longer than that of the auricular contraction.

Interpretation. The contraction due to sinus venosus is not at all observed only because firstly sinus contraction is very weak and secondly auricular contraction is much more stronger as compared to sinus contraction.

Inference. Under normal conditions the contraction of sinus venosus is generally followed by auricular contraction and ventricular contraction. The downward tracing is the contraction and upward tracing is the relaxation of the heart, while small short tracings are the pause taken up by heart during series of each changes in each cardiac cycle.

Precautions. The following precautions are to be taken :

1. Rotations of kymograph should not be fast otherwise contractions of the auricles will not be recorded.
2. The paper should be blackened well.
3. Edge of the pointer should be accurate.
4. Current flow should be uniform.
5. Heart should be kept wet in 0.6% saline solution.

EXPERIMENT No. 20

Object. *To study the effects of acetylcholine, adrenaline, hot and cold Ringer's solution, pH change, Ca, Na and K ions on heart beat of frog.*

Requirements. Same as in experiment No. 19 and the substances whose effects are to be observed.

Procedure. Set the experiment as laid down in procedure for experiment No. 19 and then go on observing the heart beat for individual substances as mentioned below :

(a) *Effect of Acetylcholine.* Take few drops of acetylcholine in a pipette and pour its 4 or 5 drops on the heart and record the resultant heart beat. Observe that the heart rate is **decreased**.

(b) *Effect of Adrenaline.* Take few drops of adrenaline in a pipette and pour 4 or 5 drops of it on the heart and record the resultant heart beat. Observe that the rate of heart beat is **increased**.

(c) *Effect of hot or cold Ringer's solution.* For this, soak cotton ball in hot or cold Ringer's solution and then squeez it on the heart. Record the resultant heart beat and observe that **hot** solution **increases** the heart rate, while **cold** solution **decreases** it.

(d) *Effect of change in pH.* For this, change the pH of Ringer's solution to slightly **acidic** and pour this solution on the heart and record the resultant heart beat. Observe that the heart rate is **decreased**. Similarly, change the pH of Ringer's solution to slightly **alkaline** and pour this solution on the heart. Record the resultant heart beat and observe that the heart rate is **increased.**

(e) *Effect of Ca ions.* For this, prepare calcium chloride solution of increasing concentrations, *viz.*, 0.01%, 0.1%, 1% and 10%. Pour these solutions on the heart in increasing concentration and record the heart beat. Observe that with increasing concentration the heat rate is gradually **decreased** and it becomes zero up to the use of 10% solution, *i.e.*, the heart stops beating. The only care to be taken in this experiment is that calcium chloride solution of different concentrations should be poured on the heart with a minimum gap of 5 minutes.

Identical results can also be obtained with Na and K ions.

RECORDING OF MUSCULAR CONTRACTION

EXPERIMENT No. 21

Object. *Recording of muscle contraction in the gastrocnemius muscle-nerve preparation of frog.*

Requirements. Sherrington's recording drum, Dubois Rejmonds induction coil, muscle chamber, make and break key, lever weight, wire, muscle, 0.65% saline water, live frog and battery.

Connections. Drum is primary circuit, movable electrodes form secondary circuit.

Strength of current. Sub-maximal induced break shock should be effective.

Preparation of gastrocnemius muscle. First of all the frog is pithed, then the skin is lifted from the urostyle. The skin covering it is cut and the incision is then completed up to the skull. The skin is cut from the urostyle up to the knee-joint. With the help of foreceps the urostyle is lifted up and a blunt pair of scissors is passed in such a manner that the scissors should cut along the body surface of the urostyle. Just below the dissection a purely white and cord-like structure will be observed. The course of the sciatic nerve can now be observed without injuring any nerve. With the help of bone cutter the pelvic girdle and the vertebral column is cut near the origin of the sciatic nerve.

On removal of the layers of the muscles in the thigh along the groove which exists there, the white cord-like sciatic nerve will be observed in that region. The cut end of the vertebral end must be held in hand and separated out downwards as far as the nerve gets bifurcated in the thigh. Throughout this procedure the nerve is not pinched. Thus, the sciatic nerve will be dissected out throughout its body length from its origin up to the knee. From the knee joint the skin is cut up to the tendon of the gastrocnemius muscle by a knife blade is inserted beneath. The tendon of the gatrocnemius muscle and with the help of needle, a thread is passed below the tendo Achillrs, in such a way that it does not but remains intact to the tendon-end, the thread is firmly tied to the tendon. The tendon is cut away from the tied end by holding the thread, the muscle is separated up to the knee joint. With the help of the bone cutter the tibia and femur are cut for about 1.25 cm away from the knee joint. Now the frog's gastrocenemius sciatic nerve muscle preparation is ready. After the dissection is completed the precaution is taken that the muscle and nerve are not allowed to dry due to the exposure to atmosphere. In order to avoid this the nerve-muscle preparation is kept moist by normal saline water which in case of frog is generally 0.65% NaCI solution.

Procedure. A nerve-muscle preparation is dissected out and is fixed in the muscle chamber containing 0.65% NaCI solution as it is isotonic with frogs muscle tissue. The apparatus is arranged for a single induction shock and the secondary induction shock will be effective. The drum is included in the primary circuit. A small weight say about 1 gm is attached to the pointer arm which keeps the pointer in a horizontal position. The strength and intensity of the current is adjusted in such a way that the muscle will be in a position to produce a simple twich about 3.75 cm amplitude. The primary as well as secondary make and break key is closed and the lever point is brought just to touch the writing surface in order to have the optimum pressure on the smoked drum. When the contact in the primary circuit is just broken due to revolving base electrodes on the Sherrington's drum the muscle will be stimulated via the sciatic nerve and it will record a single twich. As soon as the writing point has returned to the base line, the short circulating key and secondary make and break key is closed immediately. The drum is dropped with the help of stop key and the writing point is withdrawn immediately from the drum so that the tracing are not in a position to overlap one another. It will be observed that the twitching is in the form of a curve only because the drum is rotating with certain speed.

In this experiment the tension due to weight on the muscle is merely constant but the muscle fibres are free to shorten in length. Hence, the contraction is said to be isotonic (change in length of muscle fibres not in tension).

The latent period in the phase during which no apparatus changes takes place in the muscle. The latent period extends from the point at which the contracting muscle begins to raise the lever.

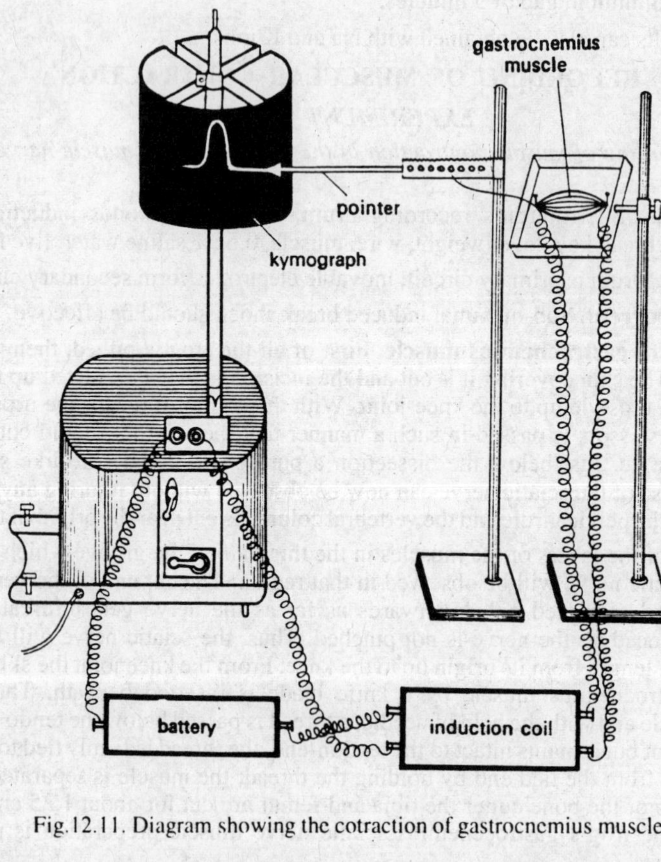

Fig.12.11. Diagram showing the cotraction of gastrocnemius muscle.

The contraction period extends from the point where the writing arm of the *"l"* shaped lever begins to rise to the point highest above the base line. The contraction period starts at the end of the latent period and is, thus, in active phase in the contraction.

The relaxation phase or period is a passive phase in the muscular contraction and extends from the highest point of the simple muscle curve up to the point at which the simple muscle curve twitch retouches the base line.

Inference. An isolated twitch consists of three phases, *viz.*, the latent phase, the contraction phase and the relaxation phase. Secondarily a simple muscle twitch represents the response to a single stimulus and a stimulus causes a sudden change in the external as well as in the internal development.

EXPERIMENT No. 22

Object. *Recording of tetanus.*

Requirements. Same as for the experiment No. 21.

(**Note:** *Connections, preparation of gastrocnemius muscle, procedure of setting the experiment same as described for experiment No. 21*).

Strength of stimulus (current). Threshold, rapid and constant.

Procedure. For recording tetanus, the muscle is stimulated so rapidly that there is no time between successive stimuli for muscle fibre to relax. The response then is smooth, sustained and maximal (Fig. 12.12). Such a response is called tetanus. Complete tetanus curve is obtained by stimulating the muscle fibre at the rate of 40 to 100 stimuli per second; if only 5 to 10 stimuli per second are given to the muscle fibre, incomplete tetanus (also referred to as tonus) is recorded. However, when the muscle fibre is stimulated constantly for a long time then finally a point is reached at which muscle response becomes zero. Now the muscle is said to be fatigued.

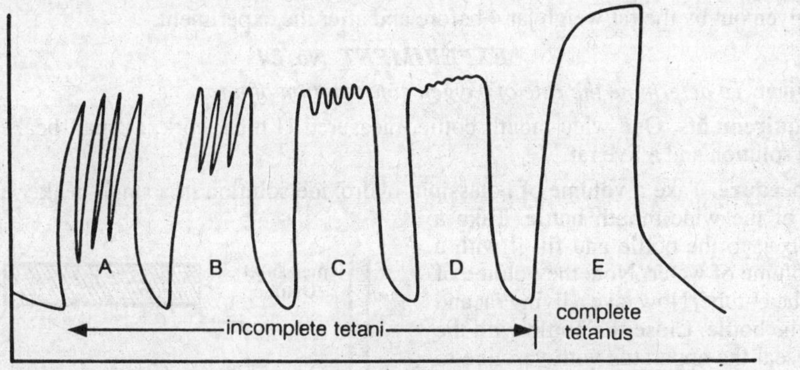

Fig.12.12. Muscle curves. A to D incomplete tetani and E complete tetanus.

RESPIRATION

EXPERIMENT No. 23

Object. *To determine the oxygen consumption of a rat.*

Requirements. Haldane's respiration apparatus, moist soda lime, pumice stone socked in sulphuric acid, and a live rat.

Procedure. Set up an apparatus as shown in Fig. 12.13. This apparatus is known as Haldane's respiration apparatus and it is used to determine the rate of oxygen consumption in small animals. Jars 1 and 4 contain moist soda lime for removing carbon dioxide, jars 2, 3 and 5 have pumice stone soaked in sulphuric acid for removing moisture. In the chamber rat is placed. M is a gas meter. J is a manometer having an inverted bell jar standing in a water trough, it prevents an excess of negative pressure and indicates the pressure actually employed. P is an aspirator for drawing air through the entire apparatus.

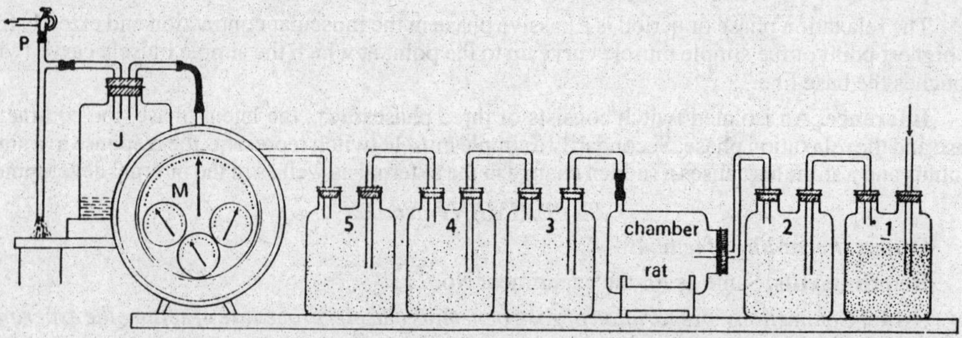

Fig. 12.13. Haldane's apparatus to determine oxygen consumption of a rat. 1 and 4 – soda lime; 2, 3 and 5 – pumice stone and sulphuric acid; rat chamber; J – manometer; M – meter; P – aspirator.

Draw a current of air through the apparatus by means of the aspirator for a specified period of time. The air entering the rat chamber has been freed of carbon dioxide and moisture by jars 1 and 2 respectively, thus, the chamber receives only oxygen. This oxygen is used for respiration needs of the rat. The rat gives out carbon dioxide and moisture into the outgoing air. Jar 3 removes the moisture, jar 4 absorbs the carbon dioxide, jar 5 is to remove any moisture gained from soda lime of jar 4. The quantity of oxygen consumed by the rat is determined indirectly. The oxygen consumed is equal to the carbon dioxide and moisture given out. Thus, weigh jars 3, 4 and 5, and the chamber including the rat both before and after the experiment. The increase in weight during a specified period of time gives the quantity of oxygen consumed by the rat during that period. To find the quantity of carbon dioxide given out by the rat weigh jar 4 before and after the experiment.

EXPERIMENT No. 24

Object. *To determine the rate of oxygen consumption of a rat.*

Requirements. One wide-mouth bottle, measured U-tube, cork, a small beaker, potassium hydoxide solution and a live rat.

Procedure. Take a volume of potassium hydroxide solution in a small beaker and keep it at the base of the wide-mouth bottle. Take a U-tube, fix it to the bottle and fill it with a known volume of water. Note the volume of water in the U-tube. Now take a living rat and put it in the bottle. Close the bottle with the cork and seal the apparatus with vaseline to make it airtight.

Observations. Note the level of water in the U-tube after an hour. It is found to be lowered. This is due to the fact that the rat respires and consumes the oxygen from the air of the bottle. The carbon dioxide released is absorbed by the potassium hydroxide solution. This reduces the pressure in the bottle and, hence, the level of water in the U-tube falls.

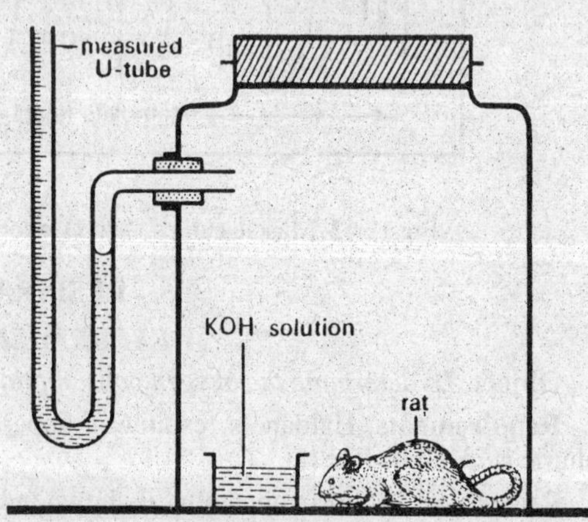

Fig.12.14. Apparatus for the demonstration of rate of respiration.

Calculations. The fall in the level of water is measured in a definite period of time that is the rate of oxygen consumption per unit time. Suppose :

(i) Level of water in the beginning = 25 ml.
(ii) Level of water after one hour = 20 ml.
(iii) Fall of water level in 1 hour = 5 ml.
(iv) Hence, the oxygen consumption by rat in one hour = 5 ml.

Precaution. The apparatus should be air-tight.

NERVOUS SYSTEM

EXPERIMENT No. 25

Object. *To study reflex action in a frog.*

Requirement. A live frog.

Principle and procedure. If a live frog is touched, it at once moves away involuntarily as a consequence of the stimulus of touching without involving the will, this type of reaction is called a **reflex action**. The course along which the impulse travelled from the point of touching the skin to the muscles which caused the animal to move away is known as a **reflex arc**. The reflex arc is the structural basis of the reflex action. The impulse of touching the skin goes to the dendrites of a sensory neuron (afferent) and then into its cell body situated in the dorsal root ganglion of the nerve concerned, and then along its axon into the gray matter of the spinal cord where its terminal axon endings form a synapsis with the dendrites of an association or internuncial neuron which lies within the gray matter of the nerve cord. The internuncial neuron forms a synapsis with the dendrites of a motor neuron (efferent) whose cell body lies in the ventral part of the gray matter of the nerve cord, then from the cell body the axon of the motor neuron conveys the impulse through the ventral root of the spinal nerve concerned to the muscles which contract and the frog moves away. This is a very simple reflex arc, it involves a sensory neuron, an association neuron, and a motor neuron; thus, in a simple reflex arc only three neurons are involved and their synapses are in the nerve cord. But in reflex arcs of mammals usually there is a chain of several association neurons lying in the central nervous system between a sensory and a motor neuron.

EXPERIMENT No. 26

Object. *To study the wiping reflex in a frog.*

Requirements. A live frog, stout pins, wire, hydrochloric acid and filter paper.

Procedure. First pith a frog by destroying the brai/—*/n or by passing a stout pin just behind the head cutting between the brain and spinal cord. In pithing the brain has been severed and destroyed but the spinal cord is intact. Hang the frog with a wire from a stand and fix the other end of the wire to the lower jaw. Prick a pin in one toe, the frog will withdraw its foot. A more violent stimulation of one foot will make both legs move as in jumping. Place a filter paper soaked in hydrochloric acid on the thigh of the frog, the leg will jerk to move the acidified paper, if it fails then the other leg will wipe the paper away by reflex action.

EXPERIMENT No. 27

Object. *To study the knee-jerk reflex in man.*

Requirements. Man and a small hammer or book.

Procedure. Make a man sit on the edge of a chair or table with one knee crossed over the other, and with the legs suspended and relaxed. Tap firmly on the ligment just below the knee cap of the upper leg with the side of the hand or with the edge of a book or with the small hammer. The leg will straighten with a jerk. This is due to sudden contraction of the quadriceps femoris muscle by reflex action whose reflex arc involves the spinal cord. The knee-jerk response will be much more pronounced if this experiment is tried with the fists of the man clenched tightly, because the degree of response depends on the condition of the body.

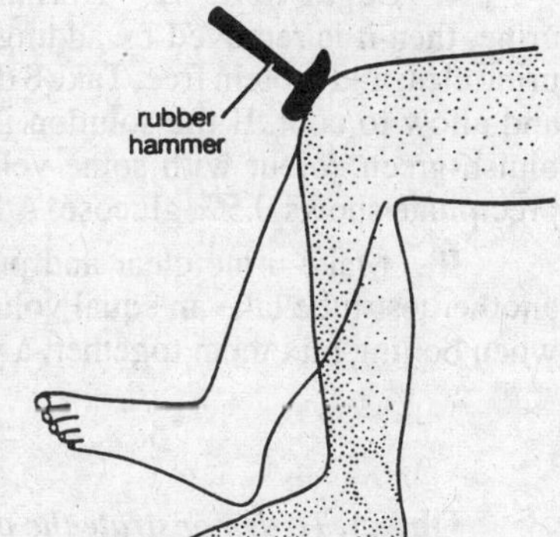

Fig.12.15. Diagram showing knee-jerk reflex in man.

Explanation. The knee-jerk is an automatic, unlearned and invoiutary response to the stimulus which involves contraction of quadriceps fempris muscle stretching the leg forward.

EXCRETORY SYSTEM

EXPERIMENT No. 28

Object. *To test the urine for urea, proteins, ketones and sugar.*

Requirements. Test-tubes, soyabean flour, litmus solution, acetic acid, phenolphthalein, sodium hydrate, copper sulphate, sulphosalicylic acid, ammonium sulphate, ammonia, sodium nitroprusside, Benedict's reagent and samples of urine.

Procedure. The following procedure is adopted for the test of urea, protein, ketone and sugar from the samples of urine.

1. Urea test. Normal urine contains 2 to 3 per cent urea. Take 2 ml of urine in a test-tube, add a pinch of soyabean flour, a little litmus solution, and a few drops of 1% acetic acid. In a few minutes the mixture becomes alkaline and turns blue because of decomposition of urea into ammonium carbonate. Alternatively, take 5 ml of urine in a test tube, add 2 or 3 drops of phenolphathalein, and 0.1 gm soyabean flour. Mix the contents and place test-tube in a water bath at 50°C. The contents will turn pink showing urea.

2. Protein test. A. Normal urine is free from proteins, but at times it may contain albumin, globulin, and other proteins. Heat urine to almost dryness by evaporation. Then heat this in a dry test-tube, fumes of ammonia are given off. Remove from flame and cool, add a few drops of sodium hydrate and a drop of copper sulphate solution. A pink colour appears showing the presence of proteins.

B. Filter urine to remove any turbidity and to make it clear. Take this clear urine in a test-tube and boil only the upper part for two minutes holding the test-tube by the bottom end. If the upper part becomes turbid and cloudy, then it may be due to albumin or phosphates. To find out add 3 to 5 drops of 5% acetic acid, if turbidity appears or persists, then it is due to the presence of albumin in urine.

C. Take some urine in a test-tube and boil the upper part only, add several drops of strong nitric acid slowly. If a flocculent precipitate appears and persists in the upper part of the tube, then it shows the presence of albumin.

D. Filter urine to remove any turbidity. Take 5 ml of filtered urine in a test-tube, add 6 drops of 20 % sulphosalicylic acid, if a cloud is formed it shows the presence of protein.

3. Ketone test. In metabolism fats and some amino acids are broken down into ketone bodies. The ketone bodies comprise acetone, aceto acetic acid, and beta hydroxybutyric acid. In a test-tube take 20 ml of urine and saturate it with ammonium sulphate crystals. Add 2 or 3 drops of ammonia and a few crystals of sodium nitroprusside, shake the tube well. If ketone bodies are present a reddish-purple colour appears in a few minutes; if, however, a brown colour appears, it shows the absence of ketones.

4. Sugar test. A. Normally there is no sugar (glucose) in urine. If there is any albumin in urine, then it is removed by adding a few drops of 5% acetic acid, then boil and filter to make the urine clear and protein free. Take 8 drops of this urine, add 5 ml of Benedict's reagent, boil the mixture and allow to cool. If the solution is a clear greenish-blue then it shows absence of glucose. But a bluish-green colour with some yellow precipitate shows 0.2% glucose. A deep yellow to orange precipitate shows 0.5% glucose. A heavy orange brown to red precipitate shows 1 to 3% glucose.

B. Make urine clear and protein-free as before. Take some of this urine in a test-tube. In another test-tube take an equal volume of Fehling's solution. Boil the two test-tubes separately and when boiling mix them together, a yellow to brick-red precipitate shows presence of glucose.

OSMOSIS

EXPERIMENT No. 29

Object. *To demonstrate the process of osmosis.*

Requirements. One big beaker or flask, one thistle funnel, strong cane sugar solution, pure water, semipermeable membrane (cellophane), one piece of gummed paper, wax, etc.

Principle. "Osmosis is a process in which water or any other solvent flows through a semipermeable membrane from a lower concentrated solute solution to one of higher concentrated solute solution". It is simply a case of diffusion, but the movement of solvent usually causes an obvious volume change, whereas the movement of solute has a negligible effect.

Procedure. Take a thistle funnel and the piece of a semipermeable membrane is tied tightly across the bottom of the thistle funnel and the joint is waxed with melted wax to make it water-tight. Now fill up the bowl of thistle funnel with strong solution of sugar and mark the upper level of the sugar solution on the outside of the thistle funnel by a piece of gummed paper stuck at that level. Now this sugar solution contained in the thistle funnel is stood in a beaker containing pure water with the help of some stand. This entire apparatus is put aside for some hours.

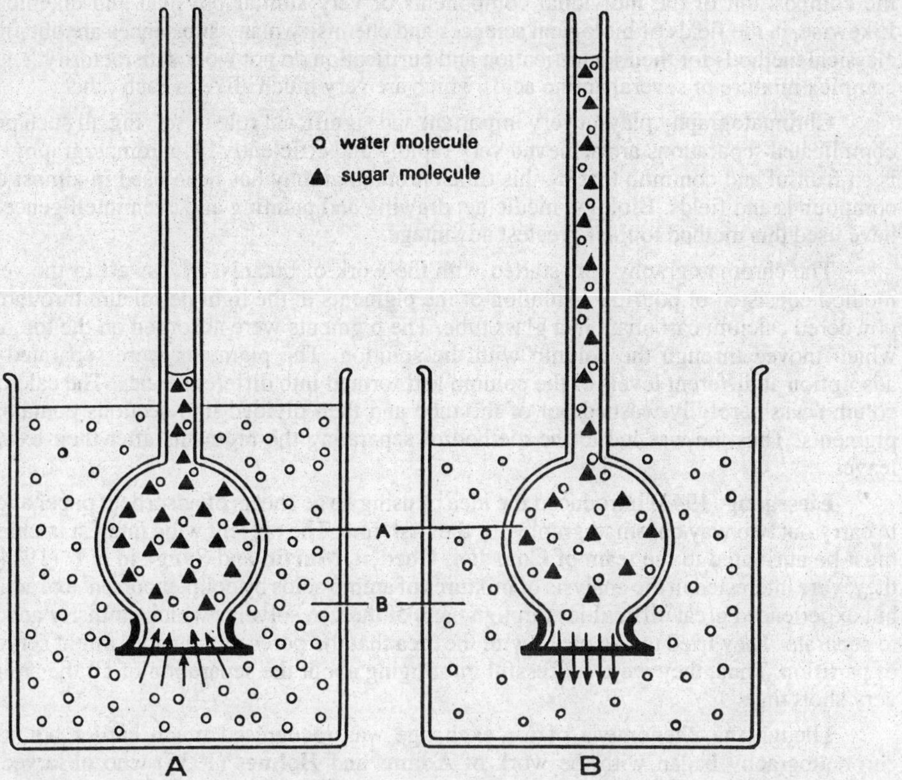

Fig. 12.16 Diagram illustrating the principle of osmosis. A – Beaker and thistle funnel which at the bottom is tied with semipermeable membrane and is filled with strong sugar solution; B – Beaker containing pure water. Arrows show direction of movement of water molecules, which can pass freely across the membrane, while the sugar molecules cannot. The right hand side diagram also expresses the state of osmotic equilibrium, at this state the same number of water molecules pass across the membrane in both directions.

Observations. After some hours it is seen that water from the beaker has entered the thistle funnel with the result the solution in the thistle funnel has increased in its volume and rises in the thistle funnel, well beyond the original mark. This level of sugar solution in thistle funnel keeps rising till the osmotic equilibrium is attained. On attaining the osmotic equilibrium the same number of water molecules pass across the membrane in both directions.

Inference. The rise in volume of sugar solution in the thistle funnel is due to passing of water molecules from the beaker containing pure water of low concentration. The sugar molecules do not pass into the beaker because of semipermeable membrane. Thus, this experiment demonstrates the process of osmosis.

Precautions. The membrane to be taken for the experiment should be semipermeable, *i.e.*, the water passes freely through it in either direction but not the solute such as sugar. If the membrane allows the solute to pass as well as water, then osmosis will not occur.

CHROMATOGRAPHY

There are many methods which have been used to obtain and identify the substances with high state of purity. Besides chemical methods, many physical methods, like fractional precipitations, distillation and crystallisation have been effectively used for separation and purification of the chemical compounds. Sometimes it becomes necessary to follow a chemical method of separation by a physical method of purification and identification. These methods were applicable quite satisfactorily in many cases but great amount of difficulty was realized with compounds which had the composition of the individual components of very similar physical and chemical properties. Likewise, in the fields of biological sciences and chemistry many substances are obtained where the classical methods for their identification and purification do not work satisfactorily, *e.g.*, protein is a complex mixture of several amino acids which are very much alike to each other.

Chromatography plays a very important and significant role in solving all such problems. The complicated separations are achieved very rapidly and efficiently by chromatography. This method is so fruitful and common that by this time chromatography has been used in almost every type of compounds and fields. Biology, medicine, drawing and painting and even intelligence departments have used this method to their greatest advantage.

The chromatography was started with the work of botanist **M.Tswett** in the year 1896. His method consisted of pouring a solution of the pigments in the light petroleum through a column of powdered calcium carbonate in a glass tube. The pigments were absorbed on the top of the column which moved through the column with the solution. The pigments were separated by selective adsorption at different levels of the column and formed into different zones. The calcium carbonate column was carefully washed out of the tube and then divided into sections containing different pigments. Thus, he was led to the method of separating the pigments after their extractions from leaves.

Liesegang (1943) introduced the idea of using large sheets of adsorbent paper and is supposed to carry out two way chromatography for the first time. The present wide interest in chromatography must be attributed to the team of **Consden, Gorden, Martin** and **Synge** in U.K.(1944). Originally they were interested in the analysis of mixtures of amino acids by counter current extraction processes but experienced great difficulties through parts of the two solvents which emulsify and was difficult to separate. They tried to use paper with the idea that the pores in the paper might serve as the basis of partition. Thus, they were successful in bringing about the separation of all the amino acids in a very short time.

Though the phenomena of ion-exchange was recognised much earlier but its use in the chromatography began with the work of **Adams** and **Holmes** (1935) who observed that certain synthetic resins were capable of exchanging ions. In 1941, **Martin** and **Synge** developed partition column chromatography and in 1944 they replaced silica gel support by the strips of paper and developed widespread use of paper chromatography. In 1952, **James** and **Martin** gave an account for the gas liquid chromatography in which mobile phase was gas rather than liquid and they got Nobel Prize in the recognition of their work. Afterwards there was rapid development in adsorption chromatography in which adsorbent in the form of thin layer on glass plate was used. In 1959, **Forth** and **Flodin** developed unique method of gel filtration in which gels of cross linked dextrans were used which separated the biological materials mostly by virtue of differences in their sizes. The wide applicability of chromatographic methods in the biological sciences is innumerable and unique.

What is chromatography ?

Chromatography is a procedure through which various substances in a mixture are separated from each other and identified. Separation involves the use of two phases, one of which is stationary and the other mobile. Separation depends upon the differential movement of the test substances between two phases. In a simple language chromatography can be defined as an analytical technique for separating compounds on the basis of differences in affinity for a stationary and mobile phase.

Chromatographic Methods

Chromatographic methods are classified into two categories, *viz.*,

1. Partition chromatography;
2. Adsorption chromatography.

1. Partition chromotography

1. Partition column chromatography;
2. Paper chromatography;
3. Thin layer chromatography;
4. Gel filtration chromatography.

} Liquid as mobile and solid as stationary phase.

5. Gas liquid chromatography or vapour phase chromatography.

} Gas as mobile and liquid as stationary phase.

2. Adsorption chromotography

1. Adsorption column chromatography.
2. Thin layer chromatography.

} Liquid as mobile and solid as stationary phase.

3. Ion exchange chromatography.

} Electrolytic solution as mobile and some ionic polymer as stationary phase.

Techniques

Following techniques are employed in chromatography :

1. Partition column chromatography. In this technique, the column is packed with a porous solid of high surface area, *e.g.*, silica gel and cellulose which is coated with water (stationary phase). The components of a mixture are separated by passing an organic solvent (mobile phase) through the column.

2. Paper chromatography. In this technique, the dissolved substances are applied as a small spot on a cellulose bound filter paper which is then kept in a container dipped in an organic solvent. The mixtures are partitioned between (stationary phase) and organic solvent (mobile phase).

3. Thin layer chromatography. In this technique, the adsorbent (stationary phase) is spread over a glass plate in the form of a thin film of even thickness. The solvent (mobile phase) moves up the plate by capillary action and, thus, effects separation.

4. Gel filtration chromatography. Sephadex contains the aqueous (stationary phase) which is distinguished from the mobile phase by its immobilization. The component of the mixture are separated according to their size, by virtue of the differential distribution between easily displaceable water present in the interbed space.

5. Gas liquid chromatography. In this technique, a column is packed with a porous inert solid coated with a thin layer of an involatile liquid as the stationary phase. Components of the mixture are separated by being partitioned between this phase and a gaseous (mobile) phase.

6. Adsorption column chromatography. In this technique, the separation of a mixture is determined by the differential adsorption of the components on an active solid (alumina, silica gel, stationary phase) as an organic solvent (mobile phase) containing them which passes over and affects separation.

7. Ion exchange chromatography. In this technique, ionised compounds are separated in aqueous solutions (mobile phase) by virtue of their differences in affinity for ionised compounds which are an integral part of the insoluble solid phase (stationary phase).

Here only **paper chromatography** will be discussed in detail because only paper chromatography is prescribed in the syllabi of under-graduate and post-graduate classes of most of the Indian Universities.

PAPER CHROMATOGRAPHY

Amongst all the chromatographic techniques known so far, paper chromatography is considered to be the simplest and most widely used. Since the outstanding work by the team of **Consden, Gordon, Martin,** and **Synge** in England in 1944, who later on were awarded Nobel Prize in recognition of their work, numerous workers have used this method so much so that at present no field in biology, biochemistry, chemistry and many other fields exists in which paper chromatography has not solved the most complicated problems.

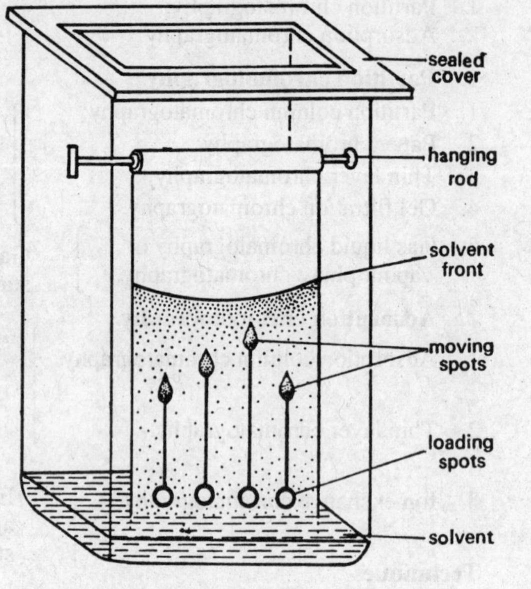

Fig.12.17. Chromatography chamber.

What is paper chromatography?

The paper chromatography is the method in which the analysis of an unknown substance is mainly done by the flow of solvents on specially designed filter paper. One of the two solvents is immiscible or partly miscible in the other solvent. The solvent rises up by the capillary action and by the adsorption on the filter paper, the separation is effected by the differential migration of the mixture of substances. This occurs due to partition coefficients.

Types of Paper Chromatography

Paper chromatography is of the following two types :

1. **Paper partition chromatography.** The standard method of analysis in which paper is used as an inter-support with one solvent as mobile phase and the other solvent as immobile phase is called paper partition chromatography.

2. **Paper adsorption chromatography.** In certain cases the paper is impregnated with an adsorbent like alumina or silica. The paper so modified, is used as the adsorbent and the single solvent is allowed to flow over the unknown components. This process is known as paper adsorption chromatography.

Techniques of Paper Chromatography

The method of paper chromatography may be **one dimensional** or **two dimensional** depending on the type of complexity involved in the analysis.

In one dimensional chromatography, a drop of the solution containing the mixture of substances is placed near one end of the filter paper. The spot is then allowed to dry. The portion of the paper with the spot is then placed in a suitable solvent. The solvent passes the spot and carries the substance along with while moving up the filter paper. Different substances move with different rates which depend on the size, nature of the substances, solvent, diffusion coefficient and many other factors. The finished paper is known as **chromatogram**. The advancing rate is determined by their R_f values. The R_f values depend on the factors as enumerated above and differs from one substances to another.

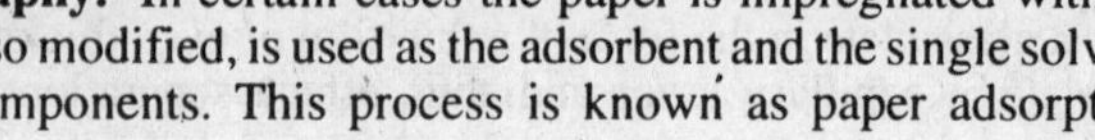

Fig.12.18. One dimensional chromatogram.

But there are many substances which cannot be separated under the above mentioned set conditions. This happens because the R_f values of many substances are very close and nearly the same. This problem can be solved by using the method of two dimensional paper chromatography.

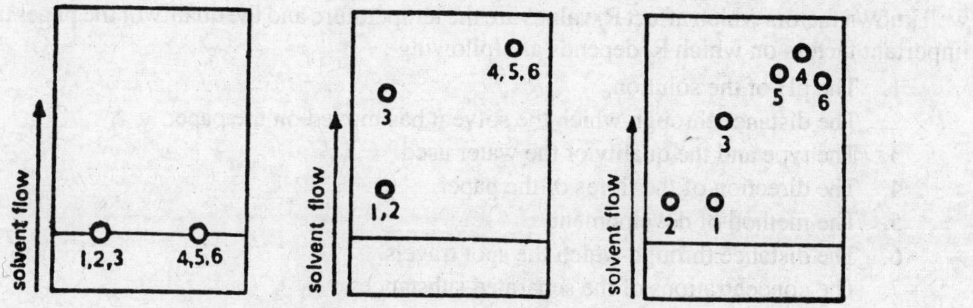

Fig.12.19. Two dimensional chromatogram.

In two dimensional chromatography; first the one dimensional chromatogram is obtained, the paper is turned at right angles to the first. The chromatography is done by using different solvents. By doing so ovelrapping R_f values are separated to a reasonable extent.

R_f value

The R_f value expresses the relative rate of movement of solute and solvent. R_f value is defined as *the ratio of the distance travelled by the compound at its maximum concentration to the distance travelled by the solvent.* Both the distances are measured from the point of application of the sample. R_f value has no unit.

$$R_f = \frac{\text{Distance travelled by the substance from origin}}{\text{Distance travelled by the solvent from origin}}$$

Thus, as shown in the Fig. 12.20 $R_f = \frac{A}{B}$

In some cases, it has been seen, that the solvent front runs off the paper. In such cases it is more convenient to express the movement of any substance just by comparing with the movement of some standard substances. This is represented by R_f as shown in the Fig. 12.21. R_f is defined as

$$R_f = \frac{\text{Distance moved by the substance}}{\text{Distance moved by the standard substance}} = \frac{C}{D}$$

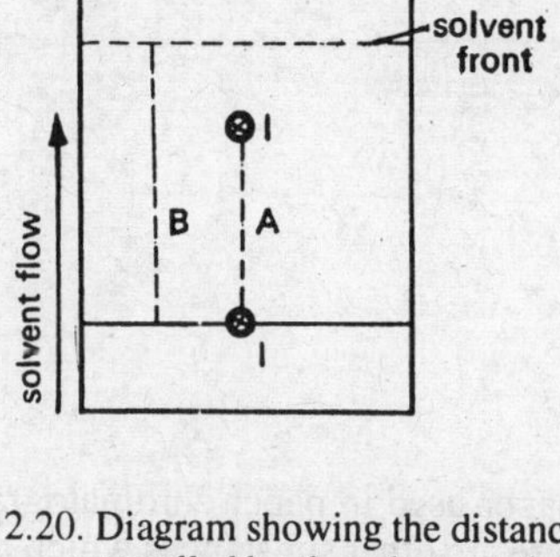

Fig. 12.20. Diagram showing the distance travelled by the substance from origin and the distance travelled by the solvent from origin.

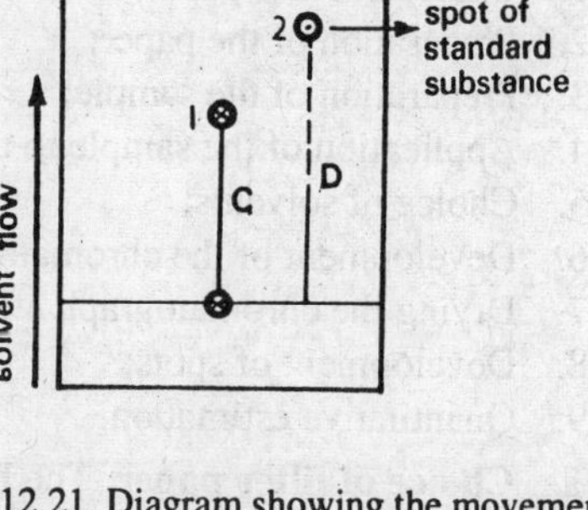

Fig.12.21. Diagram showing the movement of any substance and its comparison with the movement of some standard substance.

R_f Variation

The R_f values of different compounds are entirely different. R_f values for different compounds are given in the literature and identification is made by comparison method. Thus, it can be said that R_f is quoted as the identifying characteristic. The values vary with the solvents used, *e.g.*, two solvents will give two values. Thus, R_f value is always quoted with reference to the solvent used. Some of the well known factors which affect R_f values are the temperature and the quality of the paper used. Other important factors on which R_f depends are following :

1. The pH of the solution.
2. The distance through which the solvent has moved on the paper.
3. The type and the quality of the water used.
4. The direction of the fibres of the paper.
5. The method of development.
6. The distance through which the spot travels.
7. The concentrations of the separated substances.
8. The method of drying, locating the colourless substances, and many other operations involved in the method.
9. The impurities present on the surface of the paper.
10. Irreversible adsorption on paper.
11. Chemical reactions between substances being partitioned and the paper and ion exchange with acid groups of cellulose.

With the combined effects of 9, 10 and 11, it can be safely presumed that amongst all, the partition phenomena is most important. And if the distribution coefficient is known for the system R_f value can be very easily calculated. They are related by the following expression :

$$R_f = \frac{A_l}{(A_l + \alpha A_s)}$$

or

$$\alpha = \frac{A_l}{A_s}\left(\frac{1}{R_t} - 1\right)$$

where A_l = Cross sectional area of the mobile phase;

A_s = Cross sectional area of the immobile phase;

α = Distribution coefficient of the substance between the two solvents.

Thus, it is evident that if the distribution coefficient is known and the cross sectional areas of the two phases are measured, R_f values can be calculated.

Procedure

There are many important things which should be taken into account while doing the experiments on paper chromatography on sheets or strips of paper. The following most important operations are involved:

1. Choice of filter paper;
2. Preparation of the paper;
3. Preparation of the sample;
4. Application of the sample to the paper;
5. Choice of solvents;
6. Development of the chromatograms;
7. Drying the chromatograph after development;
8. Development of spots;
9. Quantitative estimation.

1. Choice of filter paper. The basic material of the paper used in paper chromatography is α-cellulose of high purity which is 98–99 per cent. Besides, there are other substances which are also present; like β-cellulose ether soluble matter, ammonia, nitrogen like-compounds and mineral matters. Whatman filter paper has been extensively used in chromatography. The chemical composition of Whatman paper is as follows :

(i)	α-cellulose	98	–	99	per cent
(ii)	β-cellulose	0.3	–	1.0	per cent
(iii)	Pentosans	0.4	–	0.8	per cent
(iv)	Ash	0.07	–	0.01	per cent
(v)	Ether soluble matter	0.015	–	0.1	per cent

There are various grades and types of paper, the most suitable is the one which fulfils the following requirements :

1. The paper having maximum degree and clarity of the separation.
2. The paper having negligible diffuseness of the spot.
3. The paper having the excellent rate of movement of the solvent front.

The choice of the filter paper mainly depends on the thickness, flow rate, purity and net strength.

2. Preparation of the paper. After the selection of the quality of the paper which has to be used, the shape and size of the paper is to be decided. This is particularly important in descending technique. The shape of the paper facilitates in giving better resolution of less polar components.

The shape and size of the paper depend on the type of the work involoved. Though paper of shapes like rectangular and square have been used by many workers but the most common is rectangular-shaped paper.

3. Sample preparation. In the preparation of samples for chromatographic work, the important question is to get and choose the proper solvent for making solution. The complexity arises in cases of qualitative or quantitative type of work involved therewith. For solids a weighed amount is dissolved in a volatile solvent and by careful means the minimum volume of the concentrated solution is applied on the paper. This is done simply to avoid diffusion through paper which ultimately results in larger zones. Extracts from plants and animal tissues are taken out with the help of some solvent which is then directly applied on the paper.

4. Application of sample to the paper. The amount of the sample to be applied is again an important factor. The factors which govern the amount of sample to be applied are: (i) the capacity of the solvent, (ii) the optimum concentration required for the quantitative determination, (iii) the time required for development. With the above considerations in mind the final decision is made regarding the amount of sample to be applied.

Fig.12.22. Diagram showing the marking of paper at equal distances.

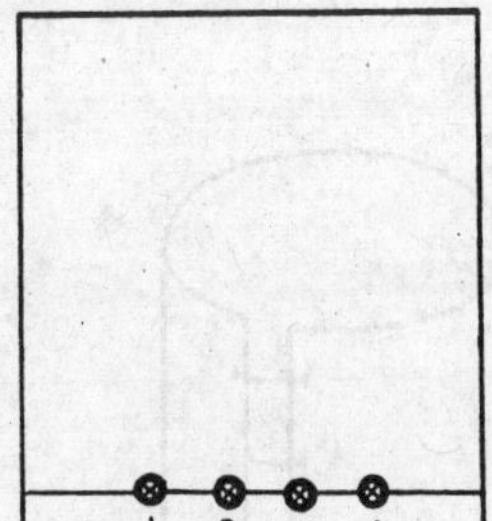

Fig. 12.23. Diagram showing the application of samples at each spot.

The samples may be applied either as spots or bands. A line is drawn with the help of pencil at a suitable distance from one end of the paper. Numerous points are marked on the line at equal distances. A drop of each sample is then put on these points with the help of capillary tubing, microsyringe or platinum loop. It is also to be noted that the spots occupy small areas which should be at the most of 5 mm in diameter. The volume of the liquid may vary from 1–2 μ or more as the case may be.

5. Choice of the solvents. There are numerous solvents which can be used in the paper chromatography. The final selection depends entirely on the nature of the substances to be separated. There are two ways to solve this problem. The first is the knowledge which one could get from literature and second is by trial and error method. The most commonly used is the polar solvent which is adsorbed on the paper. A less polar solvent, which is called the mobile phase, advances ahead by capillary action and separates the applied substances. Water miscible solvents such as furan propanol and cellosolve have been extensively used. In most cases a mixture of n-butanol, some tertiary alcohol is preferred because of the tendency of the former to esterify with organic acids. The selection of the solvents should be such so that when the components are mixed together, two definite layers result. When using solvents it is essential that they are used in the specified ratios and are shaken well before use.

In general, one-phase systems are extensively used in paper chromatography. Two-phase systems are, thus, avoided. This is simply because in that case we can get, number of possible combinations and greater variation of compositions. Since the rate of movement of the solvent is governed also by its viscosity, surface tension and density, more than one phase may also be complicated.

As far as possible minimum number of solvents are used. This facilitates in making the chamber saturated with the solvent which prevents evaporation from the paper during the development. This is difficult if number of solvents are used.

6. Development of chromatogram. Numerous apparatus have been designed to carry on chromatography. The following things are to be taken into consideration :

(i) There is sufficient amount of solvent at the bottom of the reservoir.
(ii) There is proper arrangement for the free suspension of the paper.
(iii) Large temperature variations are to be avoided.

The reservoir is made from a variety of materials. The material should be insoluble in the solvent and for that purpose glass, polythene, stainless steel and many other materials have been successfully used. The paper may be held in position by clumps, etc.

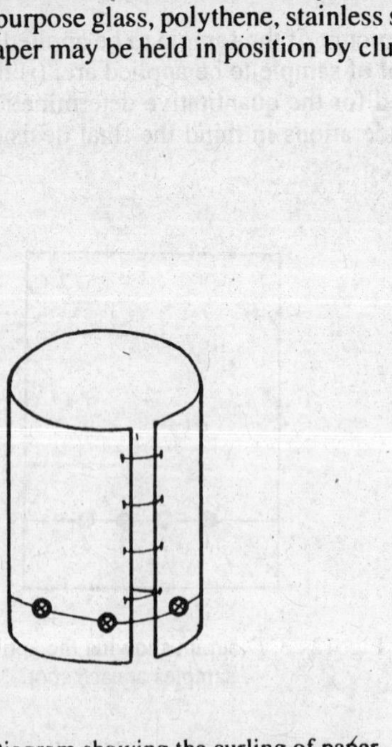

Fig. 12.24. Diagram showing the curling of paper. The two ends of the paper are curled into cylindrical form with a distance left in between them.

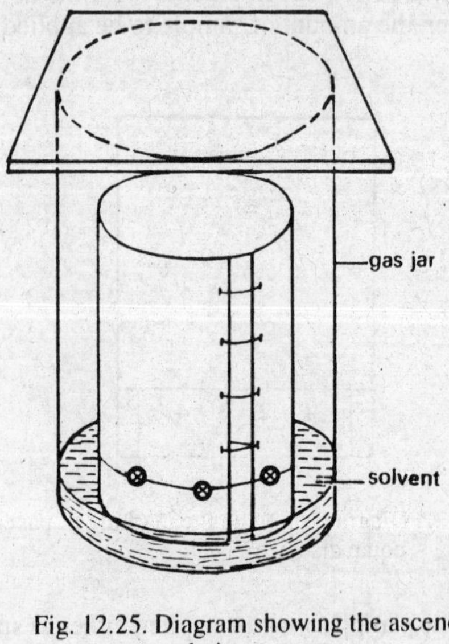

Fig. 12.25. Diagram showing the ascending technique.

The choice of the method of development depends upon the class of compounds to be investigated. The following are the types of developments :

A. Ascending development. This is the simplest of all the methods. It includes the dipping of the lower end of the paper into the solvent and allowing it to rise up the paper by capillary action. This method is preferred for quick analysis of a large number of substances. Proper care is to be taken to ensure that the lower end of the paper containing spots is so placed that it is above the solvent depth. The two ends of the paper are not allowed to touch with each other and are kept at a distance apart by clipping or by any other method.

As the solvent ascends, it does so against the gravitational force. With the increasing height the rate becomes slower. This slowness offers more time for partition equilibrium between two phases to take place.

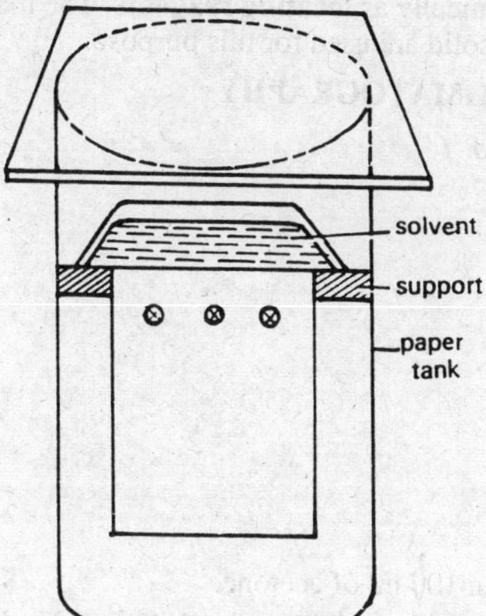

Fig. 12.26. Diagram showing the descending.

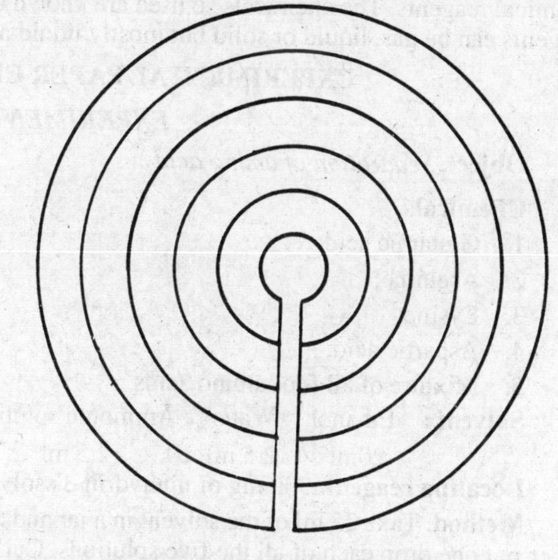

Fig. 12.27 Diagram showing radial or disc technique.

B. Descending development. This type of chromatography is done very frequently. In this method the solvent is kept in a trough at the top of the chamber and is allowed to flow down the paper. The liquid moves down by capillary action as well as by the pull of the gravity. In this case the flow is more rapid in comparison to ascending method. Due to rapid flow the chromatogram is completed comparatively in a shorter time. The apparatus required in this case is more sophisticated. The developing solvent is placed in a trough at the top which is usually made up of an inert material. The paper is then suspended in the solvent and the lid is placed at the top of the trough. By this method two dimensional chromatography cannot be done. This is because the paper must be cut to fit into the trough. Compounds with low R_f values are not completely separated by ascending method but this can be done by following descending technique.

Radial or disc development. Radial or disc development is very rarely used and that too in very special cases. In this case, a circular piece of paper is taken which has a wick cut parallel to the radius, from edge to centre. The sample is deposited at the centre of paper and at the upper end of the wick. This paper is laid on the edge of a circular disc with the wick dipping into the solvent at the bottom of the dish. The liquid ascends the wick and flow radially through the paper. While moving it carries compounds with it. As the solvent moves ahead, the radial becomes slower and takes more time for this solvent to reach the last circle.

7. Drying the chromatogram. After the development by any of the above methods, the paper is removed from the reservoir and the position of the solvent front is marked with the pencil at the two edges of the paper. The chromatogram is then dried by keeping it in an oven or over a hot plate for few minutes. This is done so that the solvents are completely dried from the paper. This drying is best done by means of a fan or hair dryer.

8. Location of spots. After drying of the developed chromatogram it is very essential to locate the separated substances. If the substances are coloured, no difficulty arises but the problem comes with the colourless substances. There are several methods which have been used to locate the spots, all of which can be grouped into the following methods: (i) physical methods and (ii) chemical methods.

(i) Physical methods. The physical methods for locating the substances are more applicable in comparison to chemical methods in the sense that the substances on the paper are not converted into other compounds. Thus, they can be removed and studied.

(ii) Chemical methods. When physical methods are not applicable, the chemical methods are used. The colourless compounds are converted into coloured compounds by reaction with some chemical reagents. The chemicals so used are known technically as **locating reagents**. The locating reagents can be gas, liquid or solid but mostly liquid and solid are used for this purpose.

EXPERIMENTAL PAPER CHROMATOGRAPHY

EXPERIMENT No. 1

Object. *Separation of amino acids.*

Chemicals.

1. Glutamic acid;
2. Arginine;
3. Lysine;
4. Aspartic acid;
5. Mixture of all four amino acids.

Solvent :	Ethanol	: Water	: Ammonia solution
	20ml	: 2.5 ml	: 2.5 ml

Locating reagent. 300 mg of ninhydrin dissolved in 100 ml of acetone.

Method. Take 25 ml of the solvent in a jar and cover with the lid. On the Whatman No.1 filter paper put one drop each of all the five solutions. Curl the filter paper into cylindrical from put into the jar so that the paper does not touch the sides of the jar. Run the chromatogram for about 3 hours. Take the paper out and mark the solvent front with the help of the pencil. When dry spray this with ninhydrin solution. Then put the filter paper into an over and heat fo 5–6 minutes at a temperature of 100–105 °C. The different amino acids appear as blue like spots. Record and compare R_f values with the mixture.

EXPERIMENT No. 2

Object. *Separation of sugars.*

Chemicals.

1. D-Glucose;
2. D-Xylose;
3. Lactose;
4. Mixture of all three.

Solvent.	Ethyl acetate	: Pyridine	: Water
	14	: 6	: 5

Locating reagent		
	1. m phenyline diamine	0.5 gm
	2. Stannous chloride	1.2 gm
	3. Acetic acid	20 ml
	4. Ethanol	80 ml

Method. Take 25 ml of the solvent in a jar and cover with the lid. On the Whatman No. 1 filter paper place 4 drops, one for each solution. Curl the paper into cylindrical form and put it into the jar so that the paper does not touch the sides of jar. Run the chromatogram for 3 hours. Take the paper out and mark the solvent front with the help of a pencil. Dry the paper and dip into the locating reagent solution. Now put this paper in the oven at 100–105°C for 5–6 minutes when different spots of sugars are observed from dark yellow to different forms in colour. Record and compare the R_f values of the mixture.

TABLE 1. R_f Values of Sugar and Sugar Derivatives 20 °C

Compounds	*Solvent*			
	1	2	3	4
D-Glucose	0.39	0.18	0.43	0.64
D-Mannose	0.45	0.20	0.46	0.69
D-Fructose	0.51	0.23	0.48	0.68
D-Arabinose	0.54	0.21	0.56	0.70
D-Ribose	0.59	0.31	—	0.76
Lactose	0.38	0.09	0.33	0.46
Sucrose	0.39	0.14	0.39	0.62
L-Ascorbic acid	0.24	0.38	0.59	0.84
Dehydroascorbic acid	0.28	0.48	—	0.88
D-Galacturonic acid	0.13	0.14	0.40	0.31
Raffinose	0.27	0.05	—	0.45

1. Water-saturated phenol + 1 % NH_3 (HCN atmosphere).
2. n-Butanol : acetic acid : water (4 : 1 : 5 V/V) descending.
3. n-Butanol : acetic acid : water (4 : 1 : 5 V/V) horizontal.
4. i-Propanol : pyridine : water : acetic acid : (8 : 8 : 4 : 1 V/V) ascending.

TABLE 2. R_f Values of Some Amino Acids

Compounds	*Solvents*							
	1	2	3	4	5	6	7	8
α-Alanine	0.34	0.10	0.39	...	0.09	0.05	0.57	0.60
Arginine	0.34	0.02	0.19	...	0.05	0.01	0.22	0.89
Cystine	0.16	0.02	0.17	...	0.01	0.00	0.24	...
Glycine	0.21	0.07	0.33	...	0.05	0.03	0.45	0.41
Diiodotyrosine	...	...	...	0.70	...	...	...	0.61
Histamine	...	...	...	0.22	...	...	...	0.95
Norvaline	...	0.23	...	0.65	0.31	0.19	...	0.80
Valine	0.50	0.18	0.56	...	0.22	0.15	0.71	0.78
Serine	0.20	0.08	0.31	...	0.05	0.02	0.45	0.36
Aspartic acid	0.18	0.01	0.33	...	0.12	0.00	0.30	0.19
Glutamic acid	0.18	0.02	0.37	...	0.01	0.01	0.36	0.31
Lysine	0.25	0.02	0.18	...	0.02	0.01	0.20	0.81
Methionine	0.55	0.27	0.51	...	0.05	0.21	0.76	0.81

1. Acetone : urea : water = 60 : 045 : 40 (V/W/V)
2. Tert. amyl alcohol saturated with water.
3. 1 Butanol : Acetic acid : Water = 250 : 60 : 250 (V/V/V)
4. 1 Butanol : Acetic acid : Water = 40 : 10 : 60.
5. 1 Butanol saturated with 2N NH_4OH.
6. 1 Butanol : Benzyl alcohol = 1 : 1, saturated with water in $(C_2H_5)_2$ NH atmosphere.
7. Pyridine : Isoamyl alcohol : Water : Diethylamine = 10 : 10 : 7 : 0.3.
8. Phenol : Water = 100 : 20 (V/V) use in presence of 1 : 1 Acetic acid.

CHAPTER 13

ECOLOGICAL EXPERIMENTS

Experimental ecology deals with the study of **biotic** and **abiotic** factors of an ecosystem, relationships between different organisms in an ecosystem and also various modifications under which an organism goes to lead a successful life in a particular environment. However, an attempt has been made in this chapter to deal with certain simple experiments relating to basic ecology.

A. Study of Physical Factors

EXPERIMENT No. 1

Object. *Recording of atmospheric temperature.*

Requirement. Maximum and minimum thermometer.

The maximum and minimum thermometer consists of a U-shaped graduated (either in Centigrade or Fahrenheit) tube. It has a small bulb B_1 at one end and a long bulb B_2 at the other end Fig. 13.1. The limb of U with bulb B_1 is the maximum thermometer and the other limb with bulb B_2

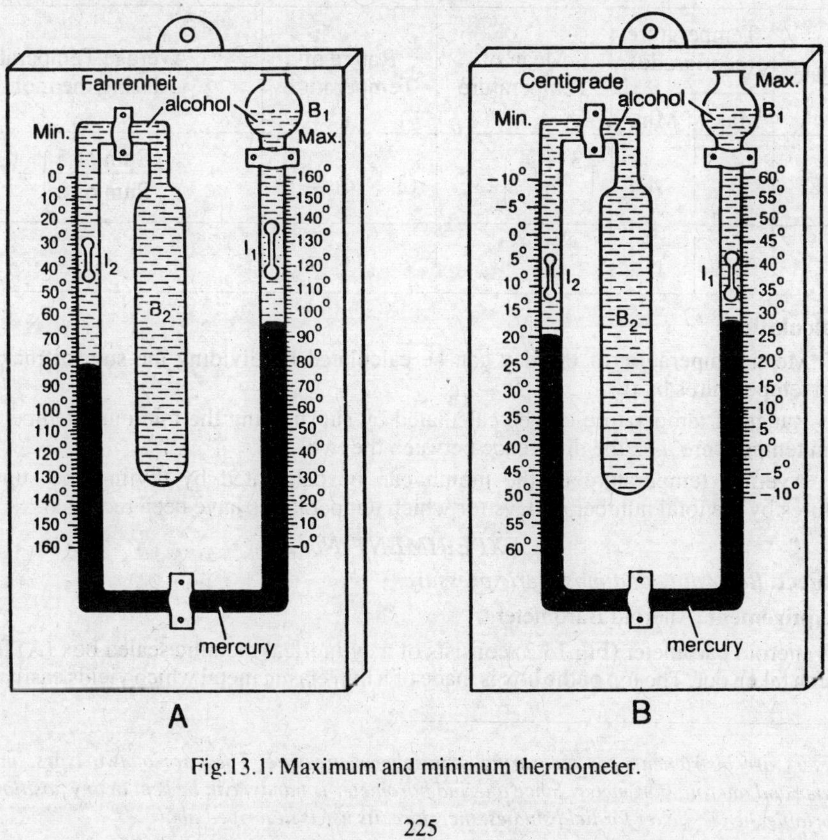

Fig.13.1. Maximum and minimum thermometer.

is the minimum thermometer. These two thermometers show the highest and the lowest temperatures reached during the day or a given period of time. The U-shaped tube contains a column of mercury as shown in the figure. The limb of U with long bulb B_2 and that with small bulb B_1 contain absolute alcohol above the level of mercury. The upper part of bulb B_1 is empty, therefore there is space for expansion of alcohol in this bulb. The limb of U with bulb B_2 contains alcohol without any empty space above the level of mercury. In addition, there is an index inside each limb of U-shaped tube lying above the mercury column. An index is a small piece of steel with a spring attached to it. Each index, marked I_1 and I_2, can be brought down into contact with the top of mercury column by means of a magnet kept outside the tube.

Working of maximum and minimum thermometer. A rise in temperature causes expansion of alcohol, while a fall in temperature causes its contraction due to which movement in mercury column occurs. Thus, when temperature rises, alcohol expands. As there is no space for expansion of alcohol in the long bulb B_2 of the maximum thermometer, it results an increase in the pressure due to which the mercury column in this limb (the limb with bulb B_2 and index I_2) is pressed down. Due to this, there occurs a rise of mercury column in the other limb with the bulb B_1 and index I_1. Therefore, the index I_1 which was in contact with the top level of mercury also moves up. When the temperature falls, alcohol contracts. This results in a downward movement of mercury column in the limb indicating maximum temperature, thus, pushing the index I_2 up which is in touch with the top level of mercury column in the limb indicating minimum temperature. The index I_2 remains in the position to which it has been pushed, even when alcohol expands again due to rise in temperature as it is checked in doing so by the spring attached to it and, thus, its lower end reads the minimum temperature. In the sameway, the index I_1 does not move down with the mercury column and its lower end reads the maximum temperature reached during the day.

Observation. Since each day has its maximum and minimum temperature. Hence, temperature record can be maintained for each day of a month as shown in the following table :

S. No.	Date	Temperature (centigrade)		Mean of Temperature	Range of Temperature	Average Temperature during the month
		Max.	Min.			
1	2	3	4	$\left(\frac{3+4}{2}\right) = 5$	$(3 - 4) = 6$	$\left(\frac{\text{Sum of 5}}{\text{Sum of 2}}\right) = 7$

Calculations

1. Mean temperature of the day can be calculated by dividing the sum of maximum and minimum temperatures by two.

2. Range of temperature can be calculated by substracting the minimum temperature from maximum temperature, *i.e.*, the difference between the two.

3. Average temperature of the month can be calculated by diviing the sum of mean temperatures by the total number of days for which temperatures have been recorded.

EXPERIMENT No. 2

Object. *Recording of atmospheric pressure.*

Requirement. Aneroid Barometer.*

An aneroid barometer (Fig.13.2) consists of a cylindrical metallic sealed box (A) from which air has been taken out. The top of the box is made of a thin elastic metal which yields easily to external

** Barometers are instruments for measuring atmospheric pressure. They are of two types: mercury type barometer and aneroid barometer. Since aneroid barometer is handy, can be kept in any position and easily transportable, hence, better for field studies. therefore, its use is described above.*

atmospheric pressure. A spring (B) is attached to the top of metallic box. The spring pulls the box upwards so that the atmospheric pressure does not disturb the box. When the atmospheric pressure increases or decreases the top of the box slightly moves downwards or upwards. The gears and levers (C) pass on the movement of the top of the metallic box to a pointer (D). The pointer moves clockwise over a scale (E) when the atmospheric pressure increases and anticlockwise when it decreases. Thus,

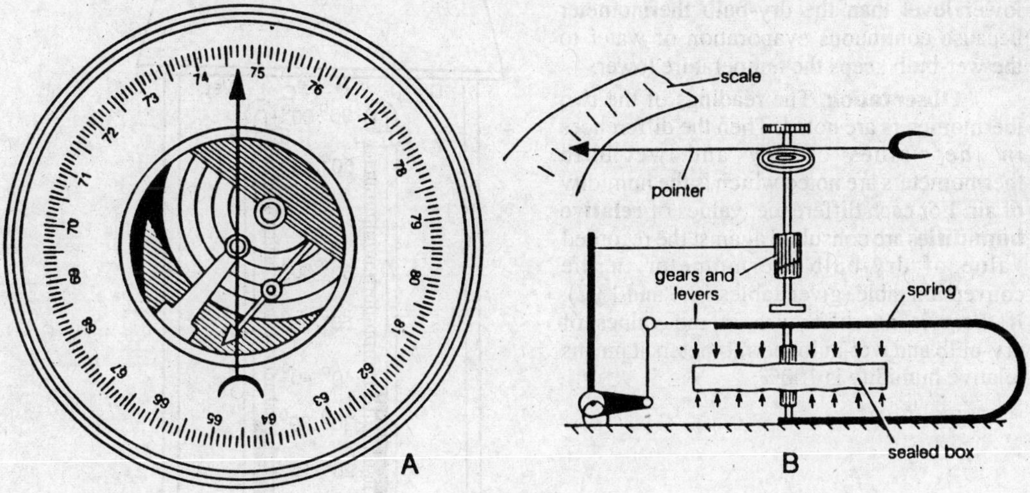

Fig.13.2. Aneroid barometer. A–Surface view; B–Diagrammatic section.

atmospheric pressure can be read on the dial of the barometer.

Observation. Since atmospheric pressure varies from time to time every day. Hence, its record can be maintained for each day as under :

S. No.	Date	Atmospheric pressure at			Average pressure	Mean pressure
		8 a.m.	12 noon	4 p.m.		
1	2	3	4	5	$\left(\frac{3+4+5}{3}\right) = 6$	$\left(\frac{\text{Sum of 6}}{\text{Sum of 2}}\right) = 7$

Calculation

1. Average pressure is calculated by dividing the sum of the columns 3,4 and 5 by 3.
2. Mean pressure for a specified period can be calculated by summing up the pressures of column 6 and dividing it by the total number of days (column 2) for a specified period.

Precaution

1. Do not forget to check the calibrations (graduations) of the dial of your aneroid barometer with a standard mercury barometer from time to time.

EXPERIMENT No. 3

Object. *Recording of relative humidity.*

Requirement. Wet and dry bulb hygrometer.

Description of hygrometer. This instrument has two mercury thermometers which are placed vertically side by side on a wooden frame. (Fig.13.3). One thermometer is marked as dry-bulb thermometer and other as wet-bulb thermometer. The lower bulb end of each thermometer is filled with mercury and is graduated in centigrade. The bulb of wet-bulb thermometer is covered with muslin cloth and the free hanging end of muslin cloth is kept dipped in water contained in a small beaker. The mercury column of the wet-bulb thermometer remains always at a lower level than the dry-bulb thermometer because continuous evaporation of water to the wet-bulb keeps the temperature lower.

Observation. The readings of the two thermometers are noted. Then the differences in the values of dry and wet-bulb thermometers are noted which is the humidity of air. For each difference, values of **relative humidities** are consulted against the recorded value of dry-bulb thermometer in the conversion table (given tables 13.1 and 13.2). If there is no difference in the values of dry-bulb and wet-bulb thermometers it means relative humidity is 100%.

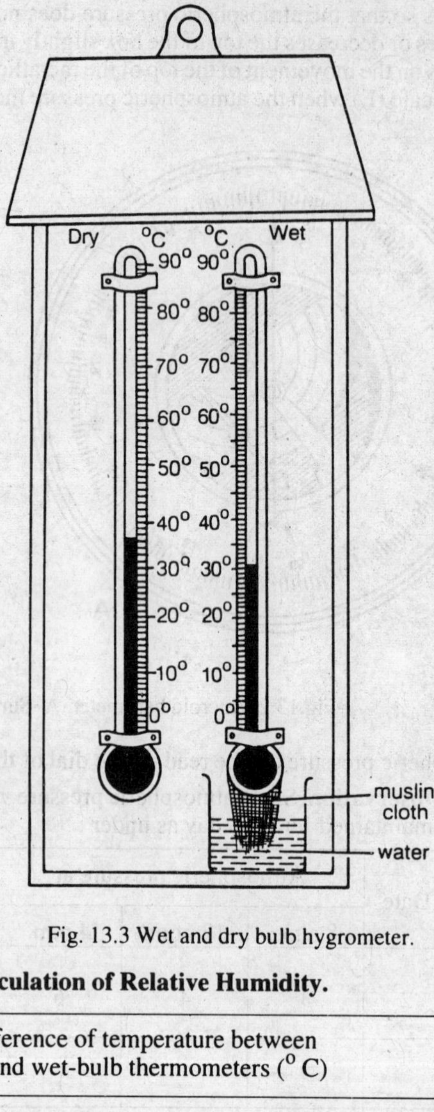

Fig. 13.3 Wet and dry bulb hygrometer.

Conversion Table 13.1 for Calculation of Relative Humidity.

Temp. indicated by dry-bulb thermometer (°C)	Vapour pressure of water (mm)	Difference of temperature between dry and wet-bulb thermometers (°C)					
		1°C	2°C	3°C	4°C	5°C	6°C
I	II	III	IV	V	VI	VII	VIII
10	9.2	8.1	7.0	6.0	5.0	4.0	3.1
11	9.8	8.7	7.6	6.5	5.5	4.5	3.2
12	10.5	9.3	8.2	7.1	6.0	5.0	4.0
13	11.2	10.0	8.9	7.6	6.5	5.5	4.5
14	11.9	10.7	9.1	8.3	7.1	6.1	5.0
15	12.7	11.4	10.1	9.0	7.8	6.6	5.5

I	II	III	IV	V	VI	VII	VIII
16	13.5	12.2	10.9	9.7	8.4	7.3	6.0
17	14.4	13.0	11.7	10.4	9.1	8.0	6.7
18	15.4	13.9	12.5	11.2	9.6	8.6	7.5
19	16.3	14.9	13.4	12.0	10.7	9.4	8.1
20	17.4	15.9	14.3	12.9	11.5	10.2	8.8

Conversion Table 13.2 for Calculation of Relative Humidities at Higher Temperatures.

Dry bulb Temp.	Wet-bulb depression (Difference in temperature shown by dry and wet-bulb thermometers)										
	1°	2°	3o	4°	5°	6°	8°	10°	12°	14°	
20° C	91	83	74	66	59	51	37	24	12		%
25° C	92	84	77	70	63	57	44	33	22	12	%
	2°	4o	6°	8°	10°	12°	14°	16°	18°	20°	
30° C	86	73	61	50	39	30	21	13	5		%
35° C	87	75	64	53	44	35	17	20	13	7	%
40° C	87	76	66	56	47	39	32				%
45° C	88	77	67	59	51	43	36				%
50° C	89	79	70	61	53	46	40				%
55° C	90	80	72	64	56	49	42				%
	2°	4°	6°	8°	10°	15°	20°	25°	30°	35°	
60° C	90	81	73	65	58	42	30	19			%
70° C	91	83	76	69	62	47	35	25	17		%
80° C	92	85	78	71	65	51	40	30	21	11	%

An Illustration for Calculation of Relative Humidity (R.H.)

Suppose, if the reading of dry-bulb thermometer is = 19° C

Suppose, if the reading of wet-bulb thermometer is = 16° C

Difference of the above two = 3° C

Then, consult column I of conversion Table 13.1 and find out 19.

Again, consult column II of the same table at the same level of 19. You will find 16.3. This refers to that at 19 °C the vapour pressure is 16.3 mm. Now, consult column V showing the difference of temperature as 3 °C and then read the figure at the level of 19 of column I. You will find that it is 12.0. This refers to that at **dew point** vapour pressure is 12.0 mm.

$$\therefore \text{ R.H.} = \frac{\text{Vapour pressure at dew point}}{\text{Vapour pressure at } 19^{\circ}\text{ C}} = \frac{12.0}{16.3} = 0.74$$

Since R.H. is expressed in %age,
Hence, R.H. = 0.74 × 100 = 74 %

(**Note:** *For higher temperatures, directly consult conversion Table 13.2 at the temperature level and difference in temperatures and note the R.H. in %. For example, if temperature is 25 °C and difference 4 °C, then R.H. will be 70%.*)

Finally, after calculating R.H. maintain your record as desired or in the form as given below :

S. No.	Date	Relative Humidity %			Average Relative Humidity %	Mean Relative Humidity %
		8 a.m.	12 noon	4 p.m.		
1	2	3	4	5	$\left(\frac{3+4+5}{3}\right) = 6$	$\left(\frac{\text{Sum of 6}}{\text{Sum of 2}}\right) = 7$

Calculation

1. Average R.H. is calculated by dividing the sum of columns 3,4 and 5 by 3.
2. Mean R.H. for a specified period can be calculated by summing up the relative humidities of column 6 and dividing it by the total number of days (column 2) for a specified period.

EXPERIMENT No. 4

Object. *Recording of light intensity.*

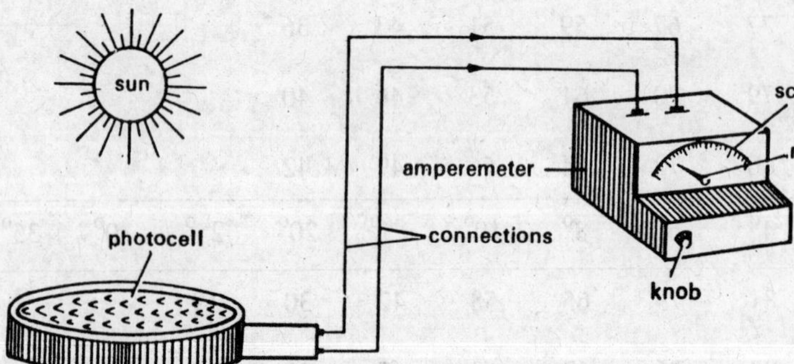

Fig. 13.4. Luxmeter (a photometer).

Requirement. Luxmeter (a photometer).

Light intensity is measured by the use of Luxmeter which is a photometer (Fig. 13.4). It consists of a photoelectric cell and an amperemeter. The photocell is exposed to light, and the value is read on the amperemeter, with deflection needle. The light intensities are measured in terms of lux units. Record the light intensities under different conditions, *i.e.*, shade, open sun, etc., and note variations, maintain your record for a desired period.

The luxmeter may be used in water for measuring light intensities. For this, photocell is sealed in a transparent water proof case and then placed at different sites and depths of pond water to find out variations in light intensities values. Keep your record as desired.

B. Study of Abiotic Factors

EXPERIMENT No. 5

Object. *To study the physical characteristics (like texture, colour and temperrature) of the soil.*

Requirements. Soil sampling tool, polythene bags for collecting samples, hand lens, sieves of different sizes, Munsell's colour chart, soil thermometer, etc.

Procedure

A. Texture and composition. For studying the texture and composition of soil proceed as under :

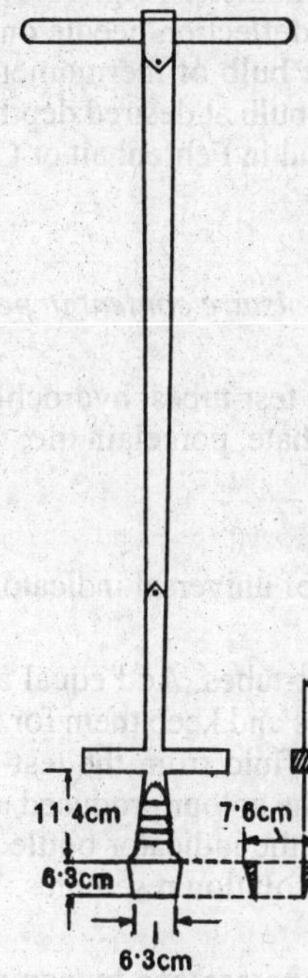

Fig. 13.5. Soil sampling tool.

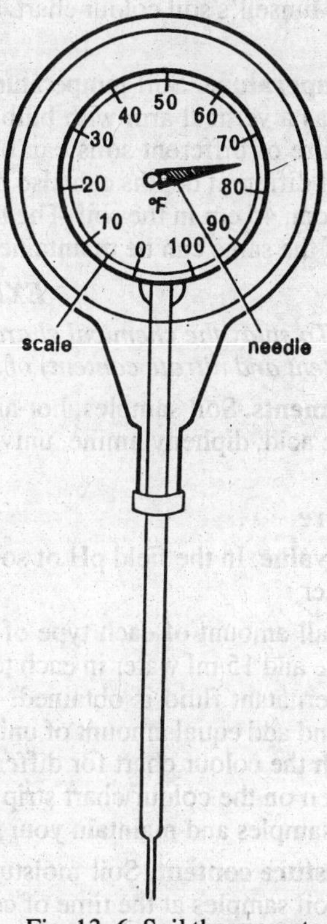

Fig. 13..6. Soil thermometer.

(i) Collect soil samples from different arears with the help of soil collecting tool (Fig. 13.5). With this tool soil from different known depths and horizons can be extracted for examination.

(ii) Examine each sample of soil under hand-lens. Also examine and observe the soil, in dry as well as wet or moist state, between thumb and fingers. Also pass the soil particles through different sized sieves. Note as under :

	Characteristic of soil	*Texture*
1.	Individual grains may be seen or felt, moist soil is squeezed. Grain size 0.02–2.0 mm.	Sandy
2.	Moist soil forms cast that does not break, dry grains fall separate	Sandy loam
3.	Dry soil is squeezed, moist soil forms cast that does not break.	Loam

	Characteristic of soil	*Texture*
4.	Dry appearing with soft feeling, dry as well as wet soil can be easily handled without breaking. Grain size 0.02–0.2 mm	Silt loam
5.	Hard when dry, breaks into lumps, moist soil is pinched forming thin ribbon which breaks easily.	Clay loam
6.	Very hard lump when dry, moist and wet soil sticky forms long, flexible ribbon, Grain size 0.002 mm.	Clay

B. Colour. For studying the colour, spread a small amount of soil sample uniformly over a white paper sheet or a cardboard. Match the colour of the particles of soil with chips of a number of colours in the Munsell's soil colour chart. Note the colour of soil sample and maintain a record of the same.

C. Temperature. Soil temperature is measured by soil thermometer (Fig. 13.6). The soil thermometer has a vertical arm with bulb at one end and a dial with deflection needle on the other end. Temperature of different soils can be measured by inserting the bulb of thermometer in soil. Temperature of different depths can also be measured by burying the bulb at desired depths like 2.5 cm, 15 cm, 30 cm, 45 cm in the soil. The temperature is read on the dial in Fehrenheit or Centigrade and a record of the same can be maintained.

EXPERIMENT No. 6

Object. *To study the chemical characteristics (like pH value, moisture content in percentage, carbonate content and nitrate content) of soil.*

Requirements. Soil samples, hot-air oven, balance, desiccator, test-tubes, hydrochloric acid, conc. sulphuric acid, diphenylamine, universal indicator, barium sulphate, porcelain tile, white spot plate, etc.

Procedure

A. pH value. In the field pH of soil is recorded with the help of universal indicator. for this, proceed as under :

Take small amount of each type of soil sample in different test-tubes. Add equal amount of barium sulphate and 15 ml water in each test-tube. Shake the test-tubes and keep them for sometime till a clear supernatant fluid is obtained. Now take few drops of the fluid from the test-tube on a porcelain tile and add equal amount of universal indicator to it. Note the colour produced and match this colour with the colour chart for different values of pH pasted on the indicator bottle. Read the pH value written on the colour chart strip with which colour of your solution resembles. Likewise, note pH of all samples and maintain your record.

B. Moisture content. Soil moisture content is determined in percentage by comparing the weight of the soil samples at the time of collection with their weight in oven-dry condition. For this proceed as under :

Take soil sample and weigh it carefully. Keep it in a hot-air oven at a temperature of 105°–110° C untill the weight of the sample becomes constant. This means that there is no moisture in it. Then, cool the soil sample in a desiccator and weigh it. The loss of weight from fresh to dry represents the moisture content of the soil. The loss in weight, expressed as a percentage, is the per cent moisture content.

Calculation

Fresh weight of soil sample $= a$ gm.

Constant weight of soil sample $= b$ gm.

Moisture present $= a - b$ gm.

Per cent moisture conte1. $= \dfrac{a-b}{b} \times 100$

In this way, calculate the moisture content of the soil samples in percentage and maintain your record according to your need.

C. Carbonate content. For measuring carbonate contents of the soil proceed as under :

Take a small but equal amount of the different soil samples in different test-tubes and then add equal amount of hydrochloric acid to each test-tube. Observe and note the degree of effervescence produced in all the test-tubes. Divide the degree of effervescence produced in all cases in different groups as +, ++, +++, ++++ or as A, B, C, D. The sample showing maximum degree of effervescence indicates the maximum amount of carbonate present. Record your observations as noted below :

Soil sample	*Degree of effervescence*
1	+ or A
2	+ + + or C
3	+ + + + or D
4	+ + or B

Result. Soil sample 3 shows maximum degree of effervescence. Hence, this soil sample contains maximum carbonate content while minimum is in soil sample 1.

D. Nitrate content. For measuring nitrogen contents of the soil proceed as under :

Make clear solutions of each soil sample in test-tubes separately by dissolving 1 part of soil sample in 5 parts of water. Take a small amount of each solution on white spot plate and add 2–3 drops of diphenylamine and 2 drops of conc. sulphuric acid (0.2%) to each. **Blue colour** appears which indicates the presence of nitrate in the sample. Observe and note the intensity of blue colour and divide the observed intensity in four arbitrary groups, as done in the case of carbonate and record your observation as noted below :

Soil sample	*Intensity of blue colour produced*
1	+ + + or C
2	+ + or B
3	+ or A
4	+ + + + or D

Result. Soil sample 4 shows maximum intensity in colour and sample 3 shows minimum intensity in colour. This suggests that sample 4 contains maximum nitrate content and that sample 3 contains minimum nitrate content.

C. Study of Water Characteristics

EXPERIMENT No. 7

Object. *To estimate pH of water sample by pH meter.*

Requirement. pH meter, buffer solution, electrodes, beaker, distilled water and water sample(s).

Procedure. For determining pH of any sample by a pH meter proceed as under :

First standardise the pH meter by immersing the glass electrodes in the buffer solution of known pH. Read the pH and adjust the pH meter so that the pH meter reads correct value of pH of the buffer solution. Then rinse the electrodes in distilled water and immerse them in water sample and let the pH meter needle settle at one point. Read the pH value and note. Take at least three readings and find out their mean value. Keep your record as suggested below :

S. No.	Date	Time	Room Temp.	Sample No.	pH	Mean pH
1	2	3	4	5	6	7

EXPERIMENT No. 8

Object. *To record temperature of water.*

Requirement. A thermistor.

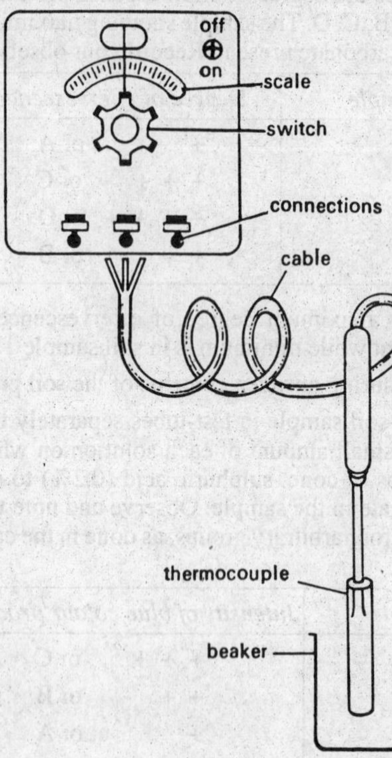

Fig. 13.7. A thermistor.

Procedure. For recording temperature of water proceed as under :

The temperature of water can be recorded with an instrument called thermistor (Fig. 13.7). This instrument works on the principles of thermocouples. To record the temperature dip the glass globule of thermistor which contains thermocouple at different depths in water with the help of an electric cord/cable. The other end of cable is connected to the potentiometer out side the water which shows the temperature value on the scale. This value can be read directly. Thus, read the temperature and maintain your record as desired.

EXPERIMENT No. 9

Object. *To estimate total hardness of a given sample of water.*

(The bicarbonates, chlorides and sulphates (mainly of Ca, Mg and Na) dissolved in water primarily account for the water hardness. The hardness caused by bicarbonates is called carbonate hardness or temporary hardness, since it can be removed by heating. The hardness caused by sulphates and chlorides of Ca and Mg is called permanent hardness; it cannot be removed by heating. The sum of these two kinds of hardness is termed as total hardness of a water sample. The salts of iron, manganese and strontium may also contribute to the water hardness.)

Requirements. Conical flask, ammonia buffer solution, Erichrome black T-indicator, burette, burette stand, beaker, EDTA* solution.

Procedure. For estimating total hardness (T.H.) of a given water sample proceed as under :

Take 100 ml of water sample in a conical flask. Add 3 ml ammonia buffer solution and 0.2 to 0.4 gm of solid Erichrome black T- indicator in it. Titrate with standard EDTA solution until the colour changes from wine red to blue and note the ml of EDTA solution used.

**EDTA – Sodium calcium adedate or calcium versenate*

Calculation

Hardness is expressed as mg/L $CaCO_3$ and may be calculated as under :

$$\text{T.H.} = \frac{\text{ml of EDTA used} \times 1000}{\text{ml of sample taken}}$$

(*When total hardness has a value more than total alkalinity, the amount of hardness equivalent to alkalinity is called* **carbonate hardness.** *The excess amount is* **non-carbonate hardness.** *When the total hardness is equal or less then total alkalnity, there is no carbonate hardness.*)

In order to estimate **non-carbonate hardness** take a known quantity of water sample and boil it for about 30 minutes, cool and filter. Repeat the procedure mentioned above and note the ml of EDTA solution used and calculate as under :

$$\text{Non-carbonate hardness} = \frac{\text{ml of EDTA used} \times 1000}{\text{ml of (boiled) sample}}$$

Carbonate hardness = T.H. – non-carbonate hardness.

EXPERIMENT No. 10

Object. *To estimate the alkalinity of a given sample of water.*

(*Alkalinity of water sample refers to its capacity to neutralize acids. It is the sum total capacity of chemical components in water that tend to raise the pH of water above 4.5. Anionic radicals such as carbonates, hydroxides and phosphates contribute to increase in alkalinity. It is expressed as mg/l of $CaCO_3$ of water. Alkalinity is important for aquatic life because it acts as a buffer, controlling pH fluctuations.*)

Requirements. Conical flasks, phenolphthalein indicator, N/50 H_2SO_4, methyl orange indicator and burette.

Procedure. For estimating alkalinity of the given water sample proceed as under :

1. **Phenolphthalein alkalinity.** Take 100 ml of given water sample in a conical flask and add 4 drops of phenolphthalein indicator in it. If colour appears in solution, it contains hydroxide or normal carbonate. If sample becomes pink then add N/50 sulphuric acid through a burette until pink colour just disappears. Note the ml. of acid used and calculate as under :

Calculation. The phenolphthalein alkalinity, expressed in ppm of $CaCO_3$, is equal to :

ml of N/50 H_2SO_4 used × 10.

2. **Methyl orange alkalinity**. Take 100 ml of given water sample in a conical flask and add 2 drops of methyl orange indicator in it. If yellow colour develops in solution, it contains carbonate or bicarbonate. Then add N/50 sulphuric acid through a burette until yellow colour shows first change from pure yellow to faint orange tint. Stop dropping H_2SO_4 and note the ml of acid used.

Calculation. The methyl orange alkalinity, expressed in ppm of $CaCO_3$, is equal to ;

ml of N/50 H_2SO_4 used × 10.

EXPERIMENT No. 11

Object. *To estimate total dissolved solids in a given sample of water by gravimatric method.*

(*Dissolved solids in water primarily include carbonates, chlorides, nitrates and sulphates (anions) of Ca, Mg, K and Na (cations). Na in irrigation water is toxic to plants and causes problems in soil structure, changing the water infiltration and permeability rates, Excess dissolved solids in drinking water can cause physiological effects, non-palatable taste, etc., and, thus, necessitate treatment of water before use.*)

Requirements. Evaporating dish (250 ml cavity), Hot-air oven, balance and dessicator.

Procedure. For estimating total solid contents in the given water sample proceed as under :

Evaporate the water sample to dryness in a cleaned, dried and weighed evaporating dish. When water is completely evaporated, place the evaporating dish in a hot air oven at 110–120 °C so that residual water, if any, is completely removed. Then cool the evaporating dish at room temperature in a dessicator. Weigh the dish with the residue and note.

Calculation

Suppose,

The weight of dry, clean, empty evaporating dish = A gm

The weight of evaporating dish with residue = B gm

The weight of residue = B – A gm

Hence, total soilds in water sample $= \frac{B - A \times 10^6}{10^2}$ ppm

$= B - A \times 10^4$ ppm.

EXPERIMENT No. 12

Object. *To estimate dissolved oxygen (D.O.) content of the given water sample by Winkler method.*

(Dissolved oxygen is a measure of the ability of a water body to support a well-balanced aquatic life. Insufficient dissolved oxygen in water leads to the onset of anaerobic conditions and subsequently to the release of noxious gases such as H_2S. So, the measurement of the dissolved oxygen in a water body is of particular importance).

Requirements. Water sampling bottles, manganous sulphate solution, alkaline potassium iodide solution, conc. H_2SO_4, pipette, burette stand, burette, conical flask, N/40 sodium thiosulphate solution, starch solution, glass rod, measuring cylinder, $K_2Cr_2O_7$ solution, potassium crystal, etc.

Procedure. For estimating dissolved oxygen in the water sample proceed as under :

Immediately after collecting water sample in water sampling bottle add 1 ml of manganous sulphate solution, with the help of pipette, well below the surface of water in bottle. Then add 1 ml of alkaline potassium iodide solution in the same manner. Replace the stopper of the bottle and mix the sample by inverting the bottle several times. Allow the precipitate to settle at the bottom of container for few minutes. Then add 1 ml of conc. H_2SO_4 with the help of a pipette. Mix well by shaking and inverting the bottle several time and then allow the sample to stand for about 10 minutes. Thus, a clear reddish brown solution is obtained. By doing so, dissolved oxygen in water sample is **fixed** and there is no chance of atmospheric air to mix with the sample.

Then take 200 ml of this sample in a conical flask and titrate rapidly with N/40 sodium thiosulphate solution until the reddish brown colour of the sample is reduced to pale straw colour. Stop adding sodium thiosulphate solution and add 1 ml of starch solution in the conical flask. A deep blue colour appears. Again continue titration rapidly and carefully till blue colour just disappears. This is the end point, stop titration and note the ml of sodium thiosulphate solution used. Avoid any reappearance of blue colour and ignore it.

Calculation

When 200 ml of sample is titrated, the amount of sodium thiosulphate solution (in ml) used is numerically equal to the dissolved oxygen in ppm and, therefore, no additional calculation is needed.

To get the result in mg/L, the value obtained above is multiplied by a constant value 0.698 (standardization value). In order to get this constant the sodium thiosulphate solution is standardized.

Standardization of sodium thiosulphate solution. Take 10 ml of $K_2Cr_2O_7$ solution in a beaker and add 1 ml conc. H_2SO_4 and approximately 1 gm of potassium iodide crystals. The reaction is exothermic, so cool the beaker and keep it in dark for 10 minutes. Then titrate it with sodium thiosulphate solution in the same way in which the main titration has been done. If the sodium thiosulphate solution has been prepared correctly, then 10 ml of this solution is needed to nutralize 10 ml of $K_2Cr_2O_7$ solution. In this case the constant value is 0.698).

EXPERIMENT No. 13

Object. *To estimate the amount of free carbon dioxide in water sample.*

Requirements. Conical flask, phenolphthalein indicator, N/44 sodium hydroxide solution, burette, burette stand, glass rod, etc.

Procedure. For estimating the amount of free CO_2 in water sample proceed as under :

Take 100 ml of water sample in a conical flask and add 10 drops of phenolphthalein indicator in it. Titrate this solution by N/44 sodium hydroxide from burette. Keep on stirring the sample with the help of glass rod gently during titration. Titrate till a faint permanent pink colour appears. Stop the titration and note the reading of burette to know the volume of sodium hydroxide used during titration.

Calculation

The amount of free CO_2 present in the sample (expressed in ppm) is obtained by multiplying the volume of N/44 sodium hydroxide used during titration by 10.

For example, if 1.7 ml of N/44 sodium hydroxide was used during titration, the amount of free CO_2 present in the sample will be $1.7 \times 10 = 17.0$ ppm.

Precaution

Free CO_2 escapes rapidly from water, hence, analysis of water for free CO_2 estimation must be carried out immediately the sample is collected.

EXPERIMENT No. 14

Object. *Qualitative study of planktons in the sample of fresh water.*

Requirements. Water sample, microscope, plain slides, dropper, watch glass, petridish, 10% formalin.

(Planktons are free floating or drifting organisms and mostly microscopic. These are grouped into phytoplanktons and zooplanktons. The movement of planktons are restricted to some extent by water current. Some zooplanktons show active swimming movement but this only helps them in maintaining their vertical position. These are found in euphotic zone, i.e., up to 600 feet depth from the surface. The producers of this zone are the phytoplanktons and zooplanktons are the consumers. The zooplanktons are important in the sense that they constitute the chief source of food for fishes.)

Procedure. For the qualitative study of planktons proceed as under :

Collect water sample from just below the surface of water source, *i.e.*, a pond, ditch or lake in wide mouth glass bottle and bring it to your laboratory. Shake the bottle well and pour some water in a petridish. Add few drops of 10% formalin in it which will cease the movement of planktons (may finally kill them). Then with the help of a dropper pour a drop of this water on a clean slide and examine under microscope. Note the number, name and type of plankton(s) present in this drop of water. Likewise, take atleast 20 observations. Record your observations in the following foim :

Water drop	Planktons	Water drop	Planktons
1st drop	*Cyclops*	11th drop	NIL
2nd drop	*Cyclops, Daphnia*	12th drop	*Cypris*
3rd drop	Alga	13th drop	*Daphnia*
4th drop	*Cypris*, Alga	14th drop	Alga, Alga
5th drop	NIL	15th drop	*Cyclops, Cyclops*
6th drop	*Cypris, Cyclops*	16th drop	Alga, Alga
7th drop	Alga	17th drop	NIL
8th drop	*Daphnia,* Alga	18th drop	*Cypris, Cyclops*
9th drop	*Cyclops, Cyclops*	19th drop	Larva, Alga
10th drop	Alga, Alga	20th drop	Alga

Conclusion. From the observations made above, following conclusions may be drawn :

1. Both zooplanktons and phytoplanktons are present in the water sample.
2. In 20 drops of water, total 27 planktons are present out of which 16 are zooplanktons.

Result. The water source from which sample was taken is dominated by zooplanktons.

EXPERIMENT No. 15

Object. *Quantitative study (total count and differential count) of planktons in the sample of water.*

Requirements. Microscope with mechanical stage, plankton counting slide (Sedgwick Rafter's counting slide), plankton collecting net, plankton collecting jar, plankton sample, plastic bucket, petridish, dropper, 10% formalin, plankton centrifuge, etc.

Description of plankton net. Plankton net (Fig.13.8) is made of a circular iron rim attached to a handle. The rim is tied with bolting silk cloth of specified number like 12,16,20, etc. In one square inch of bolting silk, there is fixed number of meshes in different types of bolting silk cloth, *e.g.*, No.12 possesses 800 meshes/sq.inch, No 16 bears 10,000 meshes/sq.inch and so on. Muslin cloth may also be used in place of bolting silk. At the other end of bolting silk cloth a plankton collecting jar of desired capacity is tied (here it is of 250 ml capacity). A thick band-shaped cloth or rubber is used to tie the plankton collecting jar with the bolting silk tightly at their junction.

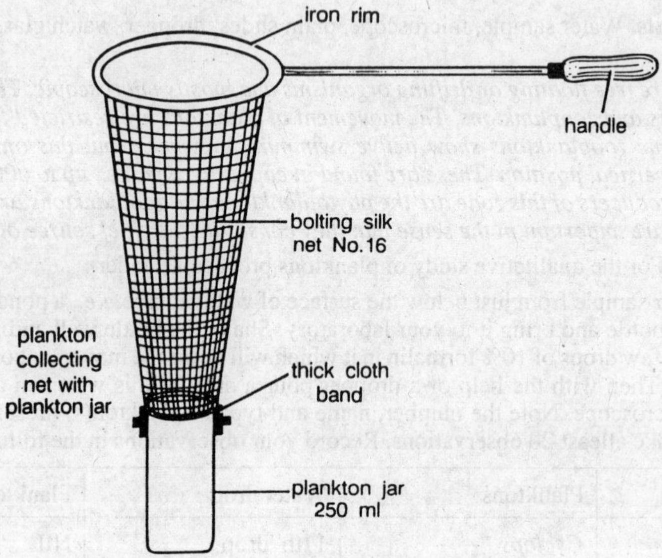

Fig.13.8. Plankton collecting net with plankton jar.

Description of Sedgwick Rafter's counting slide. Sedgwick Rafter's counting slide (Fig.13.9 A and B) is a special type of grooved slide meant for counting planktons, therefore, commonly referred to as plankton counting cell or slide. This slide has a groove of 50 mm length, 20 mm width and 1 mm depth. The groove is graduated having 50 squares along its length and 20 squares along its width; total squares being $50 \times 20 = 1{,}000$ and total area being 1000 mm^2 or 1 ml. One counting chamber (unit) is 1 mm^3.

Procedure. For quantitative study of planktons in the water sample proceed as under :

Filter about 50 litres of water from a pond through plankton net having bolting silk No.16 and collect planktons in plankton collecting jar of 250 ml capacity. Preserve the filtrate in 10% formalin and then sample is centrifuged in a plankton centrifuge.

Take formalin treated centrifuged sample in the groove of plankton counting slide (described above) and cover it with the help of a long rectangular coverglass; when there is no air bubble, the groove contains 1 ml of sample. Now, bring the slide under mechanical stage of microscope and examine for the count of planktons. Choose 2,3,4 or more random arears (hereafter called fields) containing 100 squares on the slide and proceed as under :

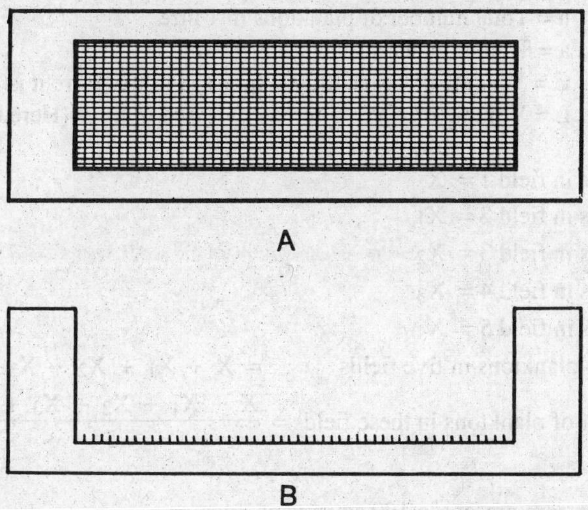

Fig. 13.9. A–Sedgwick Rafter's counting slide (surface view); B–Sedgwick Rafter's counting slide in T.S.

A. Total count. Total count of plankton means an enumeration of all planktons recognisable under the condition of examination without an attempt to distinguish between the different kinds. Each complete plankton or a part of plankton in one square (counting unit) is counted as one plankton.

As referred to, count the total number of planktons in randomly chosen area, say field 1 (see Fig. 13.10). Repeat it by choosing not less than 5 fields on the slide. Add all the number of planktons counted to get total number of planktons in the examined fields. Divide this number by the number of fields examined to get an average number.

×	✓	×	×	×	✓	✓	✓	×	✓
×	×	✓	×	×	×	✓	×	✓	×
✓	✓	×	✓	✓	✓	×	×	×	✓
✓	✓	✓	×	✓	✓	✓	×	×	✓
×	✓	×	✓	×	×	×	✓	✓	×
×	✓	✓	✓	✓	×	×	×	×	✓
✓	×	×	×	✓	✓	✓	✓	×	×
✓	×	×	✓	×	×	×	×	×	✓
×	✓	✓	×	✓	✓	✓	×	×	✓
×	✓	✓	✓	×	×	✓	×	✓	×

Field 1

Total number of planktons = X (X is the sum of all ticks (✓) examined within 100 squares showing presence of planktons)

Fig. 13.10. Hypothetical graph sheet showing total count of planktons.

Calculation

Total number of planktons present in 1 litre of water can be calculated by the following formula :

$$n = \frac{a \times 1000 \times C}{L}$$

Where n = Total number of planktons in 1 litre
a = Total number of planktons in 1 ml.
C = Volume of concentrate expressed in ml (Here it is 250 ml)
L = Volume of water filtered expressed in litres (Here it is 50 litres)

Suppose

No. of planktons in field 1 = X
No. of planktons in field 2 = X_1
No. of planktons in field 3 = X_2
No. of planktons in field 4 = X_3
No. of planktons in field 5 = X_4

Total number of planktons in five fields $= X + X_1 + X_2 + X_3 + X_4$

Average number of planktons in these fields $= \frac{X + X_1 + X_2 + X_3 + X_4}{5}$

Let it be Y

Thus, Y planktons are present in 100 squares on average.

Therefore, 1000 squares (1 ml sample) contain Y × 10

= 10 Y plankton

Now, a = 10 Y planktons in 1 ml
C = 250 ml and L = 50 litres

Putting the values in the above formula :

$$n = \frac{10\,Y \times 1000 \times 250}{50}$$

Its value will give the **total number of planktons** in 1 litre of water.

B. Differential count or Qualitative study of planktons in the water sample. This refers

D	×	×	C	D	×	C_y	A	×	V
×	×	A	×	A	×	V	D	×	C_y
C	×	D	×	V	D	C	C_y	V	×
×	×	C	A	×	×	V	×	D	×
D	C	×	C_y	×	A	×	×	×	V
A	×	V	D	×	×	C	×	C_y	×
×	×	A	×	D	×	D	×	C	C_y
V	×	C	V	×	A	×	C_y	×	C
×	D	C_y	×	V	C_y	C	×	V	C_y
C_y	×	D	C	D	×	C	×	A	×

Field 1

C = *Cyclops* = 12
C_y = *Cypris* = 11
D = *Daphnia* = 14
V = *Volvox* = 10
A = *Algae* = 09

Total = 56

Fig. 13.11. Hypothetical graph sheet showing differential count of planktons.

to the enumeration of all of the different kinds of planktons by distinguishing them qualitatively. For this, proceed as under :

First of all choose and identify all kinds of planktons to be counted. For this also prepare the sample and slide for study in the same way as described above. As mentioned earlier different random fields are chosen on the slide. Then prepare your own graph sheet having 100 squares. Make a coding for different planktons, *viz.*, C for *Cyclops*, Cy for *Cypris*, D for *Daphnia*, V for *Volvox*, A for alga, etc. Then examine the slide for these planktons and record them in the square graph sheet prepared as shown in Fig. 13.11. Court their number separately. Repeat it for the different fields choosen. Find out the total number of different planktons separately and their average value also, that too separately, to get the value of a for its use in the same formula as mentioned in the case of total count. The values of C and L to be used in the formula are the same as mentioned earlier. Calculation is to be done in the same way.

D. Study of Community

Community is a group of naturally occurring organisms both plants and animals in a common environment or a given place, *e.g.*, a grassland, a desert, a forest, etc.

EXPERIMENT No. 16

Object. *To study the community by quadrat method by determining frequency, density and abundance of different species present in the community.*

Requirements. Quadrat of definite size (50 cm × 50 cm in the present case) and meter scale

Procedure. For determining the frequency, density and abundance of different species in a community proceed as under :

Throw a quadrat of lay down a quadrat in a specified area of the grassland, forest or garden or pond or any place where community is to be studied. Note carefully each type of animals or plants and or both, as needed, occurring in the squares of the quadrat. Note down the number and names of individual types of animals and plants. Repeat it by thrrowing or laying down the quadrat randomly in the area at least 10 times. Record your data in form of the following table :

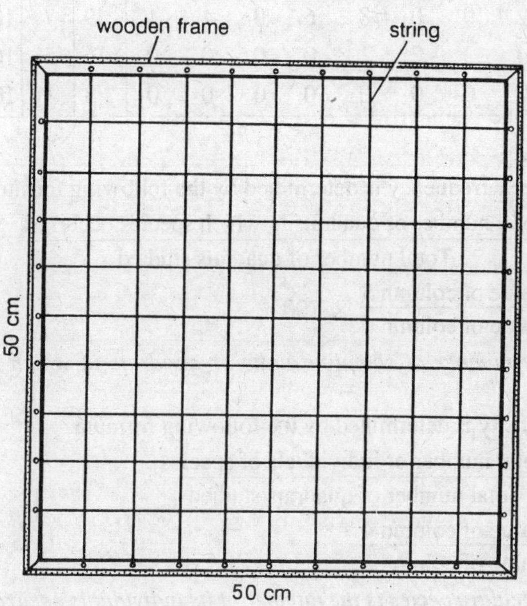

Fig. 13.12. A wooden quadrat 50cm × 50 cm.

S No.	Name of species	Number of individuals in each quadrat										Total No. of individuals of speicies	Total No. of quadrats occurence	Total No. of quadrats studied	Frequency %	Denssity (Per unit area)	Abundance
		1	2	3	4	5	6	7	8	9	10						
1	2	3										4	5	6	7	8	9
1.	A	0	4	0	0	6	0	0	0	0	0	10	2	10	20	4.0	5
2.	B	0	0	0	5	3	0	0	0	0	3	11	3	10	30	4.4	3.6
3.	C	4	0	0	0	0	0	6	0	0	0	10	2	10	20	4.0	5
4.	D	1	0	0	1	0	1	0	0	1	0	4	4	10	40	1.6	1
5.	E	2	2	2	1	3	3	1	1	1	2	18	10	10	100	7.2	1.8
6.	F	10	7	15	11	20	14	5	18	11	6	117	10	10	100	46.8	11.7
7.	G	5	0	0	4	0	0	0	6	0	0	15	3	10	30	6.0	5
8.	H	6	4	5	10	8	7	5	10	10	2	67	10	10	100	26.8	6.7
9.	I	2	3	4	2	6	0	3	2	8	2	32	9	10	90	12.8	3.5
10.	J	1	0	5	7	3	6	4	7	3	2	38	9	10	90	15.2	4.2
11.	K	6	4	0	5	3	0	2	1	7	3	31	8	10	80	12.4	3.8
12.	L	8	10	12	0	0	8	6	0	4	1	49	7	10	70	19.6	7
13.	M	0	6	4	3	2	1	0	0	0	0	16	5	10	50	6.4	3.2
14.	N	3	0	0	0	0	0	0	0	0	0	3	1	10	10	1.2	3

Calculation

A. Frequency. The % frequency is determined by the following formula :

$$\textbf{Frequency \%} = \frac{\text{Total number of quadrats in which species occurred}}{\text{Total number of quadrats studied}} \times 100$$

$$\text{or } \frac{\text{Value of column 5}}{\text{Value of column 6}} \times 100$$

The frequency is the number of sampling units or quadrats in which a particular species occurred.

B. Density. The density is determined by the following formula :

$$\textbf{Density} = \frac{\text{Total number of individuals of species}}{\text{Total number of quadrats studied}}$$

$$\text{or } \frac{\text{Value of column 4}}{\text{Value of column 6}}$$

The density of a particular species is the number of its individuals occurring per unit area.

The density values, thus, obtained from the above formula for each species, are to be expressed as individuals per unit area. For example, calculated value for species **A** according to above formula is 1. This value is in 2500 cm^2 area, *i.e.*, the area of the quadrat used 50 × 50 cm^2.

Therefore, in 2500 cm^2 area there is 1 individual.

Hence, in 1 cm^2 area there would be $\frac{1}{2500}$ individuals.

And in 100 × 100 cm^2, *i.e.*, 10000 cm^2 or (1 m × 1 m, *i.e.*, 1 m^2)

$$= \frac{10000 \times 1}{2500} = 4$$

Thus, density of species **A** is to be expressed as 4 per m^2 instead of 1.

C. Abundance. The abundance is determined by the following formula :

$$\textbf{Abundance} = \frac{\text{Total number of individuals of species}}{\text{Total number of quadrats of occurrence}}$$

$$\text{or } \frac{\text{Value of column 4}}{\text{Value of column 5}}$$

The abundance may be described as number of individuals occurring per qudrat.

EXPERIMENT No. 17

Object. *To determine the biomass of consumers of a particular area.*

(Biomass refers to the dry weight of living organisms both animals and plants per unit area.)

Requirements. Quadrat of definite size (here it is 50 × 50 cm^2), polythene bags, a weeding instrument like khurpa, balance, hot-air oven.

Procedure. Since, in the present exercise you are supposed to determine the biomass of consumers only, hence, proceed as under :

Throw or lay down quadrat at random in the specified area. Mark the boundary of the quadrat and remove it. Carefully collect different animals like earthworms, ants, insects, garden slugs, larvae, etc., in a polythene bag. Dig out the soil of this area up to 15–20 cm. and collect the animals present.

Take fresh weight of all the animals together and keep them in a hot-air oven at 70 °C for 24 hours and then weigh it again to get dry weight.

Calculation. Biomass per unit area is calculated as under :

Fresh weight = x gm

Weight after drying at 70 °C for 24 hours or dry weight = x_1 gm

Content of moisture = $x_1 - x$ gm

Since biomass is expressed in per unit area. Hence, to find out the biomass of consumers in per unit area in this case, calculate as below :

The size of quadrat used 50 × 50 cm = 2500 cm^2

Thus, 2500 cm^2 area has x_1 cm dry matter

Hence, 1 m^2 (1m × 1m or 100 cm × 100 cm) area will have,

$$\frac{x_1 \times 100 \times 100}{2500} = x_2 \text{ gm / m}^2$$

Result. The biomass of consumers = x_2 gm / m^2

(**Note:** In the same way biomass of producers can be determined. But in this case, the above ground parts of plants, fallen leaves and twigs, etc., and underground parts of plants are collected separately in separate bags. Their fresh weight is taken and then dried in oven in the same way as described above. Dry weight is taken to know the biomass and biomass per unit area is calculated similarly. Likewise, total biomass of an area including producers and consumers together can be determined. In this case also dry weights are taken separately and summed up to know the total biomass and finally multiplied by 4 to get biomass per unit area.)

E. Study of Pyramids and food Chain

Pyramids in ecology refers to a definite relationship that exists between the **numbers, biomass** and **energy** contents of the producers and consumers of different orders in any ecosystem. The diagramatic representation of these relationships is referred to as **ecological pyramids.** Thus, ecological pyramids can be drawn on numbers, biomass and energy. At present we will study all these three types of pyramids.

EXPERIMENT No. 18

Object. *To study pyramid of numbers.*

(The pyramid of numbers deals with the relationship between the numbers of primary producers and consumers of different orders in an ecosystem.)

A. *Study of pyramid of numbers of an aquatic environment.*

Requirement. Scoop sample collector, butterfly net and other accessories.

Procedure. Choose an aquatic environment like a pond or lake of which pyramid of numbers, also referred to as **food pyramid,** is to be constructed. Then by using butterfly net and scoop sample collector take separate samples from the surface, mid-water and bottom regions of choosen pond or lake. Keep these samples separately. Examine the samples carefully and note the numbers of **producers** (*i.e.*, plants, photosynthetic organisms and phytoplanktons) and **consumers** of different orders. Do not mix your data, keep them separately, *i.e.,* the organisms observed in bottom sample, in mid-water sample and in surface water sample.

You shall observe that photosynthetic organisms including phytoplanktons, from bottom sample, are maximum in number. These are **primary producers.** The sample from mid-water shows the presence of **water flea, insect larvae**, **tadople larvae,** etc., and their number would be less than primary producers. These are **primary consumers**. The sample from surface water shows two categories of organisms : (i) **small carnivorous organisms** like small fishes, water beetles, water scorpions, etc., and (ii) **large carnivorous organisms** like large fishes, turtles, etc. But the number of small carnivorous organisms would be less than primary consumers. All these are referred to as **secondary consumers** which feed on primary consumers. The large carnivorous organism are still less in number than the secondary consumers. These are called **tertiary consumers** which feed on secondary consumers.

Now construct a pyramid of number by keeping producers at the bottom then primary

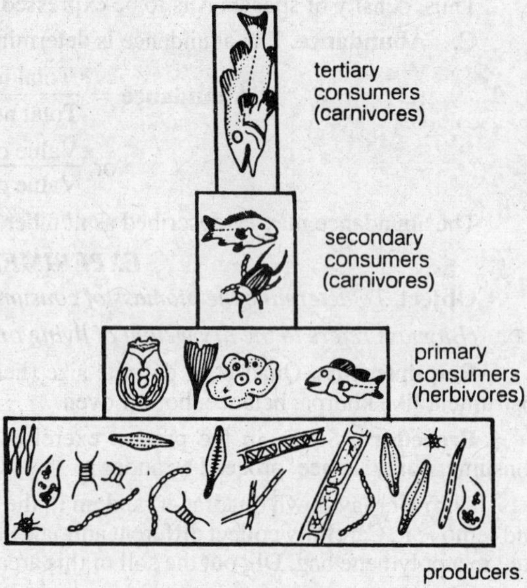

Fig. 13.12 A. Pyramid of numbers or Food pyramid of a pond.

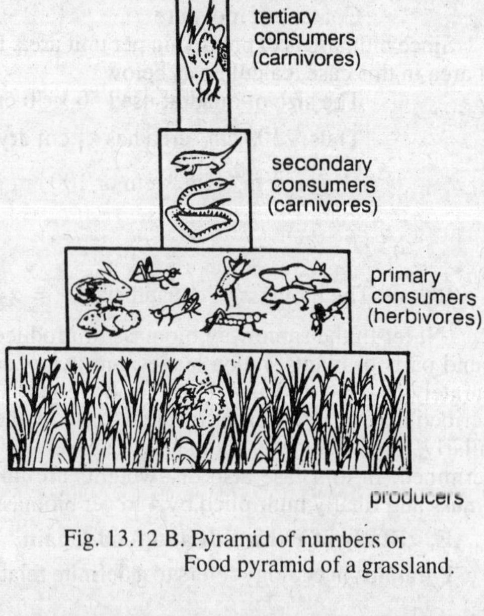

Fig. 13.12 B. Pyramid of numbers or Food pyramid of a grassland.

consumers, secondary consumers and tertiary consumers respectively as shown in Figure No. 13.12A.

B. *Study of pyramid of numbers of a grassland.*

Prodedure. Çhoose certain areas in a grassland. Collect plants and grasses from the surface of the chosen areas. Observe small insects, insect larvae, etc., feeding on plants and grasses. Collect them with the help of fine foceps or brush and count them. All these are the **primary consumers.** Note small birds and frogs, etc., feeding on them. These are the **secondary consumers.** These consumers are further eaten up by **tertiary consumers** like predatory birds.

Now, construct a payramid of number/unit in the same way as suggested above, *i.e.*, by keeping the producers at the bottom and them consumers of different grades one after the other as shown in Figure No. 13.12B.

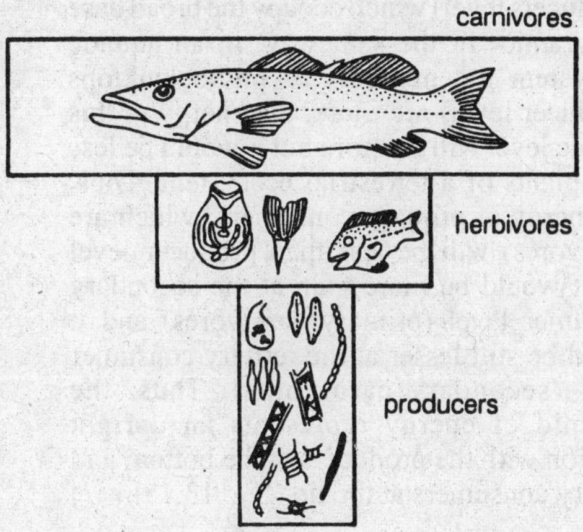

Fig. 13.13. Pyramid of biomass of a pond.

EXPERIMENT No. 19

Object. *Study of pyramid of biomass.*

(As referred to, biomass is the sum total of the dry matter of living organisms of any category per unit area.)

Prodedure. For study and construction of the pyramids of biomass, above samples described under experiment No. 18A and B are dried separately in hot air oven. Dry weight of each category is noted and a pyramid of biomass is constructed.

Pyramid of biomass, thus, constructed for **18 A** will be **inverted** (Fig. 13.13) because the biomass of producers in a pond is the least (it is because the producers in a pond are small organisms) and the biomass shows gradual increase in case of hervivores and carnivores. Hence, the pyramid is inverted.

But pyramid of biomass for **18 B** will be normal **upright** in position (Fig. 13.14) becuase the biomass of producers in a grassland is much higher (it is becuase the producers in a grassland are quite large grasses and plants) and the biomass shows gradual decrease in case of herbivores and different grades of carnivores. Hence, the pyramid is upright.

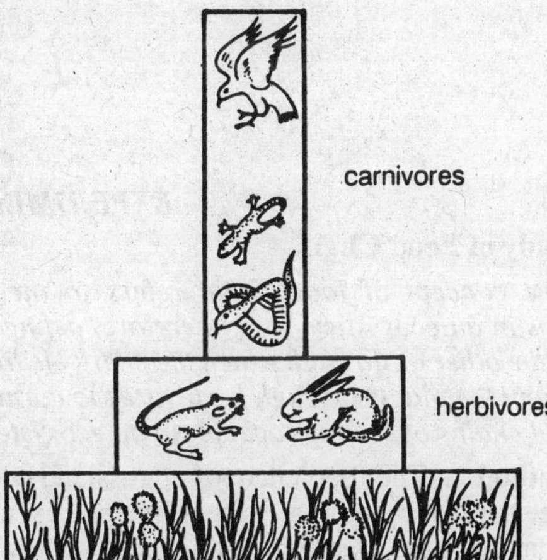

Fig. 13.14. Pyramid of biomass of a grassland.

EXPERIMENT No. 20

Object. *Study of pyramid of energy.*

(The pyramid of energy depicts the total amount of energy utilized by the organisms of different trophic levels in an unit area over a set period of time in an ecosystem; ususally square meter/year.)

The amount of energy transferred through food to sucessive higher levels becomes gradually but drasticaslly less and less. Greater amount of energy is trapped by green plants (producers level) which occupy the broad base of pyramid. In the sameway, in an aquatic ecosystem the numbers of phyloplanktons (producer level) are more, the energy at this trophic level will be more but it would be less than plants of a terrestrial ecosystem. Thus, the energy in primary consumers (which are herbivores) will be less than producer level and it would be more than at the secondary consumer level (primary carnivores) and it would be still lesser at the tertiary consumer level (secondary carnivores). Thus, the pyramid of energy represents an upright position with the producers at the bottom and tertiary consumers at the tip (Fig. 13.15).

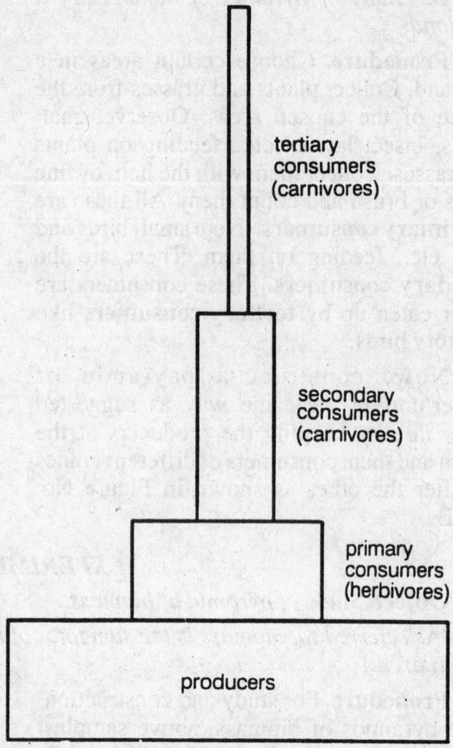

Fig 13.15. Pyramid of energy (K cal per unit area within unit time).

EXPERIMENT No. 21

Study of Food Chain

(The concept of food chain refers to the nutritional interdependence between different organisms in an ecosystem. It is, therefore, defined as a series of organisms eating one and being eaten by the other or through which the energy is transferred. Or in other words, the transfer of food energy from the plants through herbivores to carnivores is called food chain. There may be one or more food chains operating together in an ecosystem which constitute a food web.

Method 1. The food chain of a grassland ecosystem can be studied by using cages of known size. Place them at random at least at five places in the grassland. Collect all animals and vegetable debris from the cages after fumigating them with chloroform. Then sort out the animals as herbivores and carnivores and plot a food chain starting from vegetable debries. Note that in this case large carnivores may be predatory birds and, therefore, also include them at the top of the food chain.

The same method can be used in plotting food chain of a pond ecosystem. In this case your ploltting of food chain would somewhat resemble to Figure No. 13.16.

F. Study of Adaptation

EXPERIMENT No. 22

Ojbect. *To study the effect of light on colour changes in frog.*

Requirements. Live frog, pneumatic trough, source of light.

Procedure. Keep a live frog in a pneumatic trough and expose it to a strong beam of light by keeping a table lamp near the trough under dry conditions. Similarly, keep the trough containing frog in dark having humid conditions. Leave for sometime and then observe.

Observations. Observe the skin colour of frog under both the conditions first after 30 minutes then after every 15 minutes till the skin colour changes. Keep a record of your observations as under:

S. No.	Experimental condition	Change in skin colour after (in minutes)							Result
		30	45	60	75	90	105	120	
1.	Strong light dry condition								
2.	Dark humid condition								

Result. From the observations it would be apparent that skin colour of frog changes from dark to light in experimental condition number 1 and the skin colour becomes darker in shade in experimental condition number 2.

Explanation. Chromatophores, responsible for imparting colouration to the skin, are found in the upper portion of dermis usually aggregated just below the basement membrane. The concentration of chromatophores in the skin on the dorsal surface and legs of frog is always more than the ventral surface. Each pigment cell, *i.e.*, chromatophore has irregular shape with many processes and its shape is changeable. Its cytoplasm contains pigment granules. Usually pigment cells are of three types depending on the colour of pigments contained in them. These are **melanophores** containing black pigment called melanin, **xanthophores** containing yellow pigment called xanthophyll and **guanophores** containing white refractile pigment called guanin.

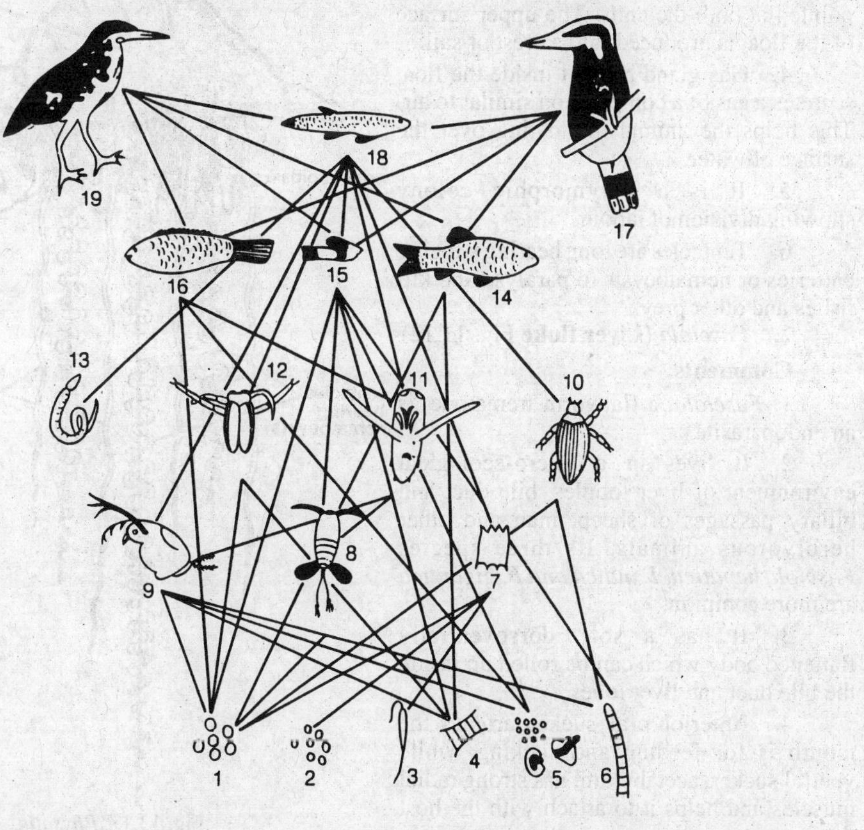

Fig. 13.16. A number of food chains operating together to form a food web in a pond water ecosystem; Diatoms 1,2,4; Algae 3,5,6; Rotifer 7; Copepods 8,9; Water beetles 10,11,12; Nematode 13; Fishes 14,15,16,18; Fish eating birds 17,19.

The olivegreen colour of the dorsal surface of the body of frog is a combined effect of all these pigments while the black patches on the skin are due to the concentration of melanin only.

Frog changes its skin colour to harmonise with its surrounding environment. This ability of animals of changing the intensity of their skin colour in relation to their surrounding is called **metachrosis.** In fact, during bright light and dry conditions the chromatophores contract and their pigment granules get concentrated in their middle region causing the skin colour light. In the sameway, during dark and humid conditions the chromatophores expand imparting skin colour darker. However, a hormone epinephrin from suprarenal gland also induces the lighening of skin colour.

EXPERIMENT No. 23

Object. *To study and comment upon the adaptive and structural modifications in some animals due to ecological conditions.*

Requirements. Animals from museum like *Physalia, Fasciola, Schistosoma, Taenia, Chaetopterus,* Stick insect, A fish, *Exocoetus, Anabas,* Flatfish, *Rhacophorus, Draco, Phrynosoma,* a bird, and Bat.

1. *Physalia* (Portuguese man of war Fig. 13.17)

Comments:

1. *Physalia* is a siphonophore coelenterate.
2. It is a marine, colonial, swimming or floating pelagic animal.
3. It has a large pneumatophore or float which is bladder-like, elongated and pointed at both the ends. The upper surface of the float is produced into a crest or sail.
4. Gas gland present inside the float secretes a gas of a composition similar to air. This helps the animal in floating over the surface of water.
5. It is a polymorphic colony showing division of labour.
6. Tentacles are long bearing stinging batteries or nematocysts to paralyse and kill fishes and other prey.

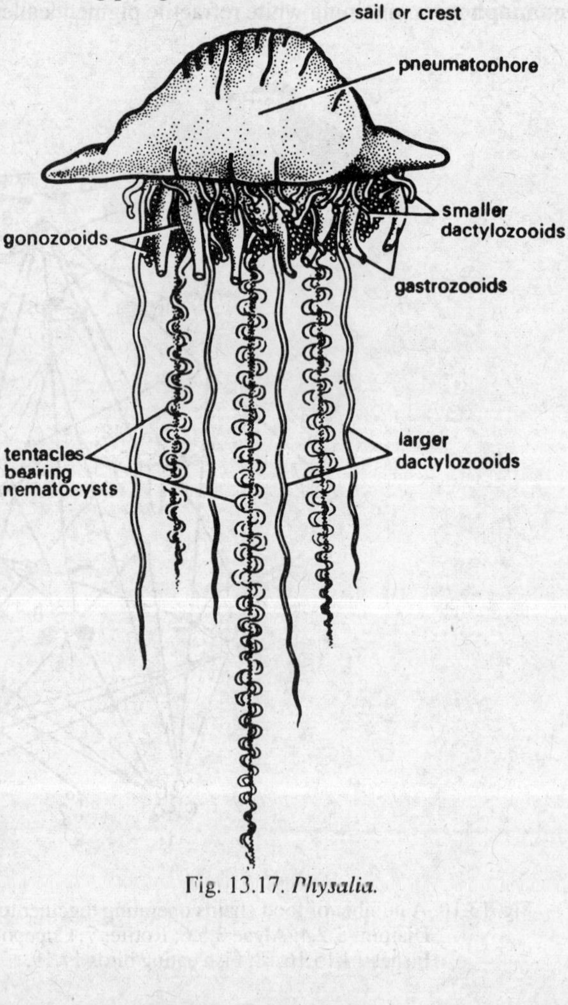

Fig. 13.17. *Physalia.*

2. *Fasciola* (Liver fluke Fig. 11.12)

Comments:

1. *Fasciola,* a flatworm trematode, is an endoparasite.
2. It lives in a micro-ecological environment of liver lobules, bile duct and biliary passages of sheep, man and other herbivorous animals. Its three species *Fasciola hepatica, F.indica and F. gigantica* are more common.
3. It has a soft dorsoventrally flattened body which can be rolled up inside the bile duct and liver lobes.
4. Anterior oral sucker around the mouth is for feeding and sucking, while ventral sucker, acetabulum, has strong radial muscles and helps it to attach with the host tissues.

3. ***Schistosoma*** **(Blood fluke** Fig. 11.13)

Comments:

1. *Schistosoma,* a flatworm trematode, is an endoparasite in the blood of man and other vertebrates.

2. Blood is nutritionally very rich micro-environment in which the schistosomes lead a very successful life. Its three species reported from man are *Schistosoma haematobium, S. mansoni and S. japonicum.*

3. Since schistosomes live in blood and blood flows in tubular blood vessels, hence, their doroso-ventrally flattened body has become tubular so that they may swim freely in the blood.

4. The outer body covering is tegument through which nutrients like sugars, amino acids and lipids from the blood are absorbed.

5. Sucker not meant for attachment.

6. Male carries the female in its gynaecophoric canal.

4. ***Taenia*** **(Tapeworm** Fig. 11.16)

Comments:

1. *Taenia,* a flatworm cestode, is an endoparasite in the small intestine of man; the small intestine is a very good micro-ecological environment in which *Taenia* lives.

2. Two species of *Taenia* are commonly found in man, *Taenia solium* (pork tapeworm) and *T. saginata* (beef tapeworm).

3. Its body has scolex, neck and long strobila being made of dorsoventrally flattened segments called proglottids.

4. Scolex bears suckers and hooks for attachment with the intestimal muscosa of the host.

5. Its body is covered with tegument through which it absorbs nutrients from the intestinal contents of the host because alimentary canal is completely absent in it.

5. ***Chaetopterus*** **(Paddle worm** Fig. 13.18)

Comments:

1. *Chaetopterus,* a polychaete annelid, is marine tube-dwelling (tubicolous) worm.

2. It lives in burrow nearly 8–9 feet deep by forming U-shaped tube opening at both the ends. To lead a successful tubicolous life several modifications occur in its body.

3. The notopodia of anterior segments are large and expanded forming collar of a funnel-like structure to generate water current inside the burrow.

4. The great wings, modified notopodia of the first segment of middle region, help in collecting and directing food towards the food cup.

5. The last three segments of middle region, modified as membranous folds forming fan-like structures, possess strong muscle fibres and act as pumping organ to maintain water current through the burrow.

6. Flow of water current through the burrow is essential as it draws food particles and oxygen from it and expells excretory wastes through it.

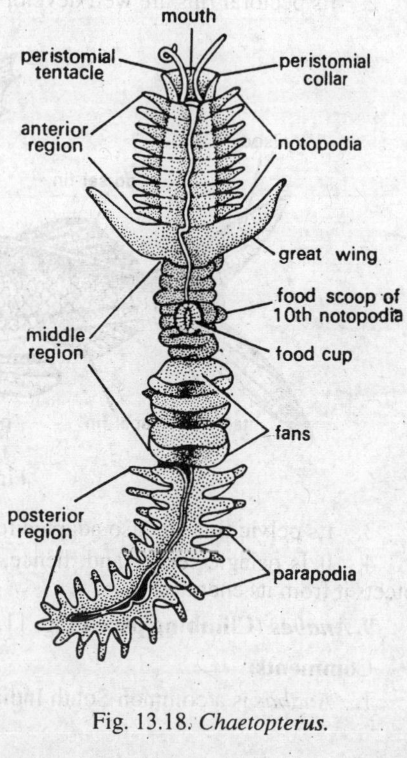

Fig. 13.18. *Chaetopterus.*

6. ***Carausius*** **(Stick insect** Fig. 13.19)

Comments:

1. *Carausius*, a pterygote insect, is leaf feeder and lives on twigs of small plants.

2. It has a slender body which resembles a twig or stick and modified greatly to mimic with the surroundings.

3. Its habit to mimic with the surrounding is of great advantage to it as it protects it from its enemies and predators.

4. It moves very slowly and poses to be dead if disturbed.

5. It is well known for its protective adaptive features.

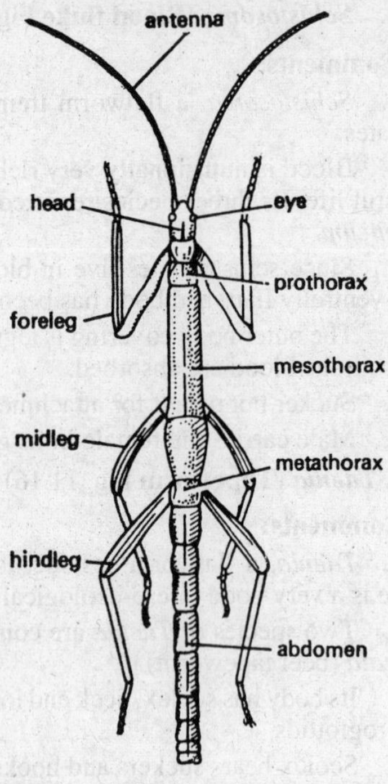

Fig. 13.19. *Carausius*.

7. *Labeo* (A fish Fig. 11.78)

Comments:

1. *Labeo*, a typical fish shows various adaptive modifications for aquatic life.

2. It has streamlined body contour which offers least resistance while swimming.

3. Its wide mouth-gap at the anterior end of body facilitates in feeding while swimming.

4. Its fins are the organs of locomotion and balance.

5. It has well developed gills for respiration with the oxygen dissolved in water.

8. *Exocoetus* (Flying fish Fig. 13.20)

Comments:

1. *Exocoetus*, a fish which do not fly as the name flying fish suggests but glides over the water surface for getting an escape from enemies.

2. Its pectoral fins are well developed, large and when spread look like wings.

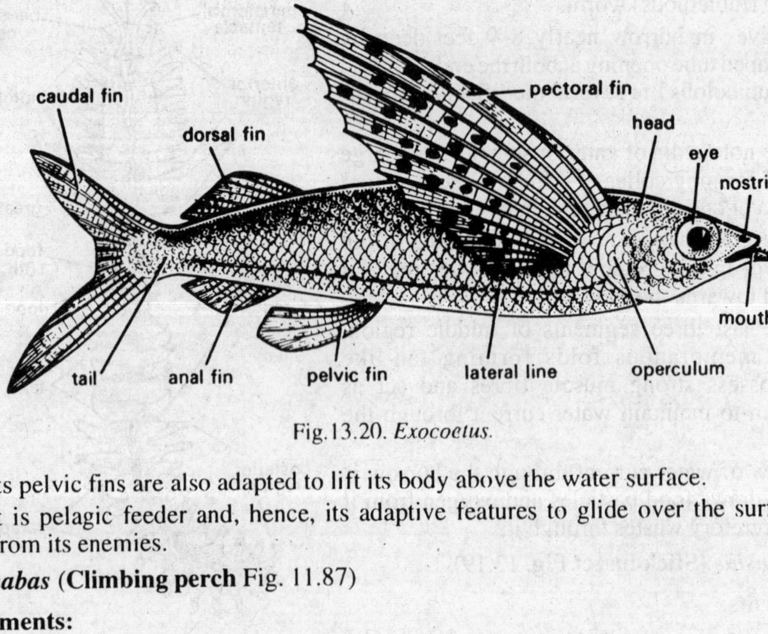

Fig.13.20. *Exocoetus*.

3. Its pelvic fins are also adapted to lift its body above the water surface.

4. It is pelagic feeder and, hence, its adaptive features to glide over the surface of water protects it from its enemies.

9. *Anabas* (Climbing perch Fig. 11.87)

Comments:

1. *Anabas* is a common South Indian freshwater fish.

2. It can live out of water for longer period as it possesses accessory respiratory organs situated over the gills.

3. It can walk on land with the help of fin spines and opercular spines.

4. It cannot climb the tree as its name climbing perch indicates. In fact, while walking on land, it is attacked by preying birds and taken to the trees as if the fish might have climbed the tree. Hence, the name climbing perch has been given to this fish.

10. *Synaptura* (Flat fish Fig. 13.21)

Comments:

1. *Synaptura*, a deep sea fish, is compressed laterally due to pressure of water, hence, its body is completely flattened.
2. Due to lateral compression of the body the laterally placed eyes have turned upwards.
3. Head is asymmetrical. Both the eyes are placed on the upper side which are reduced.
4. Dorsal and anal fins are long without spines and confluent with the caudal fin.
5. Adapted for bottom feeding.

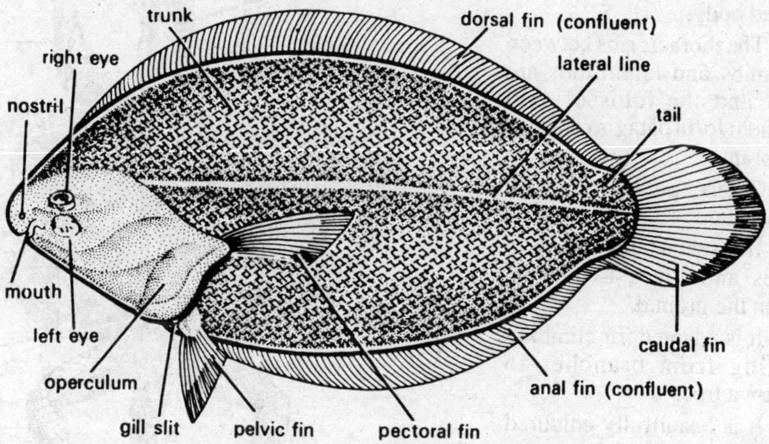

Fig. 13.21. *Synaptura.*

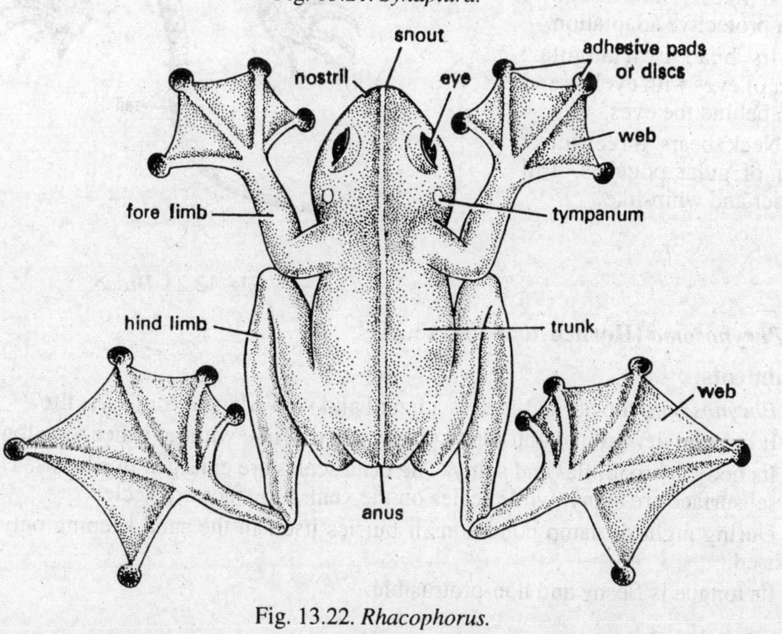

Fig. 13.22. *Rhacophorus.*

11. *Rhacophorus* (Flying frog Fig. 13.22)

Comments:

1. *Rhacophorus* is a tree-living frog and adapted to change its body colour.
2. For living on trees it has modified in different ways; its body is lightly built, limbs are long and thin with large feet, webs are developed between the digits of all the four feet and the tips of digits of all the four limbs have developed adhesive pads or discs.
3. Its large webbed feet are usually used as palms while gliding from branch to branch or tree to tree.

12. *Draco* (Flying lizard or Flying-dragon Fig. 13.23)

Comments:

1. *Draco,* a flying lizard, lives on trees and adapted for aerial and arboreal habit.
2. It has a dorso-ventrally compressed body.
3. The thoracic ribs between the fore-limbs and hind-limbs are expanded and the folds of skin between them form patagia or wing-like membrane.
4. These patagia can be folded like a fan.
5. It uses its patagia like parachutes and, thus, avoids its enemies on the ground.
6. It is adapted for climbing and gliding from branches to branches on a tree.
7. It is beautifully coloured resembling the flowers of the trees on which it lives. Thus, it shows mimicry, a protective adaptation.
8. Its head is triangular; bears a pair of eyes with eyelids and tympanum behind the eyes.
9. Neck bears three hooks and a pair of gular pouches. Tail long, slender and whip-like.

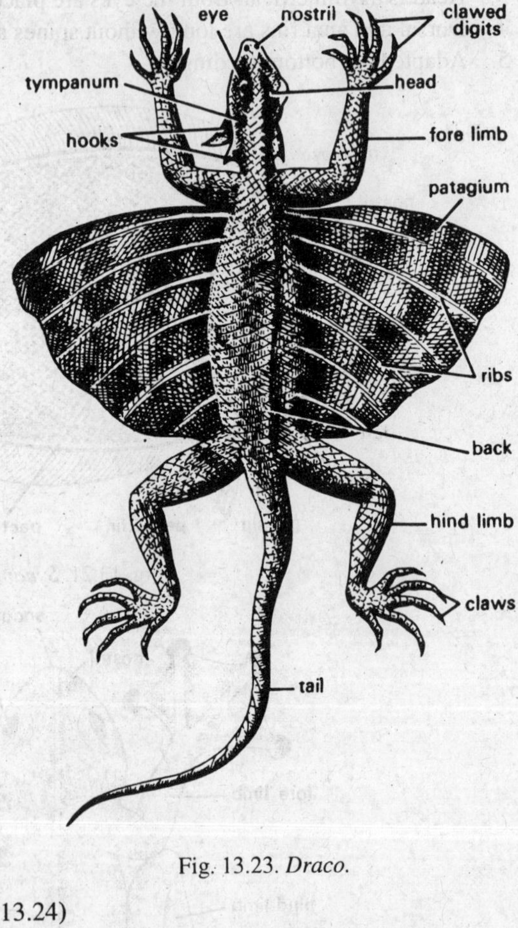

Fig. 13.23. *Draco.*

13. *Phrynosoma* (Horned-toad Fig. 13.24)

Comments:

1. *Phrynosoma,* the horned-toad is a lizard and well adapted for desert life.
2. It lives on dry ground and sandy places and can live without water for a long time.
3. Its body is broad, flat and spiny. The head scales are enlarged like horns. The body scales on the dorsal surface are spiny, while scales on the ventral surface are keeled.
4. During night or damp conditions it burries itself in the sand keeping only the head and horns exposed.
5. Its tongue is fleshy and non-protrusible.

6. Its eyes are provided with complete eyelids.
7. Its nostrils are provided with valves to prevent the entry of sand.
8. Whenever disturbed, it squirts a tiny stream of blood from the eyes which can go up to a distance of about two feet. It is a protective measure.

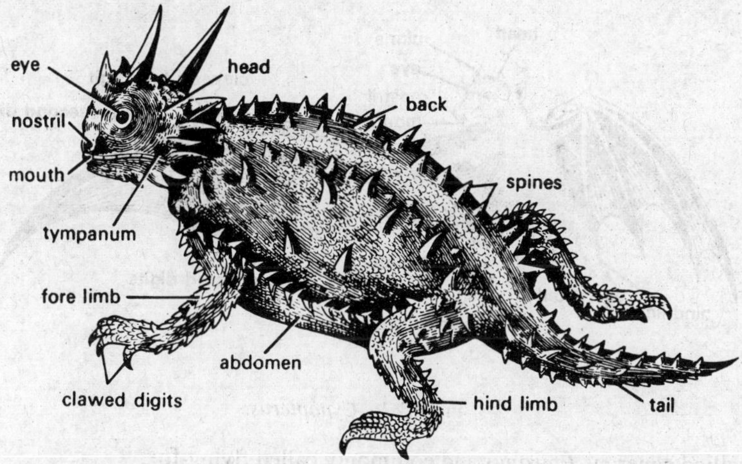

Fig. 13.24. *Phrynosoma.*

14. *Corvus* (Crow Fig. 13.25)

Comments:

1. *Corvus,* a common bird, shows several modifications for aerial mode of life.

2. It has streamlined body form which offers reatively lesser resistance during propulsion through the air.

3. Fore-limbs are modified into wings bearing feathers, hind-limbs have claws and modified for walking on the ground as well as perching on the trees.

4. Eyes large with monocular vision.

5. Tail feathers are elongated.

6. Body is lightly built because of pneumatic bones.

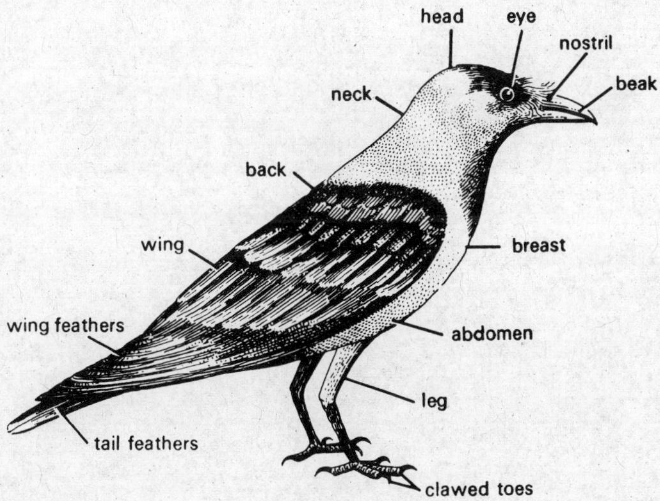

Fig. 13.25. *Corvus.*

15. *Cynopterus* (Bat Fig. 13.26)

Comments:

1. *Cynopterus*, bat, the only flying mammal is well adapted for flying mode of life.
2. Its body is well built for aerial and arboreal habit.,
3. Its fore-limbs are modified and 2nd, 3rd, 4th and 5th digits are much elongated. Skin extends from neck to digits and hind-limbs forming membranous wing or patagia.

4. Its first and second digits possess claws.

5. Its pinnae are well developed to receive the reflected back ultra-sonic sound waves after striking on the solid objects.

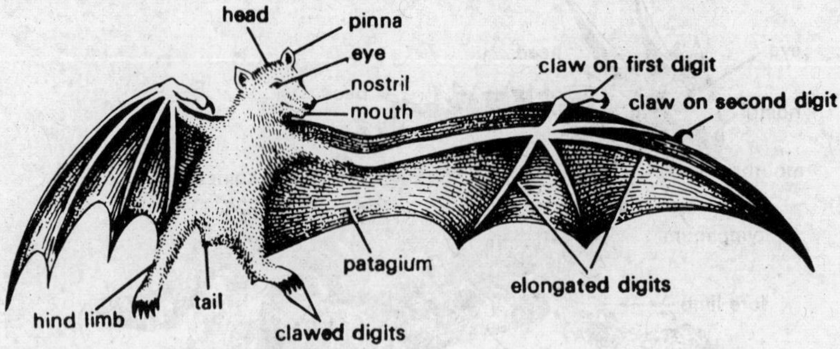

Fig. 13.26. *Cynopterus*.

6. Its allied genera is *Pteropus* and commonly called flying-fox.

7. It is frugivorous, lives in groups and often damages orchards.

8. The smaller bats like *Megaderma*, *Rinopoma*, etc., are insectivorous in food habit with more or less similar adaptive structure as described earlier.

CHAPTER 14

TOXICOLOGICAL EXPERIMENTS

Toxicology is the science of poisons. It deals with the study of economic poisons, their effects, mechanism of action and metabolism of toxicants in test animals.

EXPERIMENT No. 1

Object. *To study the bahavioural response of a fish under different doses of DDT in laboratory conditions.*

Requirements. 5 glass troughs, 5 live fishes approximately of the same weight (fish considered in the present experiment is *Clarius batrachus*), measuring cylinder, balance, stop watch, DDT powder, cotton, water, etc.

(*At present behavioural response refers to the opercular movement, related to the mechanism of respiration, of the fish when it is subjected to different doses of DDT.*)

Procedure. To study the opercular movement of the given fish under the different doses of DDT stress proceed as under :

Take 5 glass troughs. Arrange them in a line and mark them number I, II, III, IV and V. Fill the troughs with 4 liters of tap water separately. Then add 1 ppm*, 2 ppm, 4 ppm and 6 ppm DDT in trough numbers II, III, IV and V respectively. Trough number I is not added with any amount of DDT and, hence, it is the **control** trough, while the remaining 4 troughs II, III, IV and V in which DDT has been added are called **experimental** troughs. To restrict the movement of fish in the troughs, keep some cotton in each trough. Then release one fish in each trough including the control. With the help of stop watch, observe the opercular movements of the fish in these troughs per minute and keep the record in the manner suggested below. Repeat your observations atleast five times and calculate average opercular movement per minute under all experimental conditions including control.

Observations

Trough No.	*Doses of DDT*	*Opercular movement during*					*Average opercular movement/minute*
		First minute	*Second minute*	*Third minute*	*Fourth minute*	*Fifth minute*	
I	CONTROL	146	155	150	142	148	741/5 = 148.5
II	1 ppm	120	138	127	131	124	640/5 = 128
III	2 ppm	66	48	52	48	58	272/5 = 54.4
IV	4 ppm	30	28	26	32	25	141/5 = 28.2
V	6 ppm	21	20	16	18	14	89/5 = 17.8

Result. The above observations and their average values clearly indicate that with the increase of dose opercular movement decreases.

* *ppm = mg per litre*

(**Note:** The given data is hypothetical. It is only an illustration to explain that how would you record the experimental data and calculate.)

EXPERIMENT No. 2

Object. *To determine* LC_{50} *of a toxicant for Clarius batrachus under laboratory conditions.*

Requirements. 6 glass aquaria of 40 liters capacity each, 60 live fishes (*Clarius batrachus*) of approximately equal weights, measuring cylinder, balance, desired toxicant, semilogrithmic coordinate paper, stop watch, etc.

(LC_{50} is that concentration of a toxicant at which 50 % of the test animal is killed in a specified time. For example, 48 - hours LC_{50} for a toxicant refers to that concentration of the toxicant at which 50 % of the test animals are killed. LC refers to lethal concentration and 50 to 50 % population. However, lethal concentration is used for aquatic media only. Another related term LT_{50} *refers to lethal time required to kill 50 % population of the test animal.)*

Procedure. To determine LC_{50} of a toxicant for the given fish, proceed as under :

Take 6 glass aquaria. Arrange them in a line on the laboratory table and then mark them number I, II, III, IV, V and VI. Fill up each aquaria with 40 liters of tap water. Then add toxicant at the rate of 1 mg/L, 1.5 mg/L, 3 mg/L, 5 mg/L and 10 mg/L in aquaria marked number II, III, IV, V and VI respectively. Keep aquarium number I free from toxicant that works as **control**, while other aquaria are **experimental** ones. Then release 10 fishes in each aquarium including the control and observe mortality up to 96 hours. Maintain your record as suggested below:

Observations

Aquarium No.	*No. of test animals* ***Clarius batrachus***	*Concentration of toxicant mg/L*	*Mortality during*			
			48 hours		*96 hours*	
			Number	*% age*	*Number*	*% age*
I	10	CONTROL	NIL	NIL	NIL	NIL
II	10	1.0	NIL	NIL	1	10
III	10	1.5	2	20	4	40
IV	10	3.0	4	40	6	60
V	10	5.0	7	70	9	90
VI	10	10.0	10	100	10	100

Determination of LC_{50} by interpolation involves plotting the data on semilogrithmic coordinate graph paper showing concentrations of toxicants on logarithmic (Y - axis) and percentage mortality of test animals (*Clarius batrachus* in this case) on the arithmetic scale (X - axis).

Thus, plot the percentage of mortality observed in each aquarium against the concentration of toxicant in the same tank. Draw a straight line between two points representing percentage mortality at two successive concentrations. These are lethal to more than half and to less than half of the test animals. The concentration at which this line crosses 50 % lethality line will be the LC_{50} value. Refer Fig. 14.1 which illustrates the procedure described above called straight line interpolation.

Result. From graph LC_{50} determined in this case for 48 hours is 3.6 mg/L and for 96 hours is 2.3 mg/L.

(**Note:** The given data and Fig. 14.1 based on these data is hypothetical. It is only an illustration to explain that how would you record the data and proceed to determine LC_{50} value of toxicant for test animal.)

EXPERIMENT No. 3

Object. *To study the ocular effect of organophosphorus insecticide in rat.*

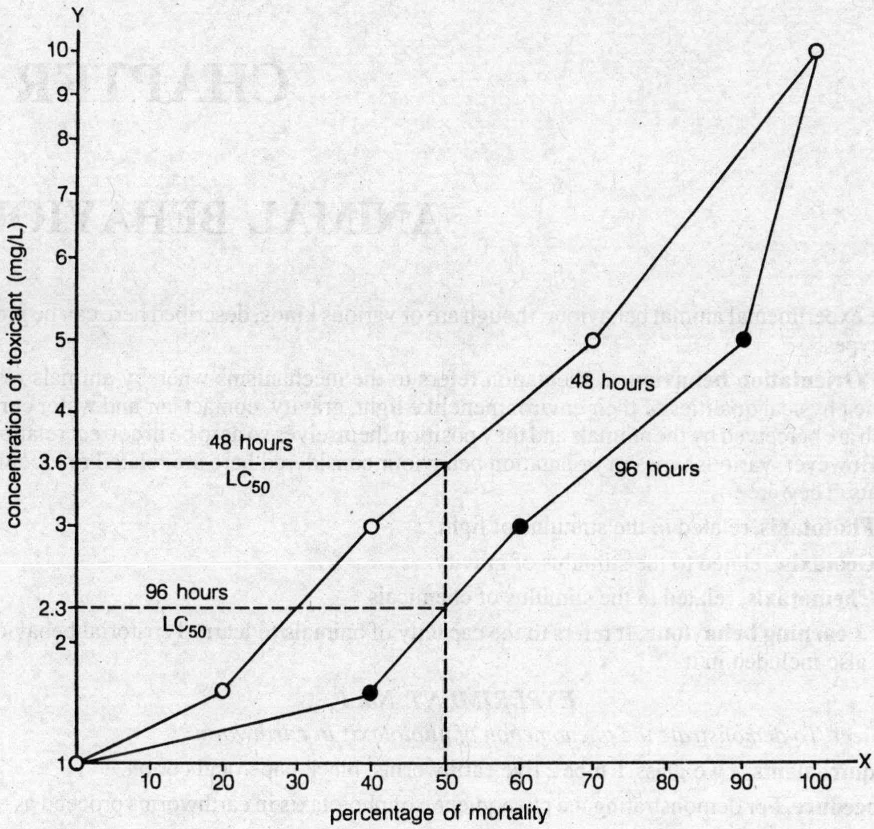

Fig. 14.1. Determination of LC_{50} by straight line graphical interpolation.

Requirements. Live albino rats, glass troughs, organophosphorus insecticides like parathion, malathion, etc.

Procedures. For testing the ocular irritancy, the standard method is that as suggested by **Draize** and **Kelly** (1952).

However, after exposing the test animals to organophosphorus insecticides for ocular effect, the eyes are externally examined for ulceration of cornea, inflammation of conjuctiva, lacrimation or discharge from eyes, etc. The ocular examinations are made up to 72 hours of the exposure of insecticides and a record of the same is maintained.

CHAPTER 15

ANIMAL BEHAVIOUR

The experimental animal behaviour, though are of various kinds, described here can be grouped into two types:

1. Orientation behaviour. Orientation refers to the mechanisms whereby animals respond to the basic physical qualities of their environment like light, gravity, contact, air and water currents, etc., which are perceived by the animals and they position themselves so as to be in correct relationship to them. However, various types of orientation behaviour considered here are related to locomotory movements. They are :

A. **Phototaxis,** related to the stimulus of light.

B. Geotaxis, related to the stimulus of gravity.

C. Chemotaxis, related to the stimulus of chemicals.

2. Learning behaviour. It refers to the capacity of animals to learn. Territorial behaviour is, however, also included in it.

EXPERIMENT No. 1

Object. *To demonstrate the phenomenon of* ***phototaxis*** *in earthworm.*

Requirements. Two glass T tubes, live earthworms, black paper and cotton.

Procedure. For demonstrating the phenomenon of phototaxis in earthworms proceed as under:

Take both the T tubes. Plug both the ends of horizontal limbs of both the T tubes with cotton. Wrap black paper on one side of the horizontal limb of both the tubes. Release few earthworms from the lower ends of vertical limb of T tubes and plug the opening with cotton. Now keep the tubes horizontally in sunlight and leave undisturbed (Fig. 15.1).

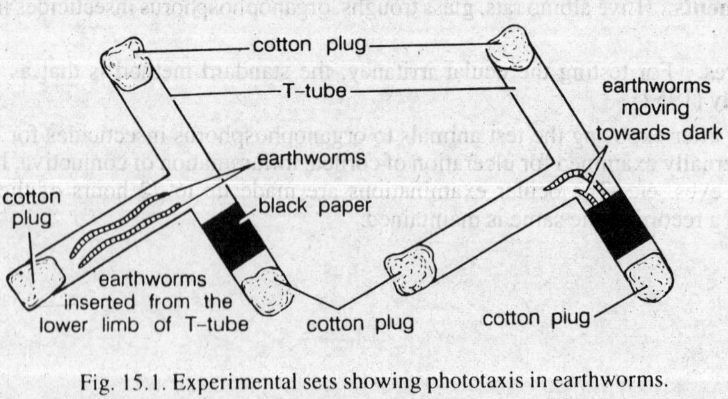

Fig. 15.1. Experimental sets showing phototaxis in earthworms.

Observation. Note that the earthworms are seen moving towards the horizontal limb of T tubes wrapped with black paper and finally settle there.

EXPERIMENT No. 2

Object. *To demonstrate the phenomenon of* ***phototaxis*** *in housefly.*

Requirements. A glass jar of about 40 cm length and 10 cm diameter, jar-lid to cover the jar, black paint or paper and 20 live houseflies.

Procedure. For demonstrating the phenomenon of phototaxis in houseflies proceed as under:

Take the glass jar and either paint or wrap black paper on its exactly lower half. Keep it in up-right position. Release 20 houseflies in the jar and cover the jar with the lid. Leave the jar undisturbed for 5 minutes and note number of flies in light and dark zones of the jar. Then disturb the jar by shaking it and again observe at described earlier. Likewise, repeat your observations for atleast 5 times.

Now invert the jar, so that upside becomes down. Leave for 5 minutes and note the number of flies in light and dark zones. In the manner described above repeat your observations for 5 times.

Then place the jar horizontally and note the number of flies in light and dark zones. Repeat the observations for 5 times. Tabulate your observations in the manner suggested below:

Observations

S. No.	*Total number of **Houseflies***	*Number of **Houseflies** in positions*					
		Upright		*Upside down*		*Horizontal*	
		Light	*Dark*	*Light*	*Dark*	*Light*	*Dark*
1	20	13	7	12	8	18	2
2	20	11	9	10	10	16	4
3	20	12	8	11	9	17	3
4	20	14	6	15	5	15	5
5	20	13	7	13	7	16	4
Average	20	12.6	7.4	12.2	7.8	16.4	3.6

Result. From the observations it is clear that in all the three positions houseflies show positive phototactic behaviour.

Note: The data given are hypothetical. They are given only to illustrate the experiment.

EXPERIMENT No. 3

Object. *To demonstrate the phenomenon of* ***geotaxis*** *in earthworm.*

Requirements. Narrow glass jars 5 in number, slightly moist soil and live earthworms.

Procedure. To demonstrate the phenomenon of geotaxis in earthworm proceed as under :

Take 5 glass jars and fill them (2/3) with the soil. Release one earthworm in each jar and keep it undisturbed.

Observation. Note that the earthworms start burrowing in the soil and within an hour they burrow completely in the soil.

Result. The experiment shows that the earthworms are positive geotactic in behaviour.

EXPERIMENT No. 4

Object. *To demonstrate the phenomenon of* ***chemotaxis*** *in Paramecium.*

Requirements. *Paramecium* culture, dropper, acetic acid, glass slide, microscope and hypodermic syringe with needle.

Procedure. To demonstrate the phenomenon of chemotaxis in *Paramecium* proceed as under:

Take a drop of *Paramecium* culture on a clean glass slide with the help of a dropper. Examine the drop with the help of microscope to be sure for the presence of paramecia in the culture. Put a cover-slip over the drop of culture, again examine under microscope and draw a diagram on white paper sheet showing relative distribution of *Paramecium* in the culture drop.

Now introduce acetic acid (Acetic acid 1 part + water 50 parts) under the cover-slip with the help of a hypodermic syringe and needle. Examine under microscope.

Observation. Note that paramecia are moving towards the acid drop under the cover-slip and aggregating there.

Result. *Paramecium* shows positive response to weak acids.

Note: Contrary to above experiment, if you observe the response of *Paramecium* for strong acids, you will find that it shows negative response.

CHAPTER 16

ENDOCRINE GLANDS

The endocrine glands are those which are without ducts and liberate their secretions, called **hormones,** directly into blood stream and for this reason they are also known as **ductless glands.** The hormones secreted by these glands guide and control the various metabolic activities, growth and differentiation of the various systems and, thus, bring about a correct physiological balance within the body. In mammals these glands which are about 10 in number are found located in different parts of the body.

At present, we are concerned only to locate them correctly in a mammalian body, to know their histological structure and some functional details.

EXERCISE No. 1

Object. *To locate the various endocrine glands in the mammal provided.*

Requirements. The given mammal (rat or rabbit), dissecting box, dissecting tray, pins or small nails, chloroform and any antiseptic solution.

Procedure. Take a freshly chloroformed rat and immerse it in some antiseptic solution. Lay down it in the dissecting tray ventro-dorsally, *i.e.*, ventral surface facing upwards. Fix it in the tray by pinning it in the limbs. Make a longitudinal incision in the skin from the pubic symphysis to the lower jaw and transversly at the anterior as well as posterior ends. Separate the skin from the muscular body wall. Reflect the flap of skin and pin them down.

Make a median longitudinal incision in the abdominal muscular wall from pubic symphysis to the xiphisternum. Cut parallel to the ribs on both the sides, reflect the muscular flaps of abdominal wall and pin them down. Now make a slight incision in the diaphragm with the tip of scissors. Thus, cut through the diaphragm and ribs, and remove them. Thus, the general viscera will be exposed.

Now proceed as described below to locate endocrine glands like thyroid, parathyroid, thymus, adrenal, panereas for islet of Langerhans, and gonads.

Thyroid gland. This gland is located in the neck at about the level of the larynx and upper part of trachea. It is the largest endocrine gland in the body. It consists of two lobes connected by a narrow bridge called isthmus across the ventral surface of the trachea.

To locate it, pick away fat from the neck and remove the lymph nodes and salivary glands. Then sternohyoid muscles are seen in the mid-ventral line. Carefully separate these where they meet in the mid-ventral line and then cut each at its attachment at the manubrium. Lift these muscles forward and remove by cutting them through their anterior attachment to the hyoid. The trachea and larynx are, thus, exposed. Note the largest gland at the anterior part of trachea or at the base of larynx. This is the thyroid gland.

Parathyroid glands. Once thyroid gland is exposed fully then try to locate parathyroid glands.

These are usually two pairs, each pair lies on either side of the midline. These are small, yellowish, flattened, rounded, ovate or pyriform in shape.

Thymus. To locate this gland, see the heart first without disturbing any tissue surrounding it. Note a flat, pinkish grey bilobed gland located at the anterior end of heart. It is easily seen in young animals but gradually it begins to atrophy.

Adrenal glands. Displace the alimentary canal particularly the intestine to one side, clamp and pin it. Try to locate the kidneys; the right kidney is more anterior to left. After locating, the kidneys locate the adrenal glands near the anteior poles of kidneys, for this reason they are also called supra real glands. These glands are oval in shape and embedded in fatty tissue.

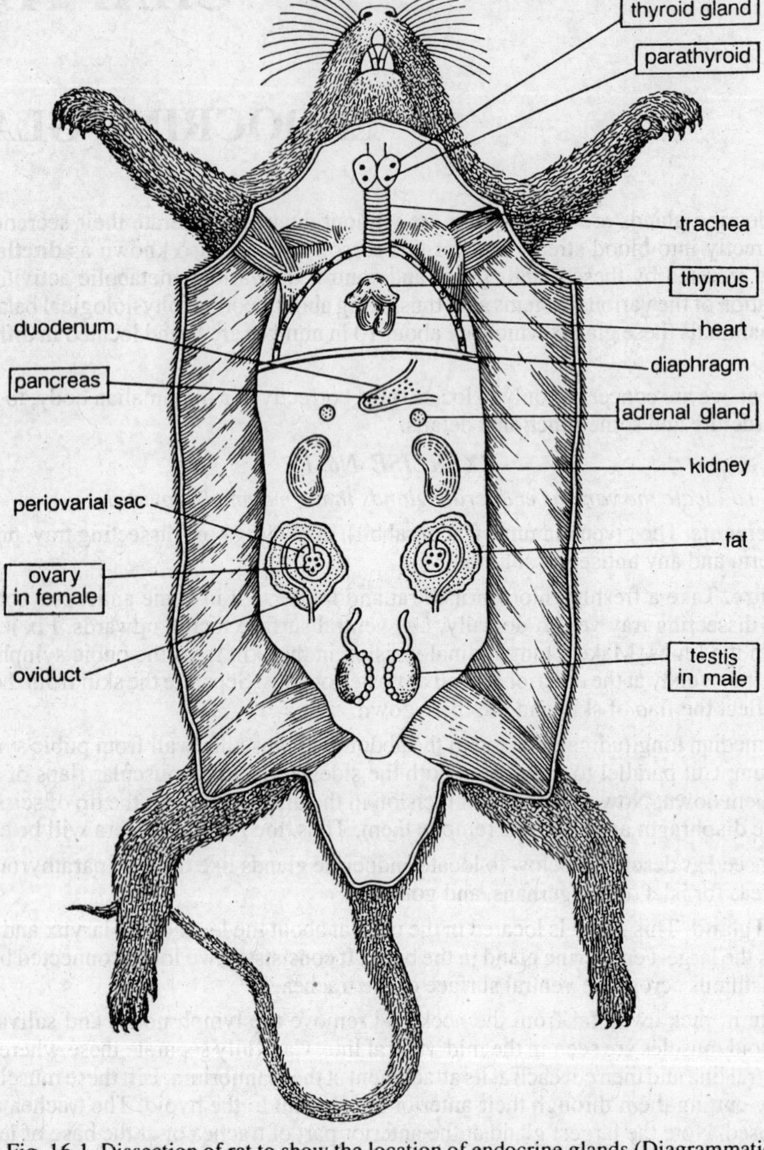

Fig. 16.1. Dissection of rat to show the location of endocrine glands (Diagrammatic).

Pancreas for islets of Langerhans. Islets of Langerhans are groups of specialized cells, which are endocrine in nature, found in the connective tissue between the pancreatic acini. Therefore, for locating islets of Langerhans, locate pancreas.

Pancreas in rat is a pink diffused mass lying between the arms of duodenum and also extends towards the stomach where it ramifies between this organ and spleen.

Gonads. If the dissected specimen is male, then locate its testes which are lying in the scrotum. Its substance contains masses of cells called **interstitial cells** or **cells of Leydig** outside the seminiferous tubules. These cells are endocrine in nature.

If the dissected specimen is female, then locate its ovaries which are lying posterior to the kidneys. The ovaries are small oval bodies attached to the dorsal body wall by folds of peritoneum called mesovarium. The ovary of mature mammal contains **corpus luteum** which is endocrine in nature.

EXERCISE No. 2

Object. *To locate pituitary gland and pineal body in the mammal provided.*

Requirement. Rat or rabbit, chloroform, any antiseptic solution, dissecting box, dissecting tray, mammalian saline, HCl and alcohol.

Procedure. Since both these glands, pituitary on the ventral surface and pineal body on the dorsal surface, are located in brain, hence, to locate them first brain is to be taken out completely from the cranium. So, for locating them proceed as under :

Take a freshly chloroformed rat and immerse its head and neck in any antiseptic solution. Then remove the head by cutting through the neck about an inch from the base of the skull. Cut away as much skin, muscle and fat as possible and then immerse it in 10% HCl and leave it for about four days until the bones have been decalicified. When you become sure that the bones are decalicified completely then take the skull out from HCl, wash well in water and then put it in alcohol for hardening for about four days. Make an incision carefully in the cranium so that alcohol may enter in it.

After about the completion of suggested period take out the skull from alcohol and cut the cranium along its mid-line by a fine tipped scissors to the nasal region. Raise the soften roof of cranium with forceps and then continue the incision along the sides of the cranium to the posterior. The roof of the cranium can now be lifted off to expose cerebral hemispheres and olfactory lobes. Now cut back in the mid-line down the posterior wall of the skull and deflect the tissue. By this, cerebellum will be exposed. Then cut away the neural arches of the anterior cervical vertebrae to free the posterior part of medulla oblongata and spinal cord. Carefully raise the spinal cord with forceps and cut the roots of cranial nerves with a fine pointed scalpel. Then turn the cranium in a petridish containing mammalian saline and shake it so that brain comes out in the petridish. If piamater is still adhered to the surface of brain then peel it off.

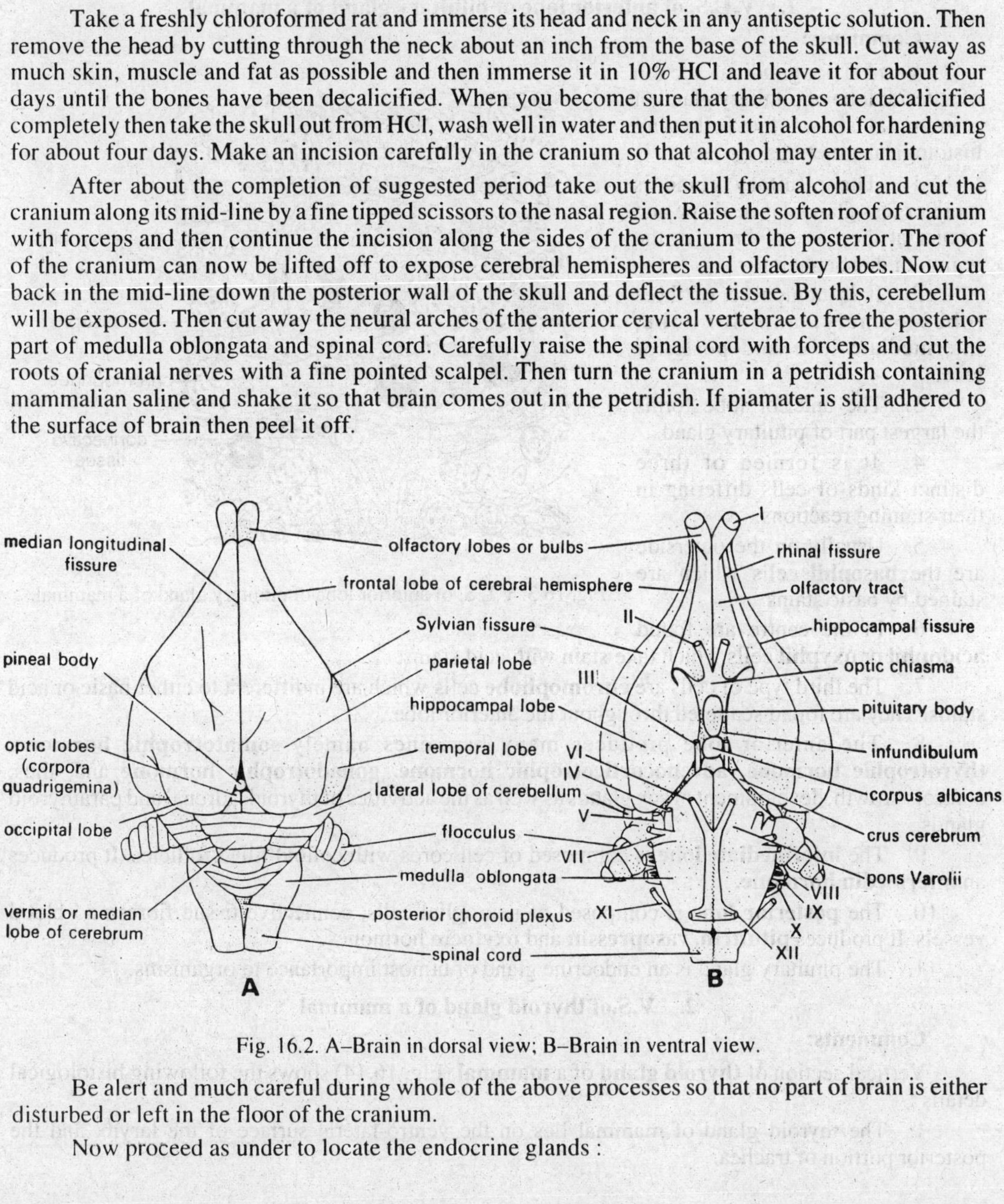

Fig. 16.2. A–Brain in dorsal view; B–Brain in ventral view.

Be alert and much careful during whole of the above processes so that no part of brain is either disturbed or left in the floor of the cranium.

Now proceed as under to locate the endocrine glands :

1. Examine the brain from dorsal view

Pineal body. It appears as a minute white mass deeply embedded between the posterior clefts of cerebral hemispheres above the anterior lobes of corpora quadrigemina (Fig. 16.2 A).

2. Examine the brain from ventral view

Pituitary gland. It appears nut-like structure situated at the base of brain (diencephalon) and is lodged in a depression of the sphenoid bone of the skull called sella turcica just at the back of optic chiasma. It is connected to the brain by a short stalk called infundibulum and is composed of three lobes; anterior lobe, intermediate lobe (pars intermedia) and posterior lobe.

EXERCISE No. 3

Object. *To comment upon the slides of endocrine glands provided.*

1. V.L.S. of anterior lobe of pituitary gland of a mammal

Comments:

Vertical longitudinal section of anterior lobe of pituitary gland of a mammal shows the following histological structures :

1. The pituitary gland is more or less glandular in shape and occurs at the base of brain in the region of diencephalon.
2. It is composed of three lobes namely, anterior lobe, intermediate lobe and posterior lobe.
3. The anterior lobe forms the largest part of pituitary gland.
4. It is formed of three distinct kinds of cells differing in their staining reactions.
5. Usually on the outerside are the **basophil** cells which are stained by basic stains.
6. In the centre are found **acidophil** or **oxyphil** cells which take stain with acid stains.

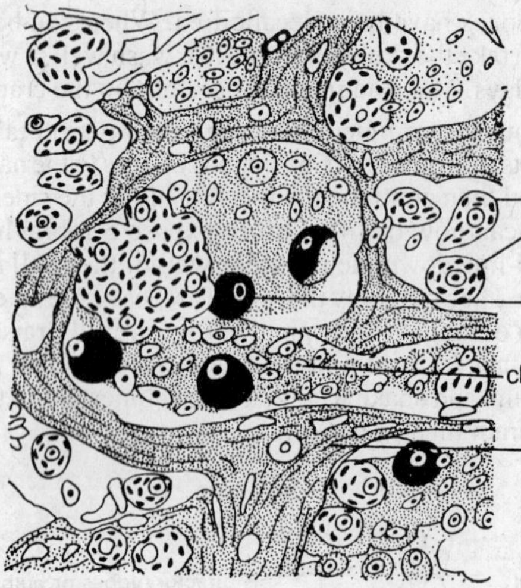

Fig. 16.3. V.L.S. of anterior lobe of pituitary gland of a mammal.

7. The third type of cells are **chromophobe** cells which are indifferent to either basic or acid stains. They are found scattered throughout the anterior lobe.
8. The anterior lobe produces many harmones namely **somatotrophic hormone, thyrotrophic hormone, adrenocorticotrophic hormone, gonadotrophic hormone** and, thus, controls growth, development of sex glands as well as the activities of thyroid, adrenal and parathyroid glands.
9. The **intermediate lobe** is composed of cell cords with colloid filled follicles. It produces an **intermedin hormone.**
10. The **posterior lobe** is composed of neuroglial cells, connective tissue fibres and blood vessels. It produces **pituitrin, vasopressin** and **oxytocin** hormones.
11. The pituitary gland is an endocrine gland of utmost importance to organisms.

2. V.S.of thyroid gland of a mammal

Comments:

Vertical section of **thyroid gland** of a **mammal** (Fig. 16.14) shows the following histological details :

1. The thyroid gland of mammal lies on the ventro-lateral surface of the larynx and the posterior portion of trachea.

2. It consists usually of two lobes connected by an isthmus.

3. Histologically it consists of an outer fibrous capsule and a number of rounded, oval or oblong thyroid follicles separated by inter-follicular tissue.

4. The fibrous capsule is composed of fibrous connective tissue containing large blood vessels and surrounds the thyroid gland.

5. Each thyroid follicle is lined with simple cuboidal epithellum.

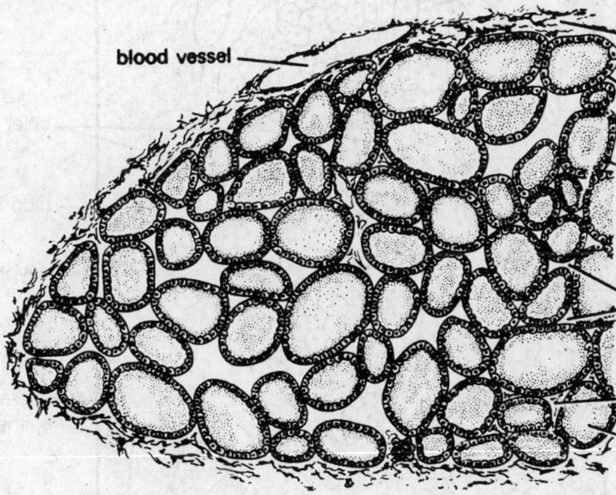

Fig. 16.4. V.S. of thyroid gland of a mammal.

6. The cells of cuboidal epithelium contains large nuclei and pour their secretion into the cavity of the follicle.

7. The cavity or lumen of each follicle is filled with **colloid.**

8. The inter-follicular tissue is composed of reticular connective tissue having numerous blood vessels and capillaries.

9. Thyroid gland secretes **thyroxine** hormone.

10. Thyroxine controls the entire metabolism of animals.

11. Thyroid is an endocrine gland.

3. T.S. of parathyroid gland

Comments:

1. Parathyroid glands are found embedded or located postero-laterally on the surface of the thyroid glands.

2. Parathyroid glands develop from the third and fourth branchial pouches.

3. Parathyroid gland (T.S. Fig. 16.5) is surrounded by inter-follicular connective tissue forming capsule around it.

4. Septa from inter-follicular connective tissue extend in its substance giving it lobulated appearance.

5. It is composed of densely packed cords of secretory epithelial cells between which are interspersed numerous small blood vessels.

6. The secretory epithelial cells are composed of two types of cells; relatively small **chief cells** and larger **oxyphil cells.**

7. The chief cells secrete **parathyroid harmone** called **parathormone** also referred to as PTH.

8. Parathormone regulates normal calcium level in the blood and body fluids. Its oversecretion causes withdrawal of calcium from the bones and, thus, increases the calcium level. Its undersecretion lowers calcium level causing severe muscular spasms and convulsive seizures, a condition called **tetany.**

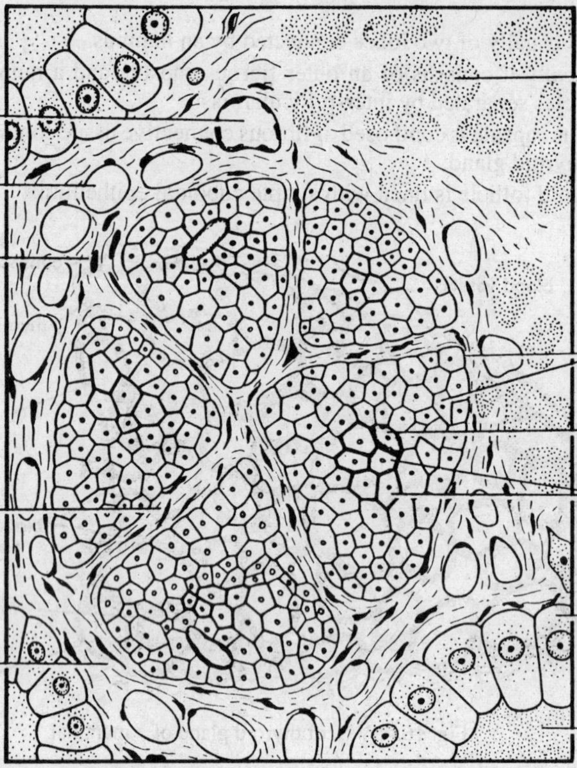

Fig. 16.5. T.S. of thyroid gland passing through parathyroid.

T.S. of adrenal gland of a mammal

Comments:

Transverse section of **adrenal gland** of a **mammal** (Fig. 16.6) shows the following histological structures :

1. The adrenal gland is composed of two distinct parts, *i.e.,* outer **cortex** and inner **medulla** surrounded by the **capsule.**

2. The **capsule** is composed of fibrous connective tissue containing blood vessels and nerves.

3. The **cortex** lies next to the capsule and is differentiated into three zones namely **zona glomerulosa**, **zona fasciculata** and **zons reticularis**.

(i) **Zona glomerulosa** is made up of columnar cells containing large nuclei. The cells are arranged in oval groups which resemble either closed or open vesicles.

(ii) **Zona fasciculata** consists of columns of large rounded cells containing nuclei. The cells are arranged radially in double rows.

(iii) **Zona reticularis** consists of networks of columnar cells containing pigment granules. Numerous blood sinusoids are found in the networks.

4. The **cortex** produces a hormone known as **cortin.** It regulates the general metabolism, controls the sodium chloride content of the blood and also promotes the breaking down of the tissue proteins to amino acids.

5. The **medulla** is the central portion, consists of networks or cords of polygonal cells and clusters of **chromaffin cells,** networks of cells contain numerous blood capillaries, sinusoids and in the centre a **central vein.**

6. Medulla secretes a hormone known as **adrenaline.** It is responsible for maintaining the blood pressure, dilation of vessels and muscles, increasing the general metabolic rate and also for hastening the coagulation of blood.

7. The adrenal glands are endocrine glands and lie just above the kidney attached to it by a fold of mesentery.

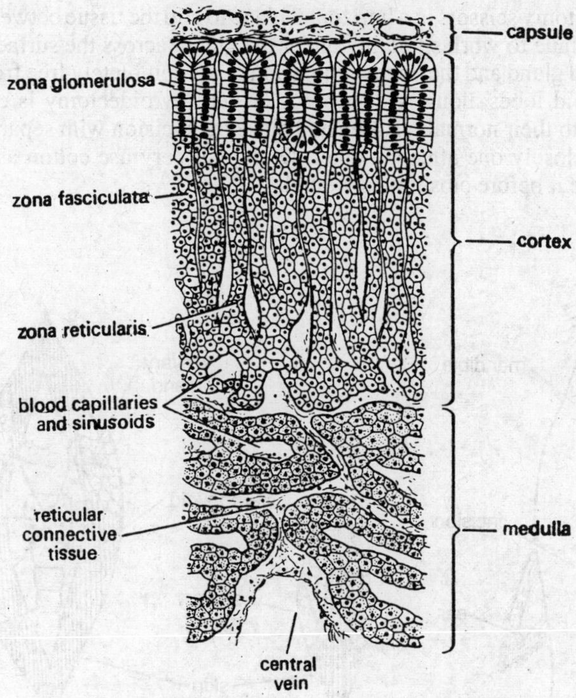

Fig. 16.6. T.S. of adrenal gland of a mammal.

EXPERIMENTS RELATING TO SURGICAL REMOVAL OF ENDOCRINE GLANDS

The various endocrine glands are removed surgically from the target animals to examine or study the impact or importance of these glands. After removing the glands post operative observations are made and their impact on the metabolism, growth and development, etc., are noted. In laboratory, however, a rat can be used for this purpose. In the following pages methods of the removal of some of these glands, *i.e.*, thyroid **(Thyroidectomy)**, adrenal **(Adrenalectomy)** and ovary **(Ovariectomy)** have been described.

EXPERIMENT No. 1

Object. *To study the impact of the removal of thyroid (**Thyroidectomy**) in rat.*

Requirements. Live rat, surgical board, dissecting apparatuses, ether, cotton, ether cone, bent clips, calcium lactate, Zephiran solution or spirit, sutures, stitching material and needle, iridectomy scissors; etc.

(*Removal of the thyroid gland from a living animal is called* ***thyroidectomy****. The surgical removal of thyroid is accompanied by the removal of parathyroid gland also. So to avoid parathormone deficiency the animal should be fed on diet mixed with calcium salts during the post operative period.*)

Procedure of surgery. Take a living rat and keep it in a small trough. Soak a cotton ball in ether and put it in the trough and make the trough air tight by placing a glass sheet over it. When rat is completely anesthetized (*i.e.*, becomes motionless) take it out and put on surgical board keeping its ventral surface upwards. Keep continuing intermittent inhalation of ether with ether cone when rat shows any type of movement during the process of surgery. Clamp the hairs of mid-ventral side of the neck and make an incision (Fig. 16.7A). Retract the skin, fascia and submaxillary glands to expose the sternomastoid and sternohyoid muscles (Fig. 16.7B). Without damaging, carefully separate the muscles along the mid-ventral line to expose the trachea and the thyroid gland (Fig. 16.7 C). Bent clips may be used if retractors are not available. Dissect carefully (without damaging nerves and blood vessels, etc.) on the lateral edge of the thyroid gland with the help of fine forceps. Lift this edge and take iridectomy scissors in a horizontal plane to clip the tissue between the thyroid gland and trachea. Now, continue to work smoothly under the gland, across the surface of trachea so that both the lobes of thyroid gland and the connecting isthmus are completely free from attached tissues. Then remove the thyroid lobes along with isthmus. When thyroidectomy is complete, allow the muscles, etc., to return to their normal position and close the incision with separate sutures with the help of needle placed closely one after the other. During surgery use cotton as sponges to reduce haemorrage and remove it before closing the wound.

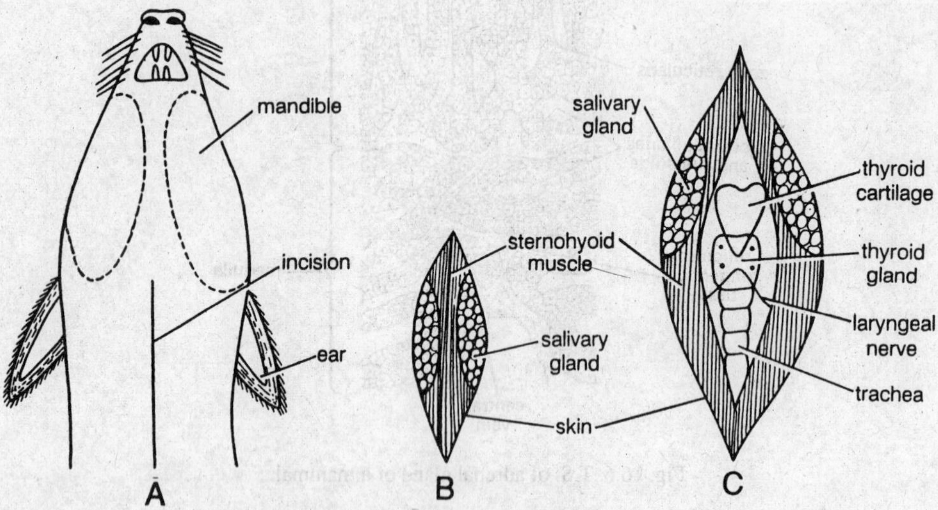

Fig. 16.7. Thyroidectomy of rat.

Immediately after the surgical process is completed, place the animal on 1% calcium lactate drinking solution and keep it in a cage for post operative care and observations.

Thyroidectomy generally effects oxygen consumption, growth and other metabolic activities.

Make your own observations and note in your practical note book.

EXPERIMENT No. 2

Object. *To study the impact of the removal of adrenal glands (**Adrenalectomy**) in rat.*

Requirements. Live rat, surgical board, dissecting apparatures, ether, cotton, ether cone, hair clamps or clips, Zephiran solution or spirit, sutures, stitching material and needle, etc.

(*Removal of adrenal glands from a living animal is called **adrenalectomy**.*)

Procedure of surgery. Anesthetize the rat with ether in the manner suggested in thyroidectomy and place it on the surgical board. Continue intermittent inhalation of ether with ether cone during the whole process of surgery when rat shows any type of movement. Clamp the hairs from the area of both sides of the back in the region of the last rib and rub this area of skin with Zephiran solution. Make an incision high on the right side at an angle between the last rib and the vertebral column (Fig. 16.8A). Once the incision through the skin is complete, make a small cut in the muscle layer and carefully insert the scissors through this cut. Make the opening wide and observe the kidney which is readily seen (Fig. 16.8 B). Adrenal gland lies slightly ahead at the anterior surface of the kidney. Remove perirenal fat carefully with forceps through the opening. Grasp the fat lying below the adrenal gland with the help of curved forceps and separate it from the kidney with the help of other forceps used for tearing the fat lying beneath the adrenal gland and kidney (Fig. 16.8 C). When adrenal gland becomes free from any attachment then take it out. Take maximum care in removing the fats and making the gland free from other tissues. Do not grasp the gland because it is fragile in nature and will be torn easily, leading to an incomplete adrenalectomy. Keep the taken out gland on a moist paper and examine it carefully to be sure that no part of the gland has been left out. When you become sure that the entire gland has been removed, close the incision in the muscle layer by placing separate sutures and then the skin by stitches.

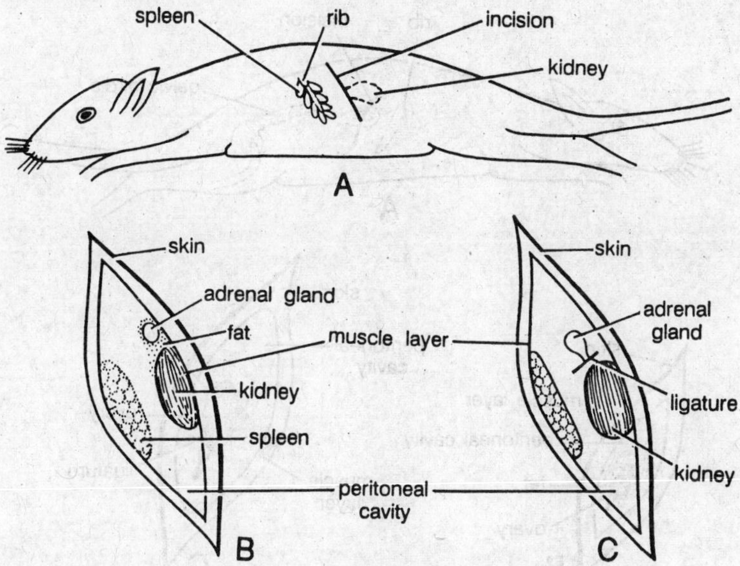

Fig. 16.8. Adrenalectomy of rat.

In the same way, proceed to remove the adrenal gland of the left side. But for the removal of left adrenal gland incision should be made slightly more posterior and central. To expose the gland of this side push the spleen to the anterior side and then the gland is seen easily.

Keep the rat for post operative care. During this period constant food supply with high carbohydrate content should be maintained and 1% NaCl drinking solution should be given. The rat should be housed at slightly higher temperature (preferably 26–29 °C) than normal and observations should be made.

The adrenal hormones are the most potent and essential for the maintenance of life. Hence, adrenalectomy causes death within few weeks. An adrenalectomized animal ceases to grow, loses weight, shows sign of muscular weakness, etc.

Make your own observations and note in your practical note book.

EXPERIMENT No. 3

Object. *To study the impact of the removal of ovaries (**Ovariectomy**) in rat.*

Requirements. Live female rat, surgical board, dissecting apparatuses, ether, cotton, ether cone, hair clamps or clips, Zephiran solution or spirit, sutures, stitching material and needle, etc.

(*Ovary also produces hormones which impart various effect in the body of female. Removal of ovaries from the body of a living rat is called **ovariectomy**.*)

Procedure of surgery. Anesthetize a female rat with ether in the manner suggested in thyroidectomy and place it on the surgical board. Continue intermittent inhalation of ether with ether cone during the whole process of surgery when rat shows any type of movement. Clamp the hairs between the last rib and the knee. Rub this area with Zephiran solution. Make an incision of about 1 to 1.5 cm long midway between the knee and last rib and about 1 cm lateral to the spinal muscles. (Fig. 16.9A). Reflect the skin and make incision in the muscle layer to open the peritoneal cavity. If the incision is correctly placed, the ovary will be seen just underneath embedded in a fatty mass (Fig.

16.9B). Remove the fat and separate the ovary from oviduct by relfecting it out and then tie off the ovary with a silk ligature (Fig. 16.9C). Now cut off the ovary and take it out and return the oviduct to the peritoneal cavity. Close the incision in the muscle with sutures and place stitches in the skin. Make similar incision on the other side and follow the same procedure to remove the ovary of opposite side.

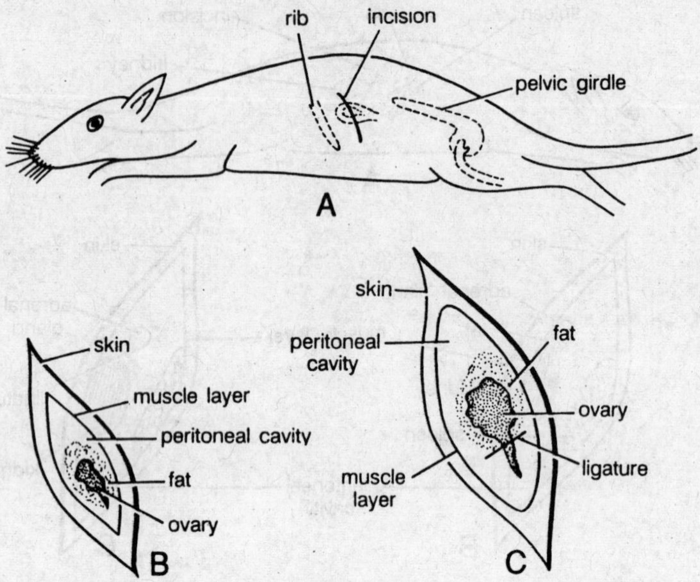

Fig. 16.9. Ovariectomy of rat.

Keep the rat under post operative care and observe the impact of the ovariectomy.

Ovariectomy leads to the atrophy of the reproductive tract and associated structures, and also disappearance of secondary sexual characteristics.

Make your own observations and note in your practical note book.

(**Note:** *Zephiran solution is suggested to be used for keeping dissecting apparatuses and cleaning the skin during all operations on rats, mice, guinea pigs and rabbits. It is prepared by dissolving benzalkonium chloride in alcohol in a proportion of 1:1000.*)

CHAPTER 17

GENETICAL EXERCISES

Genetics deals with the mechanism of heredity and causes of variations in living beings. The **heredity** or **inheritance**, in turn, deals with the process of transmission of characters from one generation to another. Several interesting experiments based on the principles and laws of heredity can be conducted in our biological laboratories. But due to several limitations our laboratories are not fully equipped for the purpose unless, otherwise, specialized. Therefore, we are left with the option of discussing the principles and laws of heredity theoretically. However, the genetical exercises have been planned in the following way to illustrate some basic principles of heredity.

A. Dominance and recessive;
B. Incomplete dominance;
C. Lethal genes;
D. Law of independent assortment;
E. Interaction of genes;
F. Multiple alleles;
G. Sex-linked inheritance.

A. Based on Dominance and Recessive

EXERCISE No. 1

Object. *Two grey Drosophila flies on breeding produce 152 grey and 49 black offsprings. Give the genotypes of the parents and justify your answer giving reasons.* (*Given that grey is dominant over black*).

Observation. The ratio of the dominant and recessive offsprings in the given problem 152:49 comes to be approximately 3:1. Hence, the parents of these offsprings will be heterozygous for grey and black traits having genotypes **Gg** and **Gg**.

Explanation. In case of a monohybrid cross when heterozygous individuals are crossed among themselves, they produce offsprings with dominant and recessive characters in a ratio of 3:1 (*See the cross given below*).

Grey Gg × Grey Gg

gametes	G	g
G	GG Grey	Gg Grey
g	gG Grey	gg Black

Grey : Black
3 : 1

Conclusions. The genotypes of the parents will be **Gg** and **Gg**. It is a case of simple Mendelian inheritance showing the phenomenon of dominance and recessive.

EXERCISE No. 2

Object. *Work out the phenotypic and genotypic ratios of F_1 and F_2 generations from a cross of a homozygous black coloured male rabbit with homozygous white coloured female rabbit considering black is dominant over white. Mention the genetical principles involved.*

Observation. The given problem is a case of monohybrid inheritance. It is related to the phenomena of dominance, recessive and segregation. Hence, it is a case which will be governed by Mendel's laws of inheritance.

Explanation. When homozygous black coloured male rabbit (**BB**) is crossed with a homozygous white coloured female rabbit (**bb**), the F_1 offsprings would be all black coloured but heterozygous with genotype **Bb**. When the individuals of F_1 are crossed among themselves then the F_2 offsprings would be phenotypically of two types in the ratio of 3:1 (3 dominant and 1 recessive) and their genotypic ratio would be 1:2:1 (*See the cross worked out below*).

Black Male Rabbit × White Female Rabbit
(Homozygous) BB ↓ (Homozygous) bb
All Black Coloured - - - F_1
(Heterozygous).Bb

F_1 Black Coloured × F_1 Black Coloured

Bb ↓ Bb

gametes ↓ gametes

gametes	B	b
B	BB Black	Bb Black
b	Black bB	white bb

------------F_2

Conclusions. F_1 phenotype is black and genotype is **Bb** (all individuals). F_2 phenotypic ratio is 3:1 (3 dominant and 1 recessive) and genotypic ratio is 1:2:1 (1 homozygous black, 2 heterozygous black and 1 homozygous white). In addition to these results something more can be concluded :

That, individuals with dominant character may be both homozygous and heterozygous.

That, individuals with recessive character will only be homozygous.

That, individuals with different genotypes may have similar phenotypes.

That, the genetical principle involved is the law of **dominance and recessive** which states that one factor (gene) in a pair may mask or prevent the expression of the other as is apparent from black heterozygous individuals in which **B** masks the expression of **b**.

That, the other genetical principle involved is the **law of segregation** or **purity of gametes** which states that during gamete formation, the two factors (genes) of each character segregate leaving one factor (gene) of a character in each gamete. It is apparent when F_1 black coloured heterozygous individuals form gametes, the factor **B** and **b** separate. Hence, gametes are always pure.

EXERCISE No. 3

Object. *Both the parents of a blue-eyed child are brown-eyed. Find out the genotypes of the parents if brown eyes (B) are dominant over blue (b) which is recessive.*

Observation. Since recessive character is only expressed in homozygous condition, hence, the genotype of the child will be **bb**. The child receives its characters from both the parents in equal amount. Therefore, its one **b** has come from one parent and the other **b** from other parent causing his eyes blue. But in the given problem both the parents are brown-eyed. Therefore, they must have been heterozygous with genotypes **Bb** and **Bb**.

Explanation. When heterozygous brown-eyed parents cross, they produce brown-eyed and blue-eyed offsprings in a ratio of 3:1. The brown-eyed offsprings having genotypes **BB** and **Bb**, while blue-eyed with **bb**. (*See the cross worked out below*).

Brown eyed father × Brown eyed mother

Bb Bb

gametes gametes

	B	b
B	BB Brown	Bb Brown
b	bB Brown	bb blue

Conclusions. The genotypes of the parents of a blue-eyed child will be **Bb**. It is a case of simple Mendelian monohybrid inheritance showing the phenomena of dominance, recessive and segregation.

Solve the following Problems

Problem 1. *If a homozygous normal man marries an albino girl, what would be the phenotypic and genotypic ratios in F_2 generation from this marriage when normal body colour is dominant over albinism. Point out genetical principles involved.*

Problem 2. *The factor for yellow seed coat in pea is dominant to its allele green. Work out the following crosses and point out the genetical principles involved* :

(i) Homozygous yellow × homozygous green;

(ii) Heterozygous yellow × homozygous green.

Problem 3. *Work out the following crosses giving gametes of each parent and the phenotypes of their progeny* :

(*i*) *Rr* × *RR*;

(*ii*) *rr* × *Rr*;

(*iii*) *RR* × *rr*;

(*iv*) *Rr* × *Rr*.

B. Based on Incomplete Dominance

EXERCISE No. 4

Object. *Explain giving reasons for the occurrence of modified 3:1 phenotypic ratio to 1:2:1 in a monohybrid cross.*

Observation. According to the phenomenon of dominance, a monohybrid cross should always result into F_2 offsprings in a phenotypic ratio of 3:1 and their genotypic ratio being 1:2:1. But sometimes, due to incomplete dominance, 3:1 phenotypic ratio comes to be 1:2:1. It means that the phenomenon of dominance shows exceptions.

Explanation. Let us consider an example to illustrate the modified 1:2:1 phenotypic ratio. In Andalusian fowl a cross between its homozygous black and white varieties results in blue hybrids. These blue hybrids when crossed among themselves, the F_2 generation offsprings are 1 black : 2 blue : 1 white (*See the cross given below*). The colour of F_1 hybrid blue is a blended character, *i.e.*, a colour midway between the two parents. It means that none of the colour of the parents are either fully dominant or recessive. Therefore, both the factors express themselves partially resulting into a blending of parental characters. Such factors are usually referred to as **intermediate factors** or **genes**.

Black Fowl × White Fowl

BB bb

♀ \ ♂	b	b
B	Bb Blue	Bb Blue
B	Bb Blue	Bb Blue

Blue Fowls ---- F_1

Cross–fertilization

♀ \ ♂	B	b
B	BB Black	Bb Blue
b	bB Blue	bb White

---- F_2

Phenotypic ratio = 1 Black : 2 Blue : 1 White.

Conclusions. The modified phenotypic 1:2:1 ratio is due to the phenomenon of incomplete dominance. As referred to, it is due to intermediate genes, hence, called intermediate inheritance.

Solve the following Problems

Problem 1. *A cross between two varieties of 4 O' clock plant (Mirabilis jalapa), pure for red* **(RR)** *and white* **(rr)** *flowers results into 12 flowers of three colours, i.e., 3 red, 6 pink and 3 white in* F_2 *generation. Explain the cause of such a result.*

Problem 2. *On crossing pink flowered pea plant, we obtained 14 red, 25 pink and 13 white flowered plants. Discuss the result and find out the genotype of the parent and progeny.*

Problem 3. *How would you produce 4 0'clock seeds all of which should yield pink-coloured plants when sown?*

C. Based on Lethal Genes

EXERCISE No. 5

Object. *In a monohybrid cross a modification of usual 3:1 phenotypic ratio comes to be 2:1. Explain giving full explanation of such a result.*

Observation. According to the laws of inheritance, a monohybrid cross should always yield a phenotypic ratio of 3 : 1 but in the given problem the phenotypic ratio 2:1 which appears to be due to some **lethal factor** causing death of one of the offsprings before birth.

Explanation. Let us consider an example to illustrate the modified 2:1 phenotypic ratio. In case of yellow race of house mouse (*Mus musculus*) though yellow colour is dominant over non-yellow yet it comes into expression only in the heterozygous individuals (*see the cross given on next page*). When two heterozygous yellow mice are mated with each other, the offsprings show a ratio of 2 yellow to 1 non-yellow instead of a usual ratio of 3 yellow (1 homozygous and 2 heterozygous) to 1 non-yellow expected for a monohybrid cross of heterozygous parents.

Heterozygous Yellow male mice Yy × Heterozygous Yellow female mice Yy

gametes Y y (female) ; gametes Y y (male)

	Y	y
Y	YY DIES	Yy Yellow
y	yY Yellow	yy Non-yellow

Yellow : Non-yellow
2 : 1

In such a cross postmortem examination (autopsis) of female mice shows that the homozygous yellow coat coloured offspring dies in the embryonic condition before birth. Thus, only the heterozygous offsprings and pure recessive offspring survive. Therefore, the normal 3:1 ratio is modified to 2:1.

Conclusions. This modification is due to some lethal genes which cause lethality showing abnormal ratio in the offsprings. In fact, various plants and animals are known to possess some genes, which when come in homozygous condition cause an early death of the offspring in embryonic condition.

Note: Lethal genes are of 3 types : **absolute lethal genes** (causing the death of the carrier in embryonic condition), **sublethal genes** (causing the death of the carrier between the birth and attainment of reproductive age), and **delayed lethal genes** (causing the death of the carrier after the attainment of reproductive age).

EXERCISE No. 6

Object. *A poultry farmer crossed two creeper birds and got 4 eggs. When he incubated the eggs, 2 creepers and 1 normal birds were obtained. One bird died during incubation. Explain the cross and point out genetic principles involved.*

Observation. It appears from the problem / object that it is a case of monohybrid inheritance yielding a phenotypic ratio of 2:1. Thus, it is a case of lethal gene action.

Explanation. From the object it is apparent that the two creeper birds considered in the cross would have been heterozygous for creeper genes. The creeper genes are lethal in birds. Hence, when they came in homozygous condition in one case, caused death of the bird in embryonic condition during incubation. Therefore, only 3 offsprings survived (2 heterozygous creepers and 1 normal). *See the cross given below :*

Heterozygous Creeper Male Cp /+ × Heterozygous Creeper Female Cp /+

gametes Cp + (female) ; gametes Cp + (male)

	Cp	+
Cp	Cp Cp DIES	Cp /+ Creeper
+	Creeper + / Cp	Normal + / +

Creeper : Normal
2 : 1

Conclusion. It is a case of absolute lethal gene action. Hence, one homozygous creeper (**Cp Cp**) died during incubation. The offsprings which survived are 2 heterozygous creepers and 1 normal.

D. Based on the Law of Independent Assortment

EXERCISE No. 7

Object. *In garden pea, tall (T) is dominant over dwarf (t) and red flower colour (R) to white (r). If pure red tall is crossed with a dwarf white, what will be (i) P_1 genotypes, (ii) the gametes of P_1, (iii) the F_1 phenotype and genotype, (iv) the gametes of F_1, and (v) F_2 phenotypic ratio, Point out the genetic principles involved.*

Observation. It appears from the object that it is a case of dihybrid cross and based on the principles of Mendelian inheritance.

Explanation. To find out the solutions of the questions asked in the object, work out the cross as suggested below:

Pure red tall means homozygous tall plants with red flowers; genotype will be **TTRR** .

Dwarf white means homozygous dwarf plants with white flowers; genotype will be **ttrr**. It will be homozygous because recessive characters are expressed only when they are in homozygous condition.

(Then workout the cross and answer the questions in conclusion)

	Phenotype	Tall red	×	Dwarf white
(P_1)	**Genotype**	TT RR		tt rr
	Gametes	TR, TR	↓	tr, tr
	Genotype	TR tr ;		TR tr
(F_1)	**Phenotype**	Tall red ;		Tall red
	Gametes	TR, Tr, tR, tr ;		TR, Tr, tR, tr
			× ↓	

(F_2) Genotypes and Phenotypes

♀ \ ♂	TR	Tr	tR	tr
TR	TR TR Tall red	TR Tr Tall red	TR tR Tall red	TR tr Tall red
Tr	Tr TR Tall red	Tr Tr Tall white	Tr tR Tall red	Tr tr Tall white
tR	tR TR Tall red	tR Tr Tall red	tR tR Dwarf red	tR tr Dwarf red
tr	tr TR Tall red	tr Tr Tall white	tr tR Dwarf red	tr tr Dwarf white

Phenotypic ratio = 9 Tall red : 3 Tall white : 3 Dwarf red : 1 Dwarf white.

Conclusions. (i) P_1 genotypes are TTRR and ttrr, (ii) the gametes of P_1 are TR and tr types, (iii) F_1 phenotype is Tall red and genotype is TRtr for all possible offsprings, (iv) the gametes of F_1 are TR, Tr, tR and tr, and (v) the F_2 phenotypic ratio is 9 : 3 : 3 : 1.

The genetic principles involved are the phenomena of dominance and recessive, law of segregation and the law of independent assortment.

EXERCISE No. 8

Object. *A plant breeder crossed a pure variety of tall tomato plant having hairy stem with a dwarf tomato plant without hairy stem (hairless) and in F_2 generation he got 4 varieties of plants numbering 619, 185, 188 and 61. Find out F_1 phenotype and genotype, F_2 phenotypic ratio and genotypic ratios if tall (T) and hairy (H) characters of tomato plant are dominant over their alleles dwarf and hairless. Point out the genetic principles involved.*

Observation. It is apparent from the object that it is a case of Mendelian dihybrid inheritance.

Explanation. A cross between pure tall hairy tomato plant with that of dwarf hairless plant yields 4 varieties of plants in F_2 generation in a ratio of 619 : 185 : 188 : 61. The numbers of 4 varieties of plants suggest an approximate ratio of 9 : 3 : 3 : 1. To find out the solutions, work out the cross as suggested below :

	Phenotype	Tall hairy	×	Dwarf hairless
(P_1)	**Genotype**	TT HH		tt hh
	Gametes	TH, TH	↓	th, th
	Genotype	THth		THth
(F_1)	**Phenotype**	Tall hairy		Tall hairy
	Gametes	TH, Th, tH, th	↓	TH, Th, tH, th

(F_2) Phenotypes and Genotypes

♀ \ ♂	TH	Th	tH	th
TH	TH TH Tall hairy	Th TH Tall hairy	tH TH Tall hairy	th TH Tall hairy
Th	TH Th Tall hairy	Th Th Tall hairless	tH Th Tall hairy	th Th Tall hairless
tH	TH tH Tall hairy	Th tH Tall hairy	tH tH Dwarf hairy	th tH Dwarf hairy
th	TH th Tall hairy	Th th Tall hairless	tH th Dwarf hairy	th th Dwarf hairless

Phenotypic ratio = Tall hairy 9, Tall hairless 3, Dwarf hairy 3, and Dwarf hairless 1.

No. of plants	Genotype	*Phenotype*
1	TTHH	Homozygous tall hairy
2	TTHh, TTHh	Heterozygous tall hairy
2	TtHH, TtHH	
4	TtHh, TtHh, TtHh, TtHh	
9	1 : 2 : 2 : 4	**Tall hairy**
1	TThh	Homozygous tall hairless
2	Tt hh, Tt hh	Heterozygous tall hairless
3	1 : 2	**Tall hairless**
1	tt HH	Homozygous dwarf hairy
2	ttHh, ttHh	Heterozygous dwarf hairy
3	1 : 2	**Dwarf hairy**
1	tthh	Homozygous **dwarf hairless**
Total F_2 offsprings **16**		**Phenotypic ratio 9 : 3 : 3 : 1**

Conclusions. F_1 phenotype is tall hairy and genotype is Tt Hh, F_2 phenotypic ratio is 9 : 3 : 3 : 1 and F_2 genotypic ratio is 1 : 2 : 2 : 4 (9 Tall hairy), 1 : 2 (3 Tall hairless), 1 : 2 (3 Dwarf hairy) and 1 dwarf hairless.

The genetic principles involved are the same as were in exercise 7.

Solve the following Problem

Problem. *In horses, black body (B) and trotting gait (T) are dominant over their alleles chestnut body (b) and pacing gait (t). If a pure black pacer is mated with a homozygous chestnut trotter, what will be (i) P_1 genotypes, (ii) F_1 phenotype and genotype, (iii) F_1 gametes, and (iv) F_2 phenotypic ratio. Explain by pointing out the genetic principles involved.*

E. Based on Interaction of Genes

EXERCISE No. 9

Object. *A poultry farmer crossed a homozygous dominant white Leghorn fowl with a homozygous recessive white Plymouth Rock fowl and then inbred the individuals of F_1 and got white feathered and coloured feathered fowls in the ratio of 13:3. Explain the principles involved.*

Observation. It appears to be a modified ratio of 9:3:3:1. The modified ratio 13:3 has come because of the interaction of genes. It is because of **epistatic** or **inhibiting genes**. When a pair of genes at one locus prevent the expression of a pair of genes at another locus, then such genes are called inhibiting genes. The phenomenon of the prevention of the expression of one pair of genes by another pair of genes is called **epistasis**. Epistasis is, in fact, inter-allelic gene suppression. Such genes are called **epistatic genes** and those which are prevented from expression are called **hypostatic genes**.

Explanation. In poultry a basic gene **C** produces colour in feathers, while the inhibiting gene **I** prevents the development of colour. Thus, gene **I** interacts with the gene **C** to suppress its expression. Because of this interaction, the Leghorn fowl is white having genotype **CCII** but the Plymouth Rock fowl is white because it has recessive genes **ccii**. However, work out the cross given in the object and notice that in F_2 offsprings, the individuals having **I** gene along with colour producing gene **C** in their genotypes produce white feathered fowls, while those without gene **I** produce coloured feathered fowls. (*See the cross given below*).

	Phenotype	Homozygous dominant White Leghorn fowl	×	Homozygous recessive White Plymouth Rock fowl.
(P_1)	**Genotype**	CCII		ccii
	Gametes	CI, CI	↓	ci, ci
	Genotype	CIci		CIci
(F_1)	**Phenotype**	Heterozygous white	×	Heterozygous white
	Gametes	CI, Ci, cI, ci	↓	CI, Ci, cI, ci

(F_2) **Offsprings**

♀ \ ♂	CI	Ci	cI	ci
CI	CICI White	CiCI White	cICI White	ciCI White
Ci	CICi White	CiCi **Coloured**	cICi White	ciCi **Coloured**
cI	CIcI White	CicI White	cIcI White	cicI White
ci	CIci White	Cici **Coloured**	cIci White	cici White

Phenotypic ratio = White : Coloured

13 : 3

Conclusions. It is a case showing **dominant epistasis**. Due to interaction of epistatic and basic genes the dihybrid ratio 9:3:3:1 is modified into 13:3. Other principles are the same as in case of a dihybrid inheritance.

EXERCISE No. 10

Object. *An animal breeder crossed a homozygous gray (BBAA) and a homozygous albino (bbaa) mice and got heterozygous gray (BbAa) mice in F_1 generation. Again he crossed two hybrids of F_1 and got gray, black and albino mice in the ratio of 9:3:4 in F_2 generation. Explain the cross pointing out genetic principles involved.*

Observation. 9:3:4 is a modified F_2 dihybrid phenotypic ratio of 9:3:3:1. The coat colour in mice is said to be governed by the principle of epistasis. Gray coat colour in mice is dominant over black coat colour. But another gene **A** is necessary to produce any pigment in mice coat. However, in homozygous recessive condition (**aa**) no pigment is produced and the mouse becomes albino irrespective to genotypes of the mouse.

Explanation. Explanations are the same as discussed in exercise 9. However, work out the cross as suggested below :

	Phenotype	Homozygous gray	×	Homozygous albino
(P_1)	**Genotype**	BBAA ·	↓	bbaa
	Gametes	BA, BA		ba, ba
	Genotype	BAba		BAba
(F_1)	**Phenotype**	Heterozygous gray	×	Heterozygous gray
	Gametes	BA, Ba, bA, ba	↓	BA, Ba, bA, ba

(F_2) **Progeny**

♀ \ ♂	BA	Ba	bA	ba
BA	BA BA Gray	Ba BA Gray	bA BA Gray	ba BA Gray
Ba	BA Ba Gray	Ba Ba **Albino**	bA Ba Gray	ba Ba **Albino**
bA	BA bA Gray	Ba bA Gray	bA bA **Black**	ba bA **Black**
ba	BA ba Gray	Ba ba **Albino**	bA ba **Black**	ba ba **Albino**

F_2 Phenotypic ratio = Gray : Albino : Black = 9 : 4 : 3.

Conclusions. The albino individuals are those which possess recessive genes for pigmentation in homozygous condition, *i.e.*, **aa**. It is a case showing **recessive epistasis** because the recessive alleles suppress the expression of a trait in the above cross. Other principles are the same as those in the case of a dihybrid inheritance.

(**Note:** *Solve the problems relating to Complementary genes, Supplementary genes, Cumulative genes and Duplicate genes.*)

F. Based on Multiple Alleles

EXERCISE No. 11

Object. *What would be the phenotypes of the progeny if the parents of following genotypes for their blood groups are mated ?*

(i) $L^A L^A \times L^A L^O$, (ii) $L^A L^O \times L^A L^O$, (iii) $L^A L^O \times L^B L^O$ (iv) $L^A L^B \times L^A L^O$, (v) $L^A L^B \times L^O L^O$.

Observation. There are 4 blood groups in human population. These are **A, B, AB** and **O**. In fact, these letters denote a protein substance called **antigen**. The persons with blood group **A** produce A antigen, those with blood group **B** produce B antigen, those with blood group **AB** produce both antigens A and B, and those with blood group **O** do not produce any of these antigens. The 4 phenotypes of blood group A, B, AB and O are produced by three alleles *A*, *B* and *O*. The gene *O* is recessive to the other two but the genes *A* and *B* are **co-dominant** and do not interfere with the expression of each other.

Explanation. Work out the crosses as given below to get the solution of the object.

(i)

Genotype $L^A L^A \times L^A L^O$

Phenotype Blood group A (Homozygous) | Blood group A (Heterozygous)

Gametes L^A, L^A | L^A, L^O

♀ \ ♂	L^A	L^O
L^A	$L^A L^A$ Group A	$L^A L^O$ Group A
L^A	$L^A L^A$ Group A	$L^A L^O$ Group A

Blood groups all A type only.

(ii)

Genotype $L^A L^O \times L^A L^O$

Phenotype Blood group A (Heterozygous) | Blood group A (Heterozygous)

Gametes L^A, L^O | L^A, L^O

♀ \ ♂	LA	L^O
L^A	$L^A L^A$ Group A	$L^A L^O$ Group A
L^O	$L^A L^O$ Group A	$L^O L^O$ Group O

Blood groups A and O types only.

(iii)

Genotype $L^A L^O \times L^B L^O$

Phenotype Blood group A (Heterozygous) | Blood group B (Heterozygous)

Gametes L^A, L^O | L^B, L^O

♀ \ ♂	L^B	L^O
L^A	$L^A L^B$ Group AB	$L^A L^O$ Group A
L^O	$L^O L^B$ Group B	$L^O L^O$ Group O

Blood groups A, B, AB and O types.

(iv)

Genotype $L^A L^B \times L^A L^O$

Phenotype Blood group AB | Blood group A

Gametes L^A, L^B | L^A, L^O

♀ \ ♂	LA	L^O
L^A	$L^A L^A$ Group A	$L^A L^O$ Group A
L^B	$L^B L^A$ Group AB	$L^B L^O$ Group B

Blood groups A, B and AB types only.

(v)

Genotype $L^A L^B \times L^O L^O$

Phenotype Blood group AB | Blood group O

Gametes L^A, L^B | L^O, L^O

♀ \ ♂	L^O	L^O
L^A	$L^A L^O$ Group A	$L^A L^O$ Group A
L^B	$L^B L^O$ Group B	$L^B L^O$ Group B

Blood group A and B types only.

Conclusions. The blood groups of the progeny in the crosses asked would be :

(i) All with blood group **A**, (ii) Blood groups **A** and **O** only in 3 : 1 ratio, (iii) All four types of blood groups **A**, **B**, **AB** and **O** in 1 : 1 : 1 : 1 ratio, (iv) Blood groups **A**, **B** and **AB** types only in 2 : 1 : 1 ratio, and (v) Blood groups **A** and **B** types only in 2 : 2 ratio.

EXERCISE No. 12

Object. *What would be the phenotypes in respect to the coat colour of the progeny if two rabbits of the following genotypes are mated ?*

(i) $CC \times Cc^{ch}$, (ii) $c^{ch}c^{ch} \times cc$, (iii) $c^{ch}c^{h} \times c^{h}c^{h}$, (iv) $C c^{ch} \times cc$.

Observation. Rabbit's coat colour has 4 phenotypes which in order of decreasing dominance are full coloured or brown (also referred to as agouti), chinchilla, Himalayan and albino. These are represented as **C**, $\mathbf{c^{ch}}$, $\mathbf{c^{h}}$ and **c** respectively. Thus, the order of dominance is $\mathbf{C > c^{ch} > c^{h} > c}$. The coat colour is governed by multiple allelism.

Explanation. Work out the crosses as given below to get the solution of the object.

(i)

Genotype	CC	×	$C c^{ch}$
Phenotype	Agouti (Homozygous)	↓	Agouti (Heterozygous)
Gametes	C, C		C, c^{ch}

♀ \ ♂	C	c^{ch}
C	C C Agouti	$C c^{ch}$ Agouti
C	C C Agouti	$C c^{ch}$ Agouti

All agouti progeny.

(ii)

Genotype	$c^{ch} c^{ch}$	×	cc
Phenotype	Chinchilla (Homozygous)	↓	Albino (Homozygous)
Gametes	c^{ch}, c^{ch}		c, c

♀ \ ♂	c	c
c^{ch}	c^{ch} c Chinchilla	c^{ch} c Chinchilla
c^{ch}	c^{ch} c Chinchilla	c^{ch} c Chinchilla

All chinchilla progeny.

(iii)

Genotype	$c^{ch} c^{h}$	×	$c^{h} c^{h}$
Phenotype	Chinchilla (Heterozygous)	↓	Himalayan (Homozygous)
Gametes	c^{ch}, c^{h}		c^{h}, c^{h}

♀ \ ♂	c^h	c^h
c^{ch}	$c^{ch}c^h$ Chinchilla	$c^{ch}c^h$ Chinchilla
c^h	c^hc^h Himalayan	c^hc^h Himalayan

Chinchilla and Himalayan both.

(iv)

Genotype Cc^{ch} × cc

Phenotype Agouti (Heterozygous) Albino (Homozygous)

Gametes C, c^{ch} c, c

♀ \ ♂	c	c
C	C c Agouti	C c Agouti
c^{ch}	c^{ch} c Chinchilla	c^{ch} c Chinchilla

Agouti and Chinchilla.

Conclusions. The phenotypes of the progeny in the crosses asked would be :

(i) All agouti, (ii) All chinchilla, (iii) Chinchilla and Himalayan in 2 : 2 ratio, and (iv) agouti and chinchilla in 2 : 2 ratio.

G. Based on Sex-linked Inheritance

EXERCISE No. 13

Object. *In man colour blindness is caused by a recessive gene* **c** *carried by the* **X** *chromosome. If a colourblind man marries a normal (homozygous) woman, what proportion of the sons and daughters will be colourblind in F_1 and F_2 generations ?*

Observation. Colourblindness is, in fact, the inability of certain human beings to distinguish red and green colour. It is a sex-linked character and is produced by a recessive gene **c**. Normal eyesight is dominant to colourblindness. Since male possesses one **X** and one **Y** sex-chromosome and the gene for colourblindness lies on **X** chromosome, hence, if his **X** chromosome carries **c** then he will be colourblind because there is no homologous gene on **Y** chromosome to check its expression. Female possesses two **X** chromosomes. So, for female to be colourblind, her both **X** chromosomes must carry a gene for colourblindness. If her one **X** chromosome carries a gene for colourblindness, its expression is checked by its dominant homologous gene present on the other **X** chromosome. It is, therefore, colourblindness is more commonly confined to males only.

Explanation. Work out the cross as given below to get the solution of questions asked in the object.

	Phenotype	Man colourblind	×	Woman normal homozygous
(P_1)	**Genotype**	X^cY		X X
	Gametes	X^c, Y	↓	X, X
	Genotype	X^cX, X^cX	×	XY, XY
(F_1)				
	Phenotype	Carrier females	↓	Normal males
	Gametes	X^c, X		X, Y

(F_2)	**Genotype**	X^cX,	X^cY,	X X,	X Y
	Phenotype	Carrier female (Normal)	Colourblind male	Normal female	Normal male

Conclusions. All F_1 females (daughters) are carrier but normal in regard to colourblindness and all males (sons) are normal. All females of F_2 are normal (though 50% carrier) but males are colourblind as well as normal in a ratio of 1 : 1.

EXERCISE No. 14

Object. *In man haemophilia is caused by a recessive gene (h) located on the X chromosome. What would be the chances of haemophilic condition in children if (i) a haemophilic man weds a normal woman, (ii) a haemophilic man weds a carrier woman, (iii) a normal man weds a haemophilic woman, and (iv) a normal man weds a carrier woman.*

Observation. Like colourblindness, haemophilia is also a character which is sex-linked. Haemophilia (bleeder's disease) is a disease of blood which prevents its clotting. It is caused by a recessive gene **h** located on **X** chromosome. Normal blood is dominant to defective blood.

Explanation. Work out the crosses asked in the object as suggested below to get their solutions.

(i)

	Phenotype	Haemophilic man	Weds	Normal woman	
(P)	**Genotype**	X^h Y	↓	X X	
	Gametes	X^h, Y		X, X	
Progeny	**Genotype**	X^h X, X^h X,		XY, XY	
	Phenotype	Normal daughters but carriers		Normal sons	

(ii)

	Phenotype		Haemophilic man	Weds	Carrier woman	
(P)	**Genotype**		X^h Y	↓	X^h X	
	Gametes		X^h, Y		X^h, X	
Progeny	**Genotype**	$X^h X^h$,	X^hX,		X^hY,	XY
	Phenotype	Haemophilic daughter	Carrier daughter		Haemophilic son	Normal son

(iii)

	Phenotypes	Normal man	Weds	Haemophilic woman	
(P)	**Genotype**	X Y	↓	$X^h X^h$	
	Gametes	X, Y		X^h, X^h	
Progeny	**Genotype**	X X^h, X X^h,		X^hY, X^hY	
	Phenotype	Normal daughters but carriers		Haemophilic sons	

(iv)

	Phenotype		Normal man	Weds	Carrier woman	
(P)	**Genotype**		X Y	↓	X^h X	
	Gametes		X, Y		X^h, X	
Progeny	**Genotype**	X X^h,	X X,		X^hY,	X Y
	Phenotype	Carrier daughter	Normal daughter		Haemophilic son	Normal son

Conclusions. (i) No haemophilic progeny will be produced, (ii) 50% of daughters and 50% of sons will be haemophilic, (iii) All daughters will be normal but 100% sons will be haemophilic, and (iv) All daughters will be normal but 50% sons will be haemophilic.

CHAPTER 18

DISSECTIONS

The meaning of **"dissection"** is to cut open the animal in order to ascertain the structure of its parts. The object of dissection is to separate the several parts from one another, so as to define their boundaries and display clearly their mutual relations. Dissection consists mainly in removing the connective tissue which binds the several parts together.

INSTRUCTIONS FOR DISSECTION

The following precautions should be followed during the dissection :

1. For every dissection, some instructions are given by the teacher. Carefully follow them.
2. Have the theoretical knowledge of the dissection which you are performing.
3. Have a well-labelled diagram of the dissection to be done.
4. Keep your instruments clean and sharp.
5. Pin down the animal firmly in the dissecting dish. Never attempt to dissect a specimen that is not fixed so.
6. In fixing animal, stick the pins obliquely not vertically so that their heads do not come in the way or obscure the dissection.
7. Never cut away any thing until you are quite certain what it is you are removing.
8. Slightly stretch the part which you are dissecting, *e.g.*, when dissecting the blood vessels or nerves of the neck, distend it by passing a small role of cotton wool down the oesophagus.
9. In cleaning blood vessels or nerves always dissect **along them** and not **across them**, and avoid laying hold of them with the forceps.
10. The organs, blood vessels and nerves are connected with each other by delicate tissues. These should be removed away with great care in order to expose the required structures.
11. Dissections should be done in water. The water must completely immerse the dissection.
12. Always keep the water clean in the dissecting dish during dissection by changing it whenever stained with blood, so that visibility remains good.
13. Display your dissection by placing black paper below the target organ or organ system exposed.
14. **Invertebrates** are always dissected from the **dorsal side.**
15. **Vertebrates** are always dissected from the **ventral side.**

GRASSHOPPER

General Anatomy

Procedure for dissection. *Take either a freshly-chloroformed or a preserved grasshopper, cut its wings and fix it in the dissecting dish* **dorso-ventrally**, *i.e., dorsal surface facing upwards. Cut the pleurites, i.e., lateral membrane between terga and sterna of the abdominal segments. Carefully remove the tergal plates and then cut and remove the tergal plates of the thoracic segments also. Study the general anatomy and draw a labelled diagram of your dissection in your practical note book.*

A. Digestive system. It consists of the following parts:

1. Mouth. Situated in the preoral cavity, bounded by the mouth parts.

2. Pharynx. Mouth leads into a short muscular **pharynx.**

3. Oesophagus and crop. The pharynx leads into a short, narrow, slender and tubular **oesophagus**. Oesophagus enlarges into a dilated conical sac-like **crop** which extends up to the posterior end of the thorax. Behind the crop is branched **salivary gland** whose ducts open into mouth.

4. Gizzard. Crop is followed by a small muscular gizzard.

5. Midgut. It starts from gizzard and includes **stomach**. Six pairs of cone-shaped **midgut** or **gastric** or **hepatic cacca** or **pouches** are found arranged around the anterior end of stomach.

6. Hindgut. Midgut leads into hindgut which consists of **ileum**, **colon** and **rectum**. Rectum opens out by anus.

Malpighian tubules, excretory organ, are yellowish thread-like tubules present at the junction of midgut and hindgut.

B. Reproductive system. Sexes are separate; females possess ovipositor at the posterior end of abdomen and, hence, can be identified from males.

Procedure for dissection. *Dissect the grasshopper in the same manner as suggested in case of general anatomy. Then carefully examine the male or female reproductive organs in the abdominal region by slightly displacing the alimentary canal. After locating the reproductive organs remove other structures so that the reproductive organs are exposed completely. Dissect male and female grasshoppers separately. Study and draw labelled diagrams in your practical note book.*

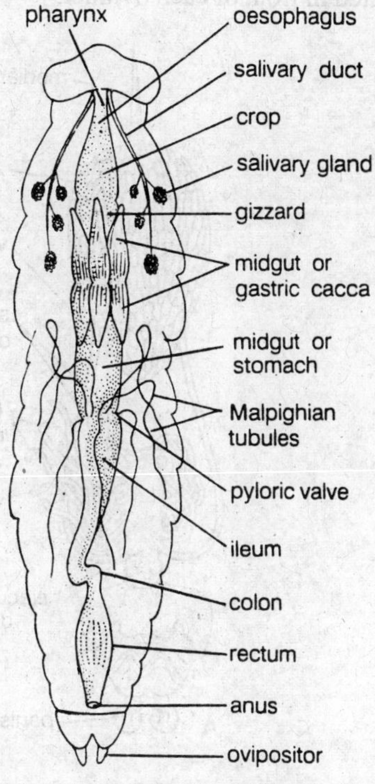

Fig. 18.1. Grasshopper. Dissected to show the digestive system.

Male reproductive system. The male reproductive organs (Fig. 18.2 A) consist of the following parts :

1. Testes. Two, each situated in 4th, 5th and 6th abdominal segments and consists of a series of slender tubules or follicles.

2. Vasa deferentia. From each testis a vas deferens or sperm duct emerges and opens in the ejaculatory duct.

3. Ejaculatory duct. It is a median wide duct and receives the sperm ducts.

4. Penis or aedeagus. The ejaculatory duct opens to the exterior through penis.

5. Accessory glands and seminal vesicle. The ejaculatory duct receives a number of accessory glands on both the sides which are in form of short and long tubules. Associated with these glands is a much longer and wider tube called seminal vesicle.

Female reproductive system. The female reproductive organs (Fig. 18.2 B) consist of the following parts :

1. Ovaries. Two, each is located over the intestine and connected to dorsal body wall by an ovarian ligament. Each ovary is formed of 8 ovarioles.

2. Oviduct. Each ovary leads into an oviduct.

3. Vagina. The oviducts of both side open into a wide muscular **vagina** that opens outside through the genital opening between the plates of the ovipositor.

4. Spermatheca or seminal receptacle. A small tubule arises from the vaginal pore called spermathecal duct which bears a short chamber called spermatheca or seminal receptacle.

5. Accessory gland. An elongated curved accessory gland, also called **collaterial gland**, is situated in front of each oviduct.

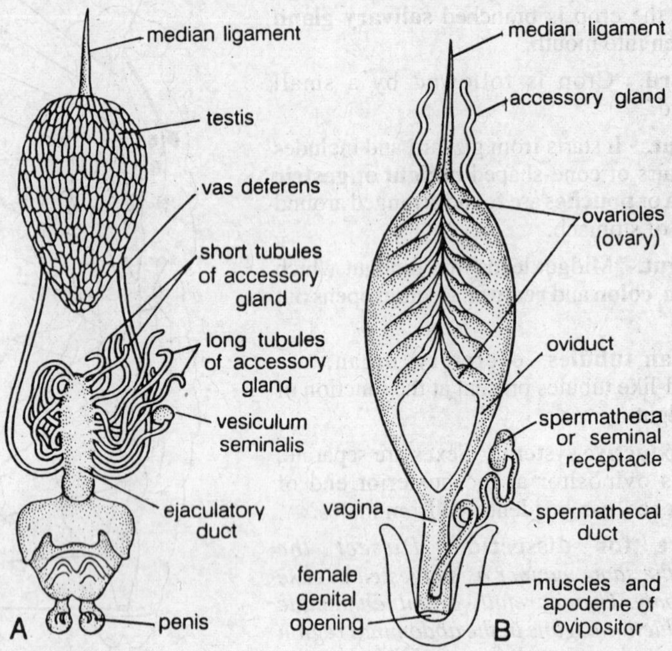

Fig. 18.2. Grasshopper. Dissected to show : A–Male reproductive system; B–Female reproductive system.

Nervous System

Procedure for dissection. *Take a preserved* **grasshopper**, *cut its wings and fix it in the dissecting dish* **dorso-ventrally**, *i.e., dorsal surface facing upwards. Either make a median longitudinal incision in the tergites on the dorsal side from the anterior end of the head to the posterior end of abdomen or cut the pleurites on both the lateral sides and remove the tergites. Now carefully remove the muscles and alimentary canal and expose the ventral nerve cord and thoracic ganglia. Expose the brain and nerves by removing the terga from their origin to their innervations.*

Study the dissection and identify the following ganglia and nerves :

1. Brain or supra-oesophageal ganglia lie dorsally in the head above the oesophagus. Brain is connected to the sub-oesophageal ganglion by two stout circum-oesophageal connectives.

2. Sub-oesophageal ganglion is situated above the mouth parts in the middle of the head slightly inclined to the posterior side.

3. Thoracic ganglia are three, one lying in each segment of the thorax, *viz.*, first thoracic ganglion, second thoracic ganglion and third thoracic ganglion.

(i)First thoracic ganglion supplies the nerves to the first pair of legs.

(ii)Second thoracic ganglion supplies nerves to the second pair of legs and fore pair of wings.

(iii)Third thoracic ganglion also supplies nerves to the hind pair of legs and hind pair of wings.

4. Ventral nerve cord is double and lies in the mid-ventral line in the abdominal cavity. It arises from the third thoracic ganglionic mass.

5. Abdominal ganglia are five pairs lying in the abdomen. Last abdominal ganglion supplies to the organs of reproduction.

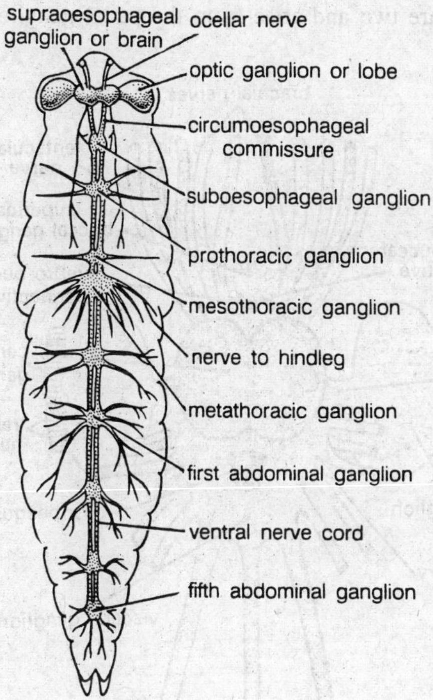

Fig. 18.3. Grasshopper. Dissected to expose the nervous system.

After completing the dissection, insert the black paper below the nerves and draw a neat and well-labelled diagram of nervous system with the help of Fig. No. 18.3.

LOLIGO (SQUID)

Nervous System

Procedure for dissection. *Take a preserved* **Loligo** *and pin down it in the dissecting dish* **dorso-ventrally**, *i.e., dorsal surface facing upwards. Make a median longitudinal incision on the dorsal side from the posterior end of the trunk to the anterior end of the collar and posteriorly in head region. Pin down the flaps of the body wall on both the lateral sides by releasing the muscles and tissues. Carefully expose the* **cerebral ganglion** *in the head region and then proceed for the* **brachial**, **tentacular** and **optic nerves**, *etc. In the trunk region expose the* **stellate ganglion** *on both the lateral sides and also trace the various nerves carefully following their course.*

Study the dissection and identify the following ganglia and nerves :

1. **Cerebral ganglion** lies in the head region in between the two eyes.

2. **Optic nerves** arise from cerebral ganglion and expand in the form of optic ganglia which gives off several retinal nerves.

3. **Visceral ganglion** is situated behind the cerebral ganglion and surrounding the oesophagus. It gives off several nerves to various organs of the body.

4. **Pallial nerves** are very long, arise from the posterior side of visceral ganglion and form the **stellate ganglia** quite posteriorly which innervate the mantle.

5. Visceral nerve arises from the posterior end of visceral ganglion and runs backwards and divides into two nerves which innervate the viscera.

6. Brachial nerves are eight, arise anteriorly from the pedal ganglionic mass and innervate the mantle.

7. Tentacular nerves are two and arise from the pedal ganglionic mass and supply the tentacles.

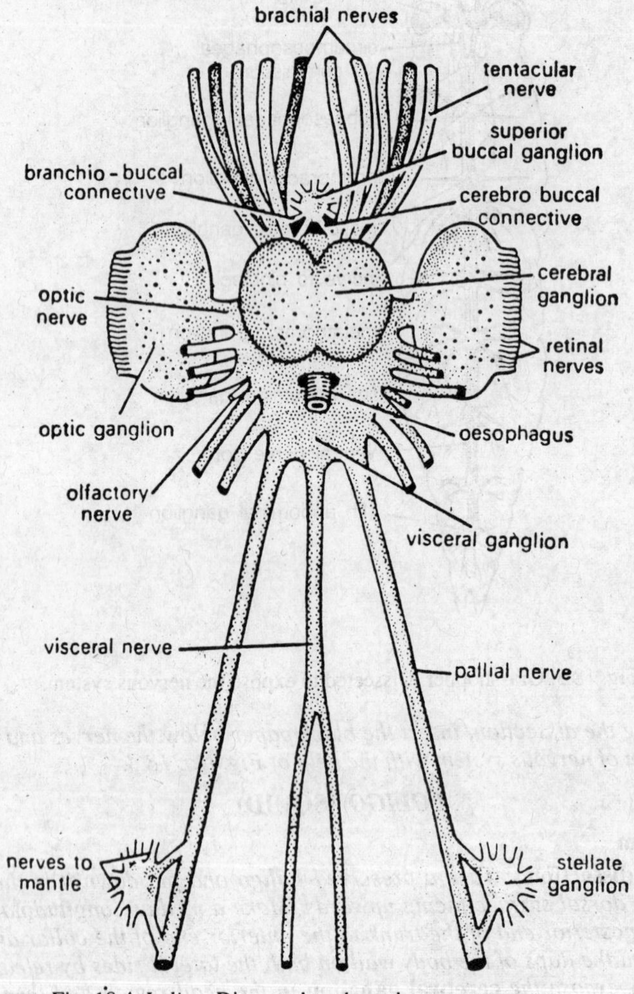

Fig. 18.4 *Loligo*. Dissected to show the nervous system.

After completing the dissection, insert the black paper below the nerves and draw a neat and well-labelled diagram of the dissection with the help of Fig No. 18.4.

LABEO (ROHU)

General Anatomy

Procedure for dissection. *Take a freshly-killed or preserved* **fish,** *wash with water and place it in a dissecting dish* **ventro-dorsally**, *i.e., ventral surface facing upwards. Pin down the fish in the dissecting dish in this position. Locate the position of pelvic and pectoral girdles by tracing them through the skin. Make a mid-ventral incision extending from pelvic to pectoral girdle. Stretch the body wall and either pin it down or remove by cutting. By doing this, visceral organs are exposed. Observe the following organs. Draw a labelled diagram of your dissection in your practical note book.*

1. Coelom. It is the body cavity in which visceral organs are lodged. It is divided into spacious **abdominal cavity** lodging the main visceral organs and a small **pericardial cavity** lodging the heart

2. Digestive system. It includes the alimentary canal and digestive glands. Note these structures:

(i) Mouth. It is a slit-like aperture surrounded by upper and lower lips.

(ii) Buccal cavity. Mouth opens posteriorly into a dorso-ventrally compressed chamber, the buccal cavity.

(iii) Pharynx. Buccal cavity opens posteriorly into a flattened chamber called pharynx. Its anterior part is perforated on the sides by gill-slits and posterior part bears pharyngeal teeth.

(iv) Oesophagus. Pharynx is followed by a narrow, short thick-walled tube called oesophagus.

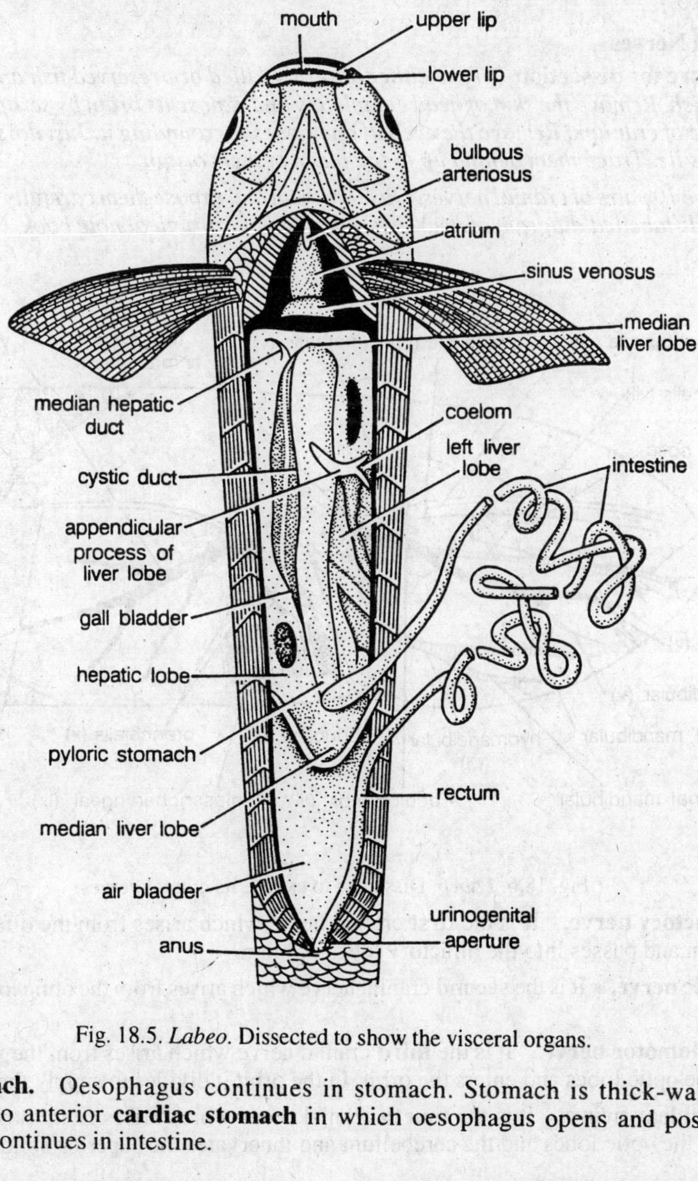

Fig. 18.5. *Labeo.* Dissected to show the visceral organs.

(v) Stomach. Oesophagus continues in stomach. Stomach is thick-walled tube and differentiated into anterior **cardiac stomach** in which oesophagus opens and posterior **pyloric stomach** which continues in intestine.

(vi) Intestine. It is a thin-walled, much coiled tube of uniform diameter and posteriorly opens into rectum.

(vii) Rectum. It is a thin-walled chamber and its diameter is more than that of intestine. It opens out by anus.

(viii) Digestive glands. These include liver and pancreas.

(a) Liver. It is a bilobed elongated dark brown coloured gland. The right lobe is longer than the left which is broader. Both the lobes are connected dorsally by a median lobe and also at different places. These is an elongated **gall bladder** lying between the right lobe of liver and cardiac stomach. Gall bladder receives three bile ducts through a **cystic duct** which finally opens into the stomach.

(b) Pancreas. It is inconspicuous diffused mass occupying almost the whole space of the abdominal cavity and is also embedded in the substance of liver and spleen.

Study the dissection and draw a neat and well-labelled diagram of general anatomy with help of Fig. No. 18.5.

Cranial Nerves

Procedure for dissection. *Take either a freshly-killed or preserved fish and wash. Keep it in a dissecting dish. Remove the skin of head completely and expose its brain by scraping the bone from the dorsal side of cranium. Remove the eyeball and bones surrounding it. Just dorsal to the orbit two slender nerves lie. Trace them behind up to the place of their origin.*

There are 10 pairs of cranial nerves. Observe them and expose them carefully and clearly. Draw a neat and well-labelled diagram of your dissection in your practical note book.

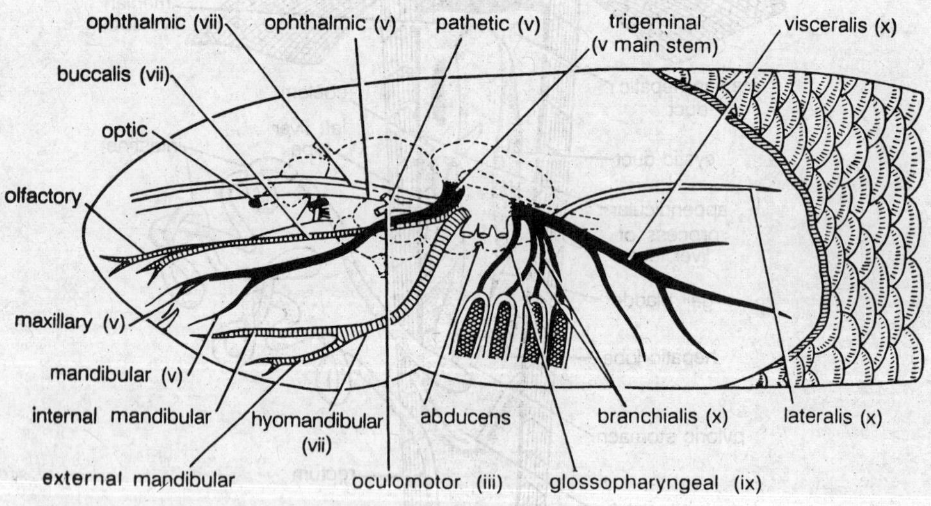

Fig. 18.6. *Labeo*. Dissected to expose its cranial nerves.

1. Olfactory nerve. It is the **first** cranial nerve which arises from the olfactory cells in the olfactory organ and passes into the olfactory lobes of brain.

2. Optic nerve. It is the **second** cranial nerve which arises from the optic lobe and innervates the retina.

3. Oculomotor nerve. It is the **third** cranial nerve which arises from the ventral side of the brain below the optic lobes and enters the orbit. In the orbit it divides to supply the eye muscles.

4. Trochlear nerve. It is the **fourth** cranial nerve which arises from the dorsal side of the brain between the optic lobes and the cerebellum and innervates the superior oblique muscles.

5. Trigeminal nerve. It is the **fifth** cranial nerve and divided into three branches :

(i) **Ophthalmic; (ii) Buccalis; (iii) Hyomandibular.**

6. Abducens. It is the **sixth** cranial nerve and innervates the eye muscles.

7. Facial nerve. It is the **seventh** cranial nerve and is divided into following branches :

(i) Ophthalmicus superficialis. It arises from anterior root and runs forward in close association with the ophthalmicus of fifth cranial nerve and innervates the skin of the dorsal surface of the snout.

(ii) Buccalis. It runs obliquely downwards and forwards across the orbit from the place of its origin and innervates the associated organs.

(iii) Hyomandibular. It runs downwards and outwards, and divides into **mandibular externus**, **mandibular internus** and **hyoidean** nerves.

(iv) Palatinus. It runs across the floor of the orbit to the palate.

8. Auditory nerve. It is the **eighth** cranial nerve which arises from medulla oblongata and innervates the internal ear.

9. Glossopharyngeal nerve. It is the **ninth** cranial nerve which arises from ventro-lateral side of medulla oblongata and innervates the first gill by its **pre**- and **post-trematic** branches.

10. Vagus nerve. It is the **tenth** cranial nerve which arises from behind the ninth cranial nerve and divides into three branches :

(i) Branchialis. It is divided into several branches to innervate 2nd, 3rd and 4th gills.

(ii) Visceralis. It has several branches to supply various visceral organs.

(iii) Lateralis. It runs backwards parallel to the lateral line and innervates the lateral region of the body.

Study the dissection and draw a neat and well-labelled diagram of cranial nerves with the help of Fig. No. 18.6.

Internal Ear

Procedure for dissection. *Take a preserved or freshly-killed fish and remove its skin from the hinder end of the cranium and expose its auditory capsule that lodges the internal ear. Carefully scrap the sides of cranium to locate a triangular area having three ridges which lodge three semicircular canals. Now carefully expose the semicircular canals with the help of scalpel and forceps, and also the other parts of the ear. When ear is completely exposed try with a forcep whether it is completely free from the auditory capsule or not ? If not then further clear the different parts of the ear. When it becomes free from auditory capsule take it out and put it in water in a watch glass.*

Study the internal ear and draw a neat and well-labelled diagram of it with the help of Fig. No. 18.7.

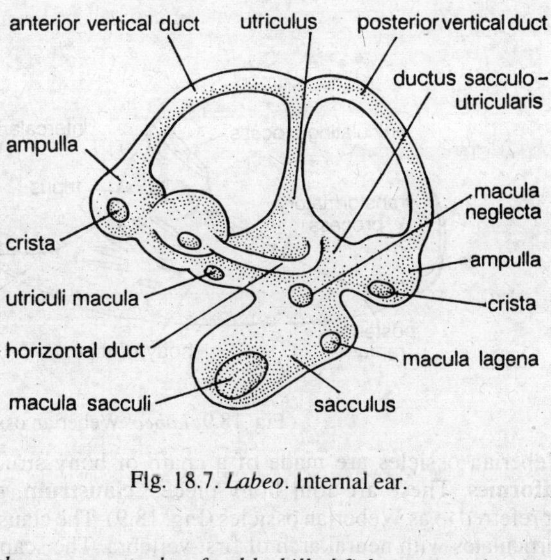

Fig. 18.7. *Labeo*. Internal ear.

Weberian Ossicles of *Labeo*

Procedure for dissection. *Take either a preserved or freshly-killed fish, remove the skin and muscles of the posterior region of the operculum. Trace out the air bladder and locate a triangular piece of bone attached to its anterior end. This triangular bone is* **tripus**, *a part of Weberian ossicles.*

Go on tracing it anteriorly and trace all the related ossicles. Take them out together and place in a watch glass containing water.

Study it and draw a well-labelled diagram of the same with the help of Fig. No. 18.8 and 18.9.

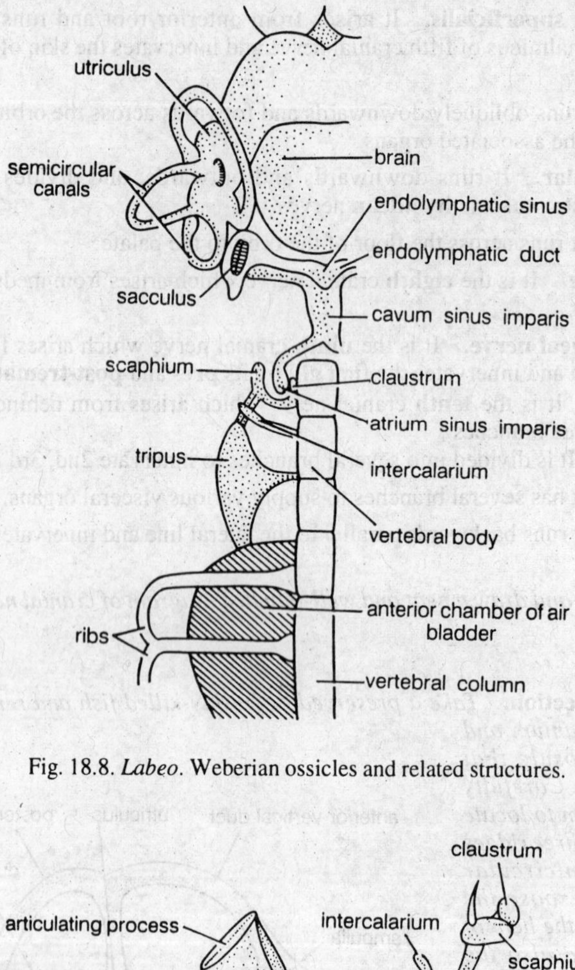

Fig. 18.8. *Labeo*. Weberian ossicles and related structures.

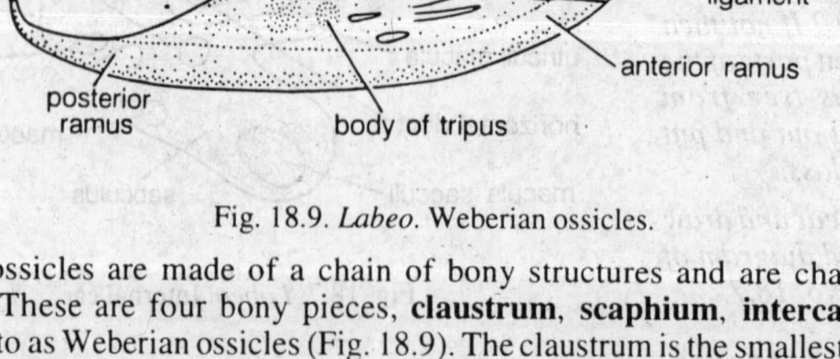

Fig. 18.9. *Labeo*. Weberian ossicles.

Weberian ossicles are made of a chain of bony structures and are characteristic of order **Cypriniformes**. These are four bony pieces, **claustrum**, **scaphium**, **intercalarium** and **tripus** together referred to as Weberian ossicles (Fig. 18.9). The claustrum is the smallest anterior most piece which articulates with neural arch of first vertebra. The scaphium is a broad and compressed bony piece which is followed by intercalarium and tripus. Tripus is the largest bony ossicle having three processes, the anterior process is connected to interossicular ligament, median process is attached to the junction of 2nd and 3rd vertebrae. The posterior process is curved and is connected with the anterior chamber wall of the air or swim bladder. Sound waves are said to travel to the internal ear through these ossicles.

WALLAGO ATTU

Cranial Nerves

Procedure for dissection. *Take either a freshly-killed or preserved fish and wash it thoroughly if the fish is preserved. Make a longitudinal incision in the integument at mid-dorsal region of the head from pectoral fin to snout. Likewise, make a transverse incision in the integument at the level of pectoral fin to the mid-ventral line of the lower jaw. Remove the integument so that the half dorsal, lateral and ventral parts of the head are fully exposed. Now, break off the cranium carefully to expose the brain. Then, trace the cranial nerves from their origin to their innervation. When your dissection is complete, display the nerves by placing black paper below them. Draw a neat and well-labelled diagram of the same in your practical record with the help of Fig. No. 18.10.*

There are 10 cranial nerves on one side, thus, there would be 10 pairs of cranial nerves, which arise from the brain and symmetrically disposed on the two sides of the head. The students are usually asked to expose V, VII, IX and X cranial nerves.

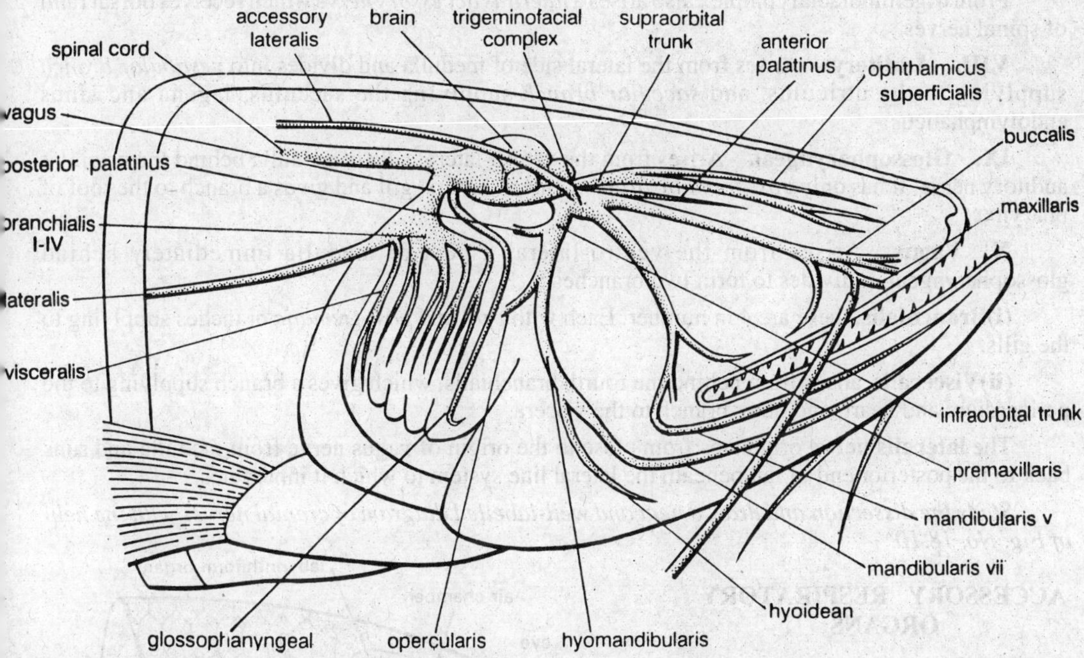

Fig. 18.10. *Wallago attu*. Cranial nerves.

The various cranial nerves are as follows:

I. Olfactory. Arises from olfactory lobe and innervates the olfactory folds.

II. Optic. Arises from the optic thalamus of diencephalon, enters the eye ball and supplies to retina.

III. Oculomotor. Arises from the ventral aspect of midbrain and supplies to eye muscles.

IV. Trochlear. Arises from the dorso-lateral aspect of the optic lobe and cerebellum, and innervates the eye muscle.

VI. Abducent. Arises from the ventral side of medulla and innervates to eye muscle.

V. Trigeminal and VII. Facial. Arise from the side of medulla, join immediately to form a **trigeminofacial complex**. The complex separates intracranially into three branches. These are :

1. Supraorbital trunk, which runs forward and forms *inner ophthalmicus superficialis* and *outer ophthalmicus superficialis*. These supply to the skin of snout.

2. Infraorbital trunk, which runs forward and outward, and divides into *maxillaris*, *buccalis* and *mandibularis*. Maxillaris supplies to maxillary barbel, buccalis supplies to the skin of snout and also to maxillary barbel, while mandibularis bifurcates to give rise *mandibularis externus* supplying to overlying skin of the anterior end of mandible, and *mandibularis internus* supplying to the mandibular barbel and anterior end of head.

The infraorbital trunk, before dividing into above 3 branches, also gives off following two branches :

(a) premaxillaris supplying to the upper jaw.

(b) palatinus anterior which runs on the roof of buccal cavity.

3. Hyomandibular trunk, which immediately gives out a stout *opercularis nerve* supplying to operculum and then divides into *hyoidean* and *mandibularis* branches. The hyoidean branch supplies to the branchiostegal membrane on under-side of the head. The mandibularis supplies to the lower lip and mandibular teeth. Hyomandibular trunk immediately after its origin from the complex also gives off a slender *palatinus posterior nerve* which also supplies to the roof of the buccal cavity.

From trigeminofacial complex also arises a *lateralis accessory nerve* which receives dorsal rami of spinal nerves.

VIII. Auditory. Arises from the lateral side of medulla and divides into *vestibular branch* supplying to the utriculus, and *saccular branch* supplying the sacculus, lagena and sinus endolymphaticus.

IX. Glossopharyngeal. Arises from the ventro-lateral side of medulla behind the origin of auditory nerve. It has only *post-trematic branch*, enters the first gill and gives a branch to the roof of pharynx.

X. Vagus. Arises from the ventro-lateral aspect of medulla immediately behind glossopharyngeal. It divides to form two branches :

(i)Branchialis, these are 4 in number. Each with *pre-* and *post-trematic* branches supplying to the gills.

(ii)Visceralis, arises from behind the fourth branchialis, which gives a branch supplying to the pericardium and heart, and other branch to the viscera.

The **lateralis nerve** originates from close to the origin of vagus nerve from medulla and runs back to the posterior end of tail beneath the lateral line system to which it innervates.

Study the dissection and draw a neat and well-labelled diagram of cranial nerves with the help of Fig. No. 18.10.

ACCESSORY RESPIRATORY ORGANS

Anabas

There are two air chambers, one on each side of the head. These are the extensions of the branchial cavities. They contain concentrically arranged wavy plates (**labyrinthiform organs**) which are outgrowths of the upper part of the first branchial arch. The plates are covered with vascular membrane. Each chamber communicates with the pharynx by the first gill-slit and with the gill cavity by an opening situated between the hyoid and the

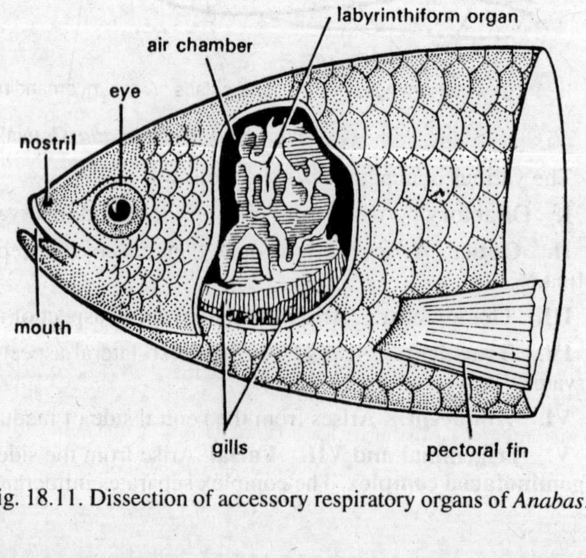

Fig. 18.11. Dissection of accessory respiratory organs of *Anabas*.

first branchial arch. Air is drawn in by the mouth into the air chamber by the first gill-slit and it passes out by the branchial aperture (Fig. 18.11).

Clarias

Clarias has air chambers which are outgrowths of the gill cavities and they enclose branched vascular processes (**dendriform** or **arborescent organs**); there are two such organs which are outgrowths of the upper parts of the second and fourth branchial arches (Fig. 18.12).

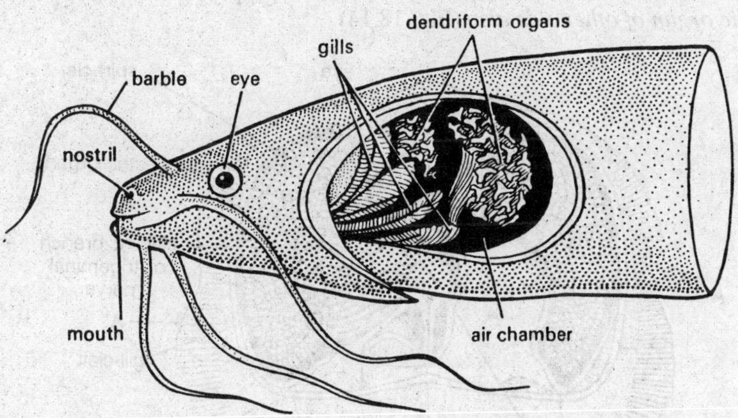

Fig. 18.12. Dissection of accessory respiratory organs of *Clarias.*

Heteropneustes* or *Saccobranchus

Heteropneustes or *Saccobranchus* has a pair of tubular **extrabranchial diverticulum** as outgrowths of the gill-chamber extending as far as the middle of the tail region. At the hind end of the extrabranchial diverticulum, there are folds forming a sort of air chamber. The air chamber communicates with the buccal cavity by a slit through which the air passes in and out (Fig. 18.13).

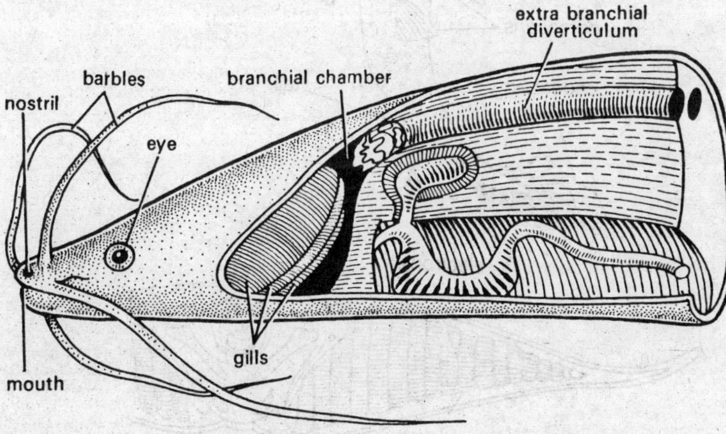

Fig. 18.13. Dissection of accessory respiratory organs of *Heteropneustes.*

ELECTRIC ORGANS

In a number of fishes **electric organs** are found which generate electric current. These organs are found both in the cartilaginous and bony fishes. Well known fishes having electric organs are **electric rays** like *Torpedo* and **skates** among cartilaginous fishes, and **electric eels, electric cat fishes** and **star-gazers** are among bony fishes.

Electric organs of electric ray (*Torpedo*) can be easily demonstrated by dissecting it in the laboratory.

Procedure for dissection. *Take either a preserved or a freshly-killed electric ray* (*Torpedo*) *and remove its skin from the middle of its head to the pectoral fin of one side on the dorsal surface. Carefully expose its chondrocranium and then the brain. At the same time, carefully remove the muscles and slowly expose a mammalian kidney-shaped structure* (*electric organ*) *between the brain and the margin of pectoral fin. Note that branches of cranial nerves VII, IX and X originating from electric lobe of medulla of brain and a branch of VI nerve supplying the electric organ. Similarly, trace the electric organ of other side also* (Fig. 18.14).

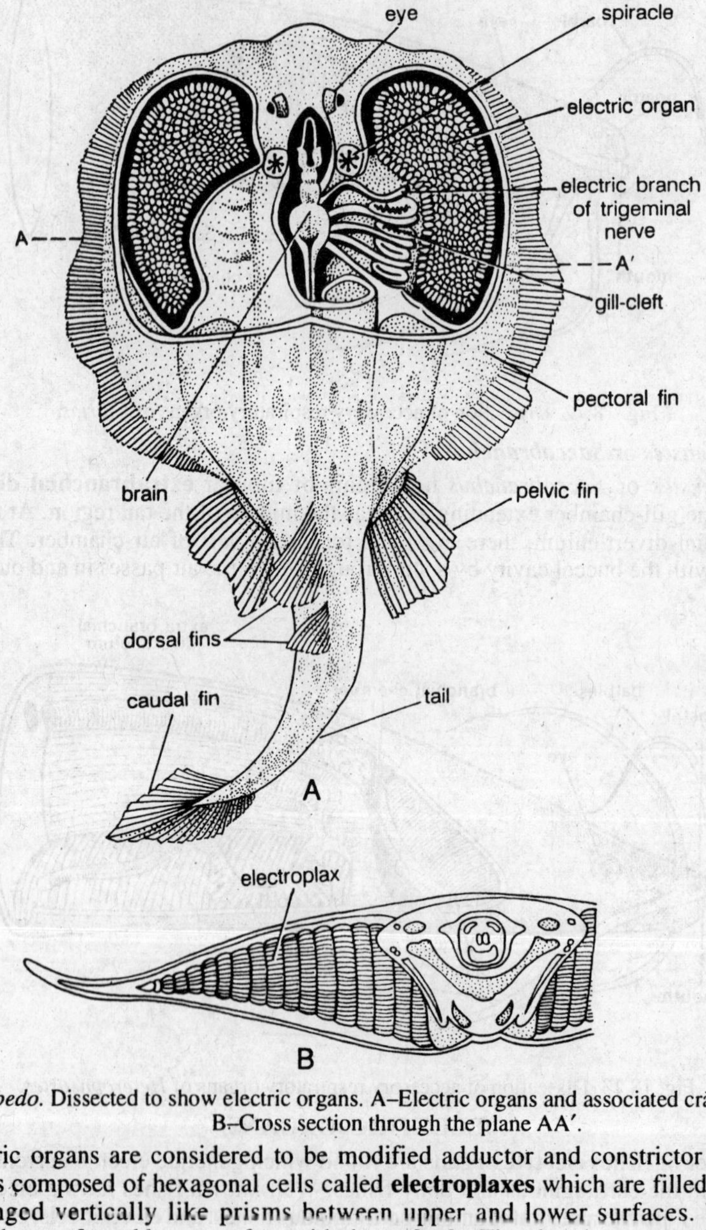

Fig. 18.14. *Torpedo.* Dissected to show electric organs. A–Electric organs and associated cranial nerves; B–Cross section through the plane AA′.

The electric organs are considered to be modified adductor and constrictor muscles. Each electric organ is composed of hexagonal cells called **electroplaxes** which are filled with jelly-like fluid and arranged vertically like prisms between upper and lower surfaces. Upper surface corresponds to the **anode** and lower surface with the **cathode**. The electric current of 50 to 60 Volts passes from anode to cathode end. Fishes use electric organs as offensive and defensive organ.

CHAPTER 19

MICROTOME AND MICROTOMY

Microtomy is the process in which **microtome** (Gr., *micro* = small, *tome* = to cut), a machine, is used for cutting the organs and tissues into very thin sections for histological studies. Apart from histological studies, histochemical, histopathological, embryological and several clinical studies also involve microtomy. **Cevalier** (1839) was the first to introduce the term microtome for the machines or instruments used in cutting sections. Sliding microtomes were developed in 1798, while rotary microtomes in 1883 and 1886. However, more efficient Spencer rotary microtome was made available in 1910. Now-a-days, different types of microtomes are available but broadly all these can be categorised into three types :

1. Sliding or celloidin microtomes;
2. Freezing microtomes;
3. Paraffin microtomes.

Most of the biological laboratories of undergraduate and postgraduate departments in our universities and colleges are provided with **paraffin microtomes.** Hence, at present we shall keep our description limited to this type of microtomes and their use in microtomy.

Paraffin Microtome

For the use of this microtome, the target organs or tissues are fixed, dehydrated, cleaned and embedded in paraffin. The paraffin blocks are then made of the embedded material. Such blocks are then mounted on a block holder which is fixed to the microtome. The block holder moves vertically up and down on a stationary knife. In the process of section cutting, the block moves forward automatically through a fixed distance, with the result that the knife cuts the material into thin sections of uniform thickness. The forward movement of block is done either by the rocking or rotary mechanism. This microtome gives us long paraffin ribbons with many sections, enabling us to study the serial sections. The sections obtained by this microtome are fine, thin and uniform and useful for various histological studies.

KNOW YOUR MICROTOME

Paraffin microtome is of two types :

1. Rocking type; 2. Rotary type.

Rocking Microtome (Fig. 19.1)

The parts of this microtome and their working are described below.

(*i*) **Heavy base.** It is long and very heavy over which all parts of the microtome are placed.

(*ii*) **Advance shaft.** It is a long wide solid cylindrical rod, almost of the same length as that of the base. It is attached by a circular disc and moves up, down and forward when the small handle is moved front and back. Block holder is attached to the surface of the front end of the advance shaft with the help of a screw.

(*iii*) **Grooved circular disc.** It is present at the posterior part of heavy base. As referred to, it is connected to the advance shaft by a spring. This disc has an adjustment lever which provides us the sections of desired thickness. A long screw is present to connect the rising shaft. There is a small handle near the circular disc.

(*iv*) **Razor holder.** There is a wide groove in the front part of heavy base, this groove is called razor holder and meant for accommodating the section cutting razor. There are screws on this holder to adjust the angle of the razor.

Working. First fix the paraffin block containing embedded material on the block holder and screw it with the advance shaft. Adjust the circular disc for getting sections of desired thickness, preferably 6 micron or 8 micron. Place and screw the razor in razor holder. Now as we move the microtome handle front and back, the advance shaft with the attached block holder, moves up and down on the razor in a vertical plane, cutting the tissue in the block into sections. During the up and down movement for cutting the sections, the advance shaft also moves forward, thus, bringing the issue in the block ahead on the razor.

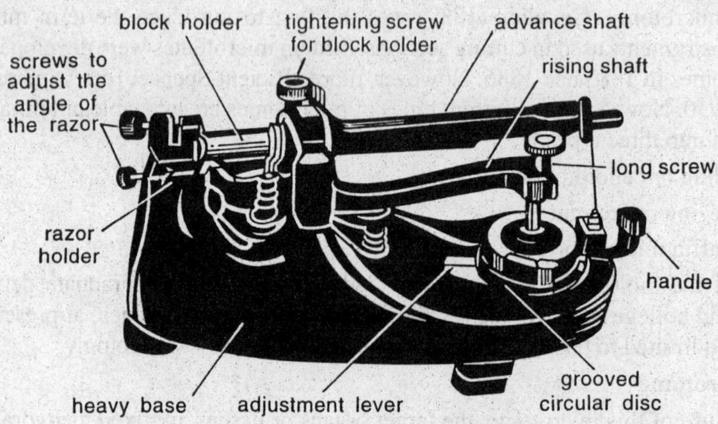

Fig. 19.1. A rocking microtome.

Rotary Microtome (Fig. 19.2 and 19.3)

The rotary microtomes are available in many models now-a-days. The basic principle is the same in these cases as well. This type of microtome has a heavy base over which all parts are placed. There is a circular wheel which has to be rotated. A device for setting the microtome to give sections of desired thickness is provided. There is a circular grooved disc at the rear end which moves anti-clockwise by the exact distance as the desired thickness of sections for which the microtome has been set. This is the advance wheel. There is a central axle to move the front portion of the microtome forwards by exactly the same distance simultaneously as the front part moves up and down. To this front part is fixed the block holder carrying the paraffin block containing embedded tissues. The razor in this case is mounted on a pair of hands with a groove and adjusting screws as in the case of a rocking microtome. This is the razor holding device which is movable forwards and backwards along two grooves on the heavy base of the microtome, so that the razor can also be brought nearer to the block if needed. Once the desired position is reached, the same can be fixed with the help of a locking lever provided.

The microtome knife is to be used most carefully as good sections can be cut only when the razor has a uniformly sharp edge on both sides and is absolutely without any nicks. After the use of knife, remove it from the microtome, clean it carefully and then keep in knife box. The knife should be well-greased, if it is not to be used for a fairly long time. When knife is to be used, clean the grease, hone and then use. Honing is done for sharpening the knife.

Cutting sections. Check all parts of the microtome and knife. Then fix the knife with the help of razor holding screws. Fix the paraffin block containing embedded tissues on the block holder and mount it on the microtome. While fixing the knife be careful, it should be neither too straight nor too inclined. Bring the razor/knife nearer to the block and see that the lower side of the latter is exactly parallel to the edge of the former. Set the microtome at whatever thickness the sections are to be cut. Then start rotating the wheel. As soon as the paraffin block makes contact with the razor, sections of paraffin will be cut. Go on rotating the wheel, a ribbon of sections is obtained. As soon as ribbon of 2.5 cm length is obtained, stop rotating the wheel and put the ribbon on a clean thinly albumen smeared slide with the help of needle and brush. Thus, go on obtaining ribbons and preserve them over slides for further processing.

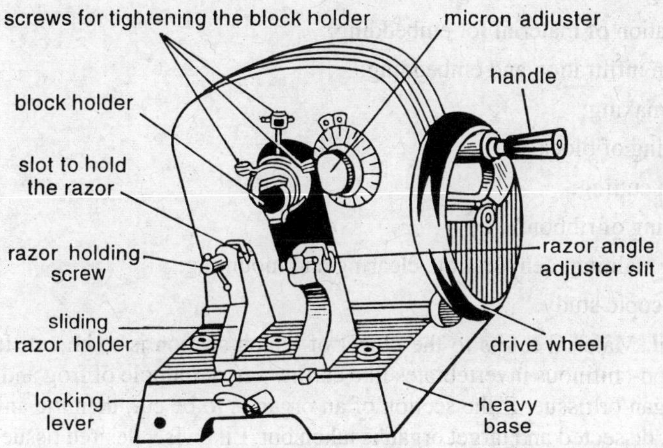

Fig. 19.2. A rotary microtome Erma type.

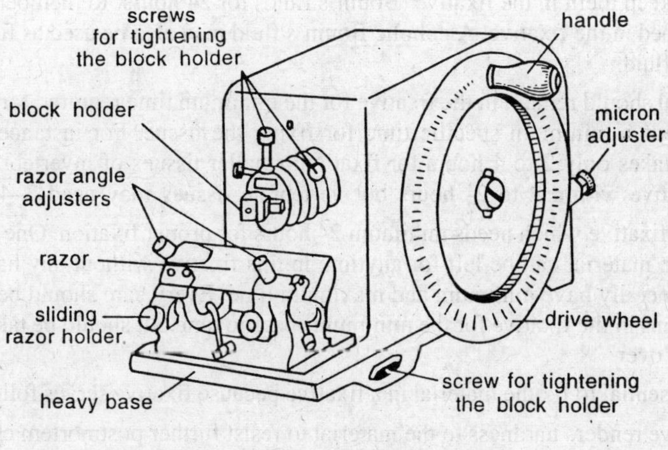

Fig. 19.3. A rotary microtome Spencer type.

MICROTOMY

As referred to, microtomy is the process of cutting thin sections of desired tissues, organs or even whole animals for histological and related studies. Microtomy, however, involves a detailed procedure that must be followed step by step with extraordinary care. As will be evident from the procedure described here under that it is a lengthy process and one step followed carelessly or omitted can spoil the result, and this would be known only when the procedure has been completed and the preparation spoilt. Therefore, whole process and each step should be followed with great care. However, microtomy includes following steps :

1. Material;
2. Fixation of material;
3. Washing of fixative from material;
4. Dehydration of material;
5. Clearing or dealcoholizing of material;
6. Preparation of material for embedding;
7. Paraffin infiltration and embedding;
8. Block making;
9. Trimming of block;
10. Section cutting;
11. Spreading of ribbon;
12. Double staining, dehydration, clearing and mounting;
13. Microscopic study.

1. Material. Material refers to the object of which section is to be cut. It may be a whole animal (if small, non-chitinous invertebrates like earthworm or tadpole of frog and other embryos of vertebrates), an organ or tissue. If the section of an organ is to be cut, then the animal concerned is first chloroformed, dissected and target organ is taken out. Likewise, desired tissue may be taken out for sectioning.

2. Fixation of material. Take cleaned 100 ml wide-mouthed glass-stoppered bottles. Keep about 50 ml of aqueous Bouin's fluid in each bottle and now transfer the materials into bottles (one bottle should be used for keeping one type of material preferably one). If the material is large, say it is intestine, liver, kidney, etc., of frog or rat then after half an hour cut the materials into two-three pieces and again keep them in the fixative (Bouin's fluid) for 24 hours. Remember that the material must be fully dipped in the fixative. Alcoholic Bouin's fluid may also be used as fixative in place of aqueous Bouin's fluid.

The material should remain in the fixative for the minimum time required for specific fixative. Each fixative needs a minimum specific time for fixing the tissue. For instance, Carnoy's fluid, another fixative, takes only 2 to 4 hours for fixation; smaller tissues of invertebrates may be well fixed in this fixative within 2 to $2\frac{1}{2}$ hours but vertebrate tissues may need 3–4 hours. Aqueous Bouin's fluid is a fixative which needs minimum 24 hours for proper fixation. One advantage of this fixative is that the material can be left for anytime in this fixative without any harm. It is because other fixatives generally have minimum and maximum time. Every care should be taken to see that the material remains in the fixative for the minimum specific time but should be taken out before the maximum time is over.

It is very essential to fix the material in a fixative because fixative serves following functions :

1. Fixative renders hardness to the material to resist further postmortem changes.
2. Fixative coagulates and render the elements of the material insoluble so that cellular substances remain intact and may not be washed away.

3. Fixative alters the refractive indices of materials and make them optically differentiated under the microscope.

3. Washing of fixative from material. After the fixation is completed (at least 24 hours in case of aqueous Bouin's fluid), take out the material from the bottle containing Bouin's fluid and keep it in a beaker containing distilled water. (Throw out the used fixative in sink and never collect it for further use as the same cannot be used again). Change the distilled water several times till yellow colour comes out from the material. In other words, keep on washing the material in distilled water as long as yellow colour comes out. In place of distilled water, running tap water may also be used for removing the fixative. Caution should be taken that only those materials fixed in aqueous fixative should be washed in distilled water or tap water. Materials fixed in 70% alcohol should be washed with 70% alcohol and those fixed in Carnoy's fluid should be washed only in absolute alcohol.

4. Dehydration of material. For dehydrating the material take a staining tube and transfer the material in it. Before transferring the material in staining tube take 30% alcohol in it and then transfer the material. Thus, starting with 30% alcohol go up to absolute alcohol through the stages of 50%, 70% and 90% alcohol. At least 25–30 minutes should be allowed in each step for vertebrate materials. If it is not possible to complete the dehydration on the same day, leave the material only in 70% alcohol. Never leave the material overnight in other grades of alcohol. At least two changes of 30 minutes duration must be given in 90% and absolute alcohol. Dehydration through grades of alcohol gradually removes water from the material to prevent putrefaction.

5. Clearing or dealcoholization of material. Clearing or dealcoholizing agents are cedar wood oil or xylene or methyl benzoate. After the dehydration has been completed, the material is treated with clearing agent which makes the material almost completely transparent. Now-a-days, methyl benzoate is preferred over other clearing agents because it needs about 24 hours to get the desired result . Another advantage of methyl benzoate is that the material can be left in it indefinitely without any ill effect. Contrary to this cedar wood oil needs longer time and it is very costly. Likewise, xylene makes the tissue brittle. Hence, methyl benzoate is advised to be used as clearing agent. Thus, for clearing the material, pour out absolute alcohol in a petridish and throw it, add methyl benzoate in the tube containing material. Leave it for 24 hours.

6. Preparation of the material for embedding. After clearing the material, remove methyl benzoate from the tube and pour benzene in it. This step is for removal of methyl benzoate from the material. Keep the material for about an hour or an hour and half an hour; this time is sufficient for complete removal of methyl benzoate from the materials belonging to vertebrate tissues. But, it is advised to give at least 3 changes of benzene, each of about 20–25 minutes to ensure complete removal of methyl benzoate. Now the material is ready for infiltration and embedding; the step to be followed. Therefore, after making all necessary preparations of the step to be followed, then start the step of removing methyl benzoate by using benzene. Because, if material remains for more than $1\frac{1}{2}$ hours in benzene it becomes brittle and useless for microtomy.

7. Paraffin infiltration and embedding. Depending on the melting point of the wax to be used, set your oven. Paraffin wax is generally used in our laboratories, its melting point ranges from 60 – 62°C. Hence, set your oven at 62°C. Take a beaker of 100 – 200 ml, scrap the wax with the help of a scalpel and put these scrappings in beaker. Now keep the beaker in the oven till the wax melts completely. Do not melt the wax on a flame. After the wax has melted completely, filter it if any dirt appears, keep this molten wax in oven for 24 hours and then this molten clean wax can be used for infiltrating the material. For infiltrating purpose, take three small porcelain cups and pour molten wax in these cups and keep them in oven in a serial. Now take out the material from benzene and place it in the first porcelain cup, after 45 minutes transfer the material in the next cup and then again after 45 minutes in the third cup. Leave the material in the third cup for 45 minutes and in the mean time all preparations for block making should be started. (**Note.** There is no need of keeping the material in a mixture of wax and benzene as is done when xylene is used for removing the clearing agent).

8. Block making. For making blocks of paraffin wax and embedding the material into it, either metal 'L'-shaped angles or paper boats or cavity blocks are taken. Preference should be given to 'L' angles. For making blocks with 'L' angles, take a glass plate and apply glycerine over it, also apply glycerine at the inner surface of the 2 'L' angles. Now arrange the 'L' angles on the glass plate in such a way that a rectangle is formed. Thus, in between the arms of two 'L' angles a rectangular cavity is formed.

Now, when infiltrating time in third cup is over, pour molten clean wax (not from the cups used for infiltration) into the rectangular cavity in 'L' angles. Fill it about 4/5th of the total height of the cavity. Then take out the third infiltrating cup and take out the material from it with the help of warm forceps and place the material gently into the rectangular cavity of 'L' angles which has just been filled with wax. Be quick at this stage so that the upper layer of wax may not solidify, otherwise proper embedding of material will not take place and block will not be formed properly. Care should also be taken that the material should be completely dipped in the wax. While placing the material in the wax, it is also important to orient the material in a proper manner so that sections can later be cut in the desired plane. Then allow the wax to solidify. After about 30 minutes, when the wax has solidified completely, lift up the glass plate along with 'L' angles and place them very gently in an enamel tray containing water. Leave the 'L' angles as such for about another 20–25 minutes. After this time solidification will be completed and most likely wax block will come up on the water surface. It is advised to leave the block in water for further few minutes. Every precaution should be taken that during block making no air bubbles should come, and if any air bubble arise remove it by using hot needle or spatula.

9. Trimming of block. Trimming of wax block is necessary to remove excess of wax. For it, scrap the excess wax with the help of a scalpel. Scrapping should be done in a way that the block becomes rectangular in shape and the material is slightly visible. At least 1–2 mm wax should be left in front and on all sides of the material, and at least 6–8 mm wax should be left behind the material to be cut. After proper trimming the block becomes more or less uniformly translucent, if every step was followed carefully keeping in view all precautions.

10. Section cutting. The blocks are sectioned either by rocking or rotary microtomes. Now apply nearly half an inch of wax on block holder. Fix up the block by melting the wax in its centre on the block holder. Once the block is set on block holder, adjust the distance of which thickness sectioning is to be done and fix up the block holder on the microtome as described while describing the microtomes. Then start cutting sections. Before starting section cutting, take few slides and apply very thin layer of Mayer's egg albumen over them. Now put ribbons on the albumen smeared slides in the manner already described.

11. Spreading of ribbons. Warm the hot plate nearly up to 40 – 45°C. Keep slides containing ribbons on the hot plate and pour few drops of water below ribbons. As the water is heated, the wax around sections start melting and it causes some expansions in the ribbons. Thus, when all sections and ribbons become flattened, drain off water from the slide and allow the slide to cool at room temperature overnight for drying the ribbons.

12. Staining, dehydration, clearing and mounting. Double staining technique is used; the stains used are generally Haematoxylin (nuclear stain) and Eosin (cytoplasmic stain). Before starting this step, dewaxing of the sections of the slide is done. Xylene is used for this purpose. Pass individual slide through two changes of xylene (keep 15 minutes in the first and 5 minutes in the second), so that the wax is completely dissolved and removed, and only the section of material remains affixed to the slide. Now pass the slide in **descending series** of alcohols : absolute → 90% → 70% → 50% → 30% → distilled water. Time recommended for keeping the slides in these series are 8–10 minutes in absolute, 5–8 minutes in 90%, and 5 minutes in remaining grades of alcohol. Give two changes in distilled water, each of 5 minutes.

Then stain the sections in aqueous Haematoxylin for 2–5 minutes. After staining dip the slide in distilled water and in tap water. The sections turn blue because of alkalinity of water. Examine the slide under microscope. If stain appears to be dark, then dip the slide in acid water and immediately dip it in distilled water. By doing this excess of stain from the sections are removed. Then dehydrate the sections through **ascending series** of alcohols → 30% → 50% → 70% → 90% (give 5 minutes, 10 minutes, 10 minutes and 15 minutes respectively in these grades). Then stain the sections in alcoholic eosin by immersing the slide in a tube containing eosin for about a minute. Wash excess eosin again in 90% alcohol and then transfer the slide in absolute alcohol for 10 minutes. Give two changes of absolute alcohol of 10 minutes each. Then transfer the slide in a tube containing absolute alcohol and xylene (1 : 1) for 5 minutes and finally transfer the slide in xylene for 15 minutes. After it, place few drops of either Canada balsam or D.P.X. on the slide with the help of glass rod. Place a rectangular coverslip on the slide in such a way that air bubbles do not get trapped below it. Keep such slides for drying and then store in slide box.

Microscopic study. Thus, prepared slides are ready to be examined under microscope. Microphotographs of these slides can be taken if necessary.

(**Note :** The process of microtomy described above is general for histological studies but for specialized studies some modifications in the steps are made. These modifications you will come across when you will proceed for specialized studies).

CHAPTER 20

BIOSTATISTICS

Statistics has been defined in various ways by different authors, but the most feasible one appears to that as proposed by **Lovitt**. According to him, *"statistics deals with the collection, classification, tabulation, analysis and interpretation of numerical data or facts, from any field of investigation, as the basis of explanation, description and comparison of phenomena."* In fact, various aspects of statistics provide methods for making reasonable and wise decisions in the face of uncertainity. When the aspects of statistics are applied to the study of organisms or living systems, the study is called **biostatistics** or **biometry** or **biometrics.**

In recent years, biostatistics has gained much importance, as in all biological experiments, observations are systematically arranged and the results are analysed by using the aspects of statistics and interpreted in such a way that reasonable conclusions are drawn at the end. However, some aspects which are normally considered and used during statistical analysis of data, consist of frequency tables, measures of central tendency, measures of dispersion, correlation and regression, etc., and also some tests of significance. The other aspect of biostatistics relates to the presentation of data with the help of graphs. These aspects are, however, illustrated in this chapter.

PRESENTATION OF DATA IN THE FORM OF FREQUENCY TABLES

In experiments, when we record data, sometimes the data become so large that it becomes difficult to present the data in original form. Therefore, methods have been evolved for the presentation of data in the form of frequency tables.

In statistical classification, data is divided into different groups or classes or facts, called **variables** or **variates** (represented by the symbol **y**). In fact, variables represent the quantitative characters which differ in value from one organism to another, from one sample to another. The

Table 1. Data on coat colour of 20 mice involving albino and black colour.

Mice no.	Coat colour	Mice no.	Coat colour
1	Albino	11	Albino
2	Albino	12	Albino
3	Black	13	Albino
4	Albino	14	Black
5	Black	15	Albino
6	Albino	16	Albino
7	Albino	17	Black
8	Albino	18	Albino
9	Black	19	Black
10	Albino	20	Black

variables may be **discrete** or **discontinuous** and **continuous**. A set of data involving units which are not further divided and always represented as integers, constitute **discrete** or **discontinuous variables**, while those where the measurements are made over a continuous scale or any value within a certain range, constitute **continuous variables**. The number of times that a variable occurs within a group or in other words, its rate of occurrence is called **frequency** (represented by the symbol **f**). The arrangement of data with a set of values of a variable into classes with associated frequencies is known as **frequency distribution**, and such tables are referred to as **frequency tables**.

1. Discrete variable. In the data for coat colour in 20 mice presented in Table 1, the results can be presented in the form of a frequency table by counting the number of mice of each type as done in Table 2.

Table 2. Frequency distribution of coat colour from data in Table 1.

Coat colour	Frequency
Albino	13
Black	7

Table 3. A frequency distribution table for number of children in 100 families of a town.

No. of children Variable (y)	Frequency (f)	f × y
0	2	0
1	4	4
2	25	50
3	24	72
4	27	108
5	11	55
6	5	30
7	2	14
Total	$\Sigma f = 100$	$\Sigma f y = 333$

In all the above cases, the frequencies are shown for discrete variable. No class intervals are needed.

2. Continuous variable. In continuous variable, **class intervals** are to be prepared. When one class ends, the next class begins, so that the continuity of variable is maintained, *e.g.*, 0 to 10, 11 to 20, 21 to 30, 31 to 40 and so on or 0 to 5, 6 to 10, 11 to 15, 16 to 20, 21 to 25 and so on. In these examples the class interval is 10 and 5 respectively. **Tallies** denote frequencies; since the values are so many on a continuous scale that the frequency for each value of the variable will give a table which may be too big to use. For example, if we weigh 50 mice from laboratory culture, a frequency table with class interval 10 (Table 5) can be prepared from the data given in Table 4.

Table 4. Data on weight (in gm) of 50 laboratory mice.

Mice no.	*Weight in gm*	*Mice no.*	*Weight in gm*	*Mice no.*	*Weight in gm*	*Mice no.*	*Weight in gm*	*Mice no.*	*Weight in gm*
1	92	11	102	21	114	31	93	41	102
2	105	12	116	22	96	32	109	42	109
3	112	13	97	23	119	33	91	43	94
4	107	14	103	24	103	34	108	44	95
5	109	15	94	25	120	35	115	45	111
6	95	16	107	26	102	36	110	46	103
7	108	17	98	27	101	37	105	47	96
8	113	18	106	28	99	38	94	48	102
9	118	19	109	29	112	39	104	49	102
10	117	20	100	30	108	40	109	50	98

Table 5. Frequency distribution table prepared from the data on weight of mice given in Table 4, using a class interval of 10.

Class interval (Weight in gm) From–To	**Tally**	**Frequency**	**Relative frequency**	**Percentage frequency**
91–100	卌 卌 卌	15	0.30	30 per cent
101–110	卌 卌 卌 卌 ||||	24	0.48	48 per cent
111–120	卌 卌 |	11	0.22	22 per cent

A better spread frequency distribution table may be prepared from the same data by using class interval of 5, as has been done in Table 6. (**Note.** It is always better to have more than five class intervals).

Table 6. Frequency distribution table prepared from the data on weight of mice given in Table 4, using a class interval of 5.

Class interval (Weight in gm) From–To	**Tally**	**Frequency**	**Relative frequency**	**Percentage frequency**
91 – 95	卌 |||	8	0.16	16 per cent
96 – 100	卌 ||	7	0.14	14 per cent
101 – 105	卌 卌 ||	12	0.24	24 per cent
106 – 110	卌 卌 ||	12	0.24	24 per cent
111 – 115	卌 |	6	0.12	12 per cent
116 – 120	卌	5	0.10	10 per cent

Table 5 and 6 depict that how class intervals are prepared and make use of the tally of the values that fall in each group. In the process of making tally, each group of five is represented by four vertical lines crossed by an oblique fifth line to make the counting easy and convenient. From the frequency distribution, **relative frequency distribution** can be obtained by dividing the frequency of each class by the total number of observations (total number of observations in above cases are 50). Likewise, on multiplying relative frequency by 100, **percentage frequency** can be obtained in each case.

It is also useful to prepare **cumulative frequency distribution** to obtain informations regarding observations that are "less than" or "not more than" a specific value. Table 7 represents such a table based on the data given in Table 4.

Table 7. A cumulative frequency distribution table from the data given in Table 4.

Class	Cumulative frequency of mice	Relative cumulative frequency	Percentage cumulative frequency
95 or less	8	0.16	16 per cent
100 or less	15	0.30	30 per cent
105 or less	27	0.54	54 per cent
110 or less	39	0.78	78 per cent
115 or less	45	0.90	90 per cent
120 or less	50	1.00	100 per cent

PRESENTATION OF DATA IN THE FORM OF GRAPHS

It is also useful to represent the data with the help of graphs, as they are more easily understandable. The frequency distribution tables can also be represented in the from of graphs, which are meant to depict the essential characteristics of the frequency distribution in a comprehensive manner. However, these may be in the following forms :

1. Pie charts. These are most often used for percentage distributions, where the number of components is usually kept less than six. Fig. 20.1 represents a pie chart prepared from the data given in Table 6, where the whole circle represents 100% and the sum of angles at the centre is 360°. Therefore, 360° is equated with 100% and the angles for different per cent frequencies can be calculated and depicted in the pie chart.

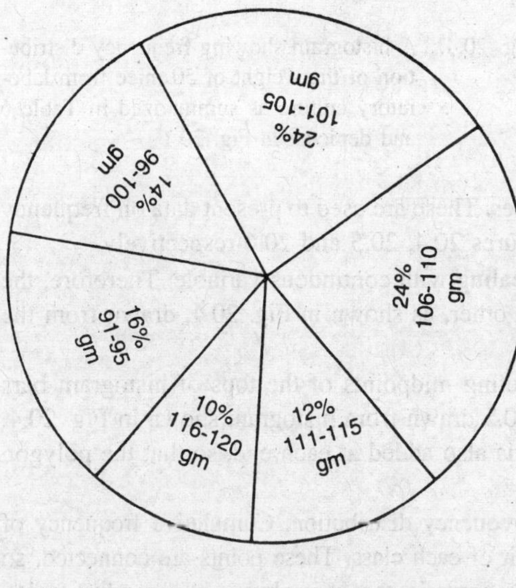

Fig. 20.1. A pie chart drawn from data in Table 6 depicting frequency distribution of weight of 50 mice.

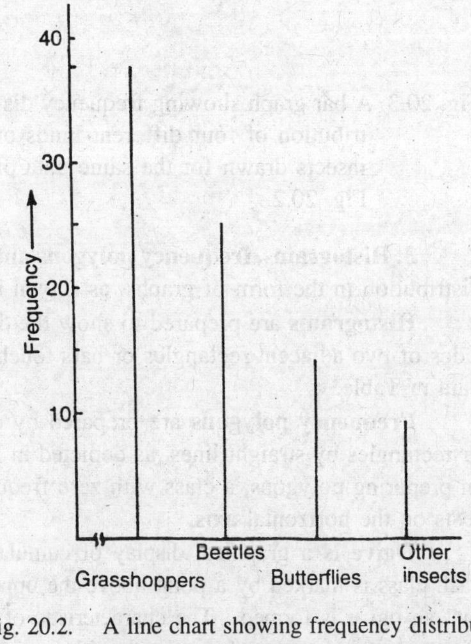

Fig. 20.2. A line chart showing frequency distribution of four different kinds of insects trapped from a field.

2. Line charts and bar graphs. These are prepared to depict discrete variables in which lines or bars are proportional to the frequencies. Figures 20.2 and 20.3 show line charts and bar graphs for the hypothetical data given below :

During an experimental survey, following different kinds of insects were trapped in a specific period from a field ; grasshoppers 37, beetles 25, butterflies 14 and other insects 9.

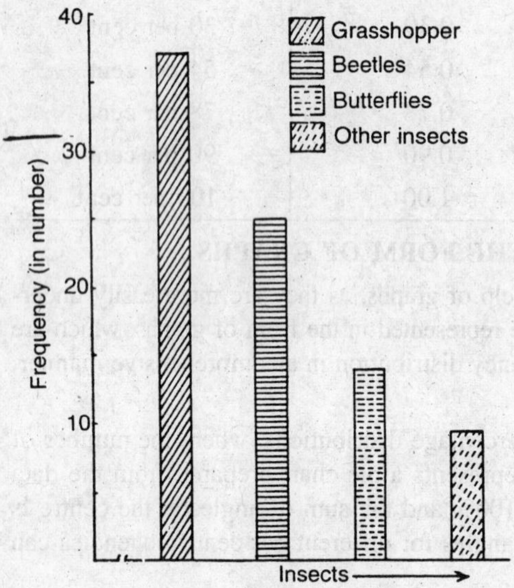

Fig. 20.3. A bar graph showing frequency distribution of four different kinds of insects drawn for the same data of Fig. 20.2

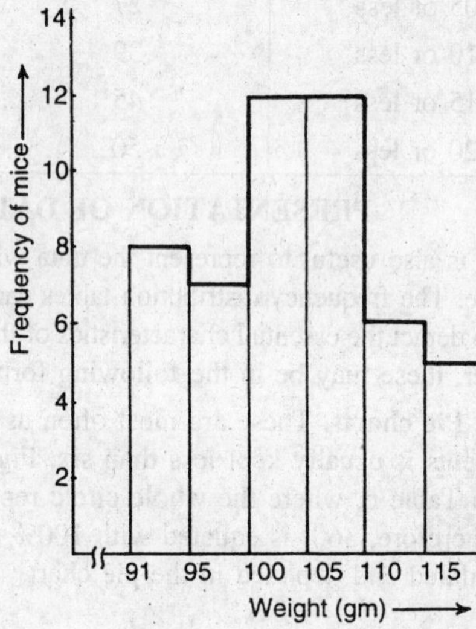

Fig. 20.4. A histogram showing frequency distribution of the weight of 50 mice from laboratory culture as summarized in Table 6 and depicted in Fig. 20.1

3. Histograms, frequency polygons and ogives. These are used to present data on frequency distribution in the form of graphs, as shown in Figures 20.4, 20.5 and 20.6 respectively.

Histograms are prepared to show the data dealing with continuous variable. Therefore, the sides of two adjacent rectangles or bars touch each other, as shown in Fig. 20.4, drawn from the data in Table. 6

Frequency polygons are prepared by connecting midpoints of the tops of histogram bars or rectangles by straight lines, as depicted in Fig. 20.5 drawn from histogram shown in Fig. 20.4. In preparing polygons, a class with zero frequency is also added at each end, so that the polygon rests on the horizontal axis.

Ogive is a graphical display of cumulative frequency distribution. Cumulative frequency of each class is marked by a point above the upper limit of each class. These points are connected, so that an ogive is formed . The characteristic of such a curve is, that it is always in ascending order. Figure 20.6 is an ogive from the data shown in Table 7.

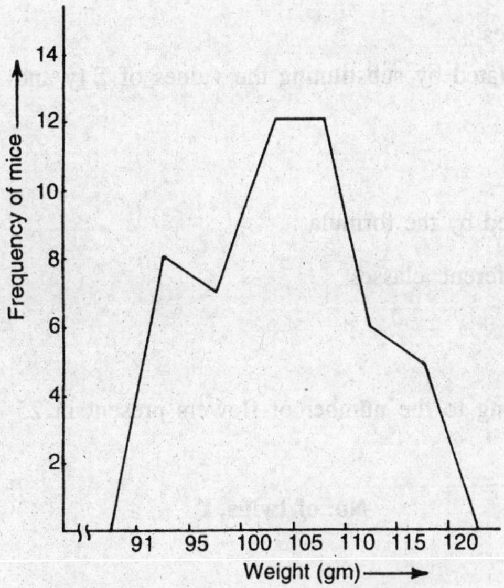

Fig. 20.5. A frequency polygon drawn from data summarized in Table 6 and depicted in Figures 20.1 and 20.4

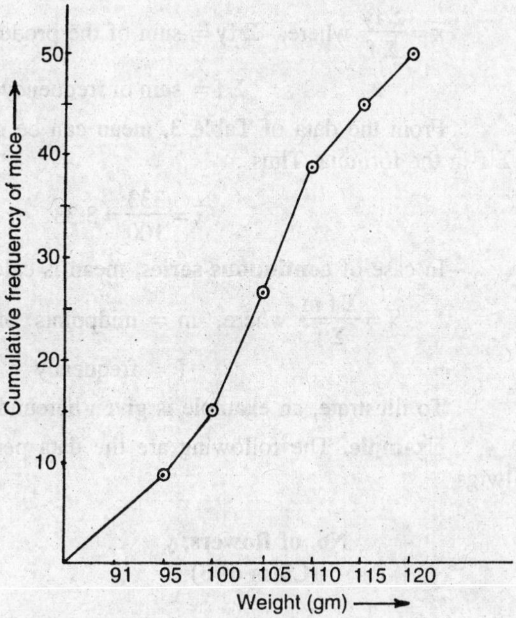

Fig. 20.6. An ogive of the cumulative frequency distribution of weights in gm as given in Table 7 and depicted in Figures 20.1, 20.4 and 20.5

MEASURES OF CENTRAL TENDENCIES

It includes mean, mode and median.

1. Mean. Arithmetic mean, usually referred to as mean, is the average of all observation values. When observations are denoted by x, values showing $x_1, x_2, x_3, \ldots x_n$; then the mean $\overline{x}$ (pronounced as x bar) for a number of observations is calculated by summing up the observations and dividing the sum by total number of observations (n). Thus :

$$\overline{x} = \frac{x_1 + x_2 + x_3 + \ldots + x_n}{n}$$

For convenience, all these summed up values ($x_1 + x_2 + x_3 + \ldots . x_n$) can be represented as Σ **x** (= summation of all x values), where Σ (pronounced as sigma) stands for summation. Then, the formula for calculating mean is written as :

$$\overline{x} = \frac{\Sigma x}{n}$$

For example, if the following data represent the number of flowers per plant 5, 7, 6, 8 4, 8, 7, 6, 5 and 4, then mean may be calculated by summing up all these values and dividing that summed up value by 10 (as 10 represent n ; number of plants taken into consideration). Thus :

$$\overline{x} = \frac{\Sigma x}{n}, \Sigma x = (5 + 7 + 6 + \ldots . + 4) = 60; n = 10$$

$$\overline{x} = \frac{60}{10} = 6$$

In case of **discrete series**, mean is calculated by the formula :

$\bar{x} = \frac{\Sigma fy}{\Sigma f}$ where, Σfy = sum of the products of frequency and variable (y).

Σf = sum of frequency values

From the data of Table 3, mean can be calculated by substituting the values of Σfy and Σf in the formula. Thus :

$$\bar{x} = \frac{333}{100} = 3.33$$

In case of **continuous series**, mean is calculated by the formula :

$\bar{x} = \frac{\Sigma fm}{\Sigma f}$ where, m = midpoints of different classes

f = frequency

To illustrate, an example is given hereunder :

Example. The following are the data pertaining to the number of flowers present in 25 twigs.

No. of flowers, y (Class of 5)	**No. of twigs, f**
5–9	3
10–14	5
15–19	6
20–24	4
25–29	3
30–34	4

From the above data, the midpoint for each class is determined and arranged in the following form :

No. of flowers *(x)*	**Mid point within class** *(m)*	**No. of twigs** *(f)*	$f \times m$
5–9	7	3	21
10–14	12	5	60
15–19	17	6	102
20–24	22	4	88
25–29	27	3	81
30–34	32	4	128
		$\Sigma f = 25$	$\Sigma fm = 480$

Now, substituting the values of Σfm and Σf in the formula,

$$\bar{x} = \frac{\Sigma fm}{\Sigma f} = \frac{480}{25} = 19.2$$

2. Median. If the items of the data are arranged in ascending or descending order, then the value of the middle term is called the **median** of data. Thus, median is the value of the variable which divides the data into two equal parts, one part consisting of all values greater and the other all values lesser than the median.

For calculating median, the raw data are arranged in ascending and descending order and the middle item is computed from the following formula :

Median $= \frac{n+1}{2}$, where n is the total number of items or observations. Here, $\frac{n+1}{2}$, will give rise the item or observation number and the numerical values of this item will denote the median.

To illustrate, if median number of flowers are to be obtained from following data :

Plant no.	1	2	3	4	5	6	7	
No. of flowers	20	17	25	18	23	21	16	(n = 7)

Then, arrange these data in **ascending** and **descending** orders as under :

Item no. or Observation no.	**No. of flowers (Ascending order)**	**No. of flowers (Descending order)**
1	16	25
2	17	23
3	18	21
4	20	20
5	21	18
6	23	17
7	25	16

From formula $\frac{n+1}{2}$, the median item comes to be $\frac{7+1}{2} = \frac{8}{2} = 4$. Now the fourth item has a value of 20 both in ascending and descending orders. Hence, the median number of flowers will be 20. The answer remains the same whether the data are arranged in ascending or in descending order. This was an example showing an **odd** number of items or observations.

When the observation/item numbers are **even**, then the value of $\frac{n+1}{2}$ will be a fraction and the average of two central values may be taken as median.

To illustrate, if median number of flowers are to be obtained from following data :

Plant no.	1	2	3	4	5	6	7	8	
No. of flowers	23	17	24	20	18	16	25	21	(n = 8)

Then, after arranging these data in **ascending** and **descending** orders, we get

Item no. or Obervation no.	**No. of flowers (Ascending order)**	**No. of flowers (Descending order)**
1	16	25
2	17	24
3	18	23
4	20	21
5	21	20
6	23	18
7	24	17
8	25	16

Using the formula $\frac{n+1}{2}$, the median item / observation comes to be $\frac{8+1}{2}=\frac{9}{2}=4.5$. Now, 4.5 is the average of items 4 and 5. Hence, median item will be the average of 4th and 5th items. The values against these items are 20 and 21. The average of these values will be the median, which comes to be $\frac{20+21}{2}=\frac{41}{2}=20.5$.

3. Mode. The value of variable which has maximum frequency is called the **mode.** In fact, in case of several observations taken during an experiment, a specific observation or observations may come several times repeatedly. The observation which occurs the largest number of times represents the mode of the observations. When concentration of data occurs at two or more points on a scale of values, such experiment is referred to as **bimodal** or **multimodal** depending upon the number of such values. This can be illustrated by the following examples :

Original data	Modes
2, 2, 3, 3, 5, 5, 6, 6, 7, 7, 8, 8	No mode
2, 3, 6, 6, 6, 6, 4, 7, 7, 5, 8, 9	One mode, *i.e.*, 6
2, 3, 4, 4, 4, 5, 7, 7, 7, 6, 8, 9	Two modes, *i.e.*, 4 and 7 (Bimodal)
2, 2, 3, 3, 4, 4, 5, 6, 6, 7, 8, 8	Five modes, *i.e.*, 2, 3, 4, 6 and 8 (Multimodal)

Since modal values are of very little statistical significance, hence, modes are of little use in statistical computation.

MEASURES OF DISPERSION

Although the measures of central tendency give us an idea of central value around which most values are clustered, but it says nothing about how they are clustered. Therefore, we need numerical measures to describe or depict the **dispersion** or the spread of data around central value. Such measures of dispersion, which are most often used, are expressed as range, mean deviation, variance and standard deviation.

1. Range. In the process of measurement, we can notice a range of spread of observations showing the deviations between the lowest and the highest values of the variable. This gives us an idea of the amount of variability existing in the observations as is evident from the following example :

For data 1, 2, 3, 4, 5, 6, 7. ; ($\Sigma x = 28$; n 7)

$$\bar{x}\text{ (mean)} = \frac{\Sigma x}{n} = \frac{28}{7} = 4$$

Range = 1 to 7, *i.e.*, 6

The range when given along with mean provides useful informations, usually when the data is a normal distribution.

2. Mean deviation. The mean deviation is an easier way of measuring dispersion with little statistical implications. For calculating mean deviation, first of all mean, of all observations in consideration, is calculated. Then deviation for each observation from mean value is computed and all such values are summed up. While computing deviation from mean value, the values may be positive or negative, depending upon whether the individual observation is greater or lesser than the mean, but they are not taken into account and mean deviation is calculated by using the formula $\Sigma\ |d|/\ n$ or $\Sigma f|d|/\Sigma f$; where $|\ d\ |$ represents that all values of deviation from mean, *i.e.*, **d** will be positive.

An **example** is given below to illustrate the computation of mean deviation when only observation data (x) is known :

Calculation of mean deviation from a given sample :

Observation no.	Observation (x)	Deviation $\lvert x-\bar{x} \rvert = \lvert d \rvert$	Mean deviation
1	3	3	
2	4	2	
3	5	1	$\frac{\Sigma \lvert d \rvert}{n} = \frac{12}{7}$
4	6	0	$= 1.71$
5	7	1	
6	8	2	
7	9	3	
n = 7	$\Sigma x = 42$; Mean $= \frac{42}{7} = 6$ ($\bar{x}$)	$\Sigma \lvert d \rvert = 12$	

(**Note.** Parallel lines on either sides show that all algebraic signs treated as positive).

An **other example** is given below to illustrate the computation of mean deviation when observation data (x) and their respective frequencies are known :

Observation (x)	Frequency (f)	fx	Deviation $\lvert x-\bar{x} \rvert = \lvert d \rvert$	$f \lvert d \rvert$
15	1	15	3	3
16	2	32	2	4
17	2	34	1	2
18	2	36	0	0
19	2	38	1	2
20	2	40	2	4
21	2	42	3	6
	$\Sigma f = 13$	$\Sigma fx = 237$		$\Sigma f \lvert d \rvert = 21$

$$\text{Mean } (\bar{x}) = \frac{\Sigma fx}{\Sigma f} = \frac{237}{13} = 18.23 \text{ ; say } 18$$

$$\text{Mean deviation} = \frac{\Sigma f \lvert d \rvert}{\Sigma f} = \frac{21}{13} = 1.61$$

3. Variance. Variance also depends on the deviations where squared deviations are summed up. To calculate variance, first the deviations, are squared and then summed up. This sum is divided by the number of observations to get the variance of the sample. Since the squaring makes all deviations (+ or –) positive, hence, the sum of squares of deviations is always positive. Therefore, *variance is defined as the mean of squares of deviations.* The sample variance, s^2 may be calculated as :

$$s^2 = \frac{\Sigma (x-\bar{x})^2}{n} \text{ or } \frac{\Sigma d^2}{n}$$

where, $\bar{x}$ = mean of the sample

n = number of observations.

When the data are grouped in the form of a **frequency distribution,** the variance, $s^2 = \Sigma f d^2 / \Sigma f$.

An illustration to compute variance from the same data which are considered in calculating mean deviation is given below :

Observation (x)	**Frequency** (f)	**Deviation** (d)	**Deviation²** (d^2)	$f \times d^2$
15	1	3	9	9
16	2	2	4	8
17	2	1	1	2
18	2	0	0	0
19	2	1	1	2
20	2	2	4	8
21	2	3	9	18
	$\Sigma f = 13$			$\Sigma f d^2 = 47$

Mean $\bar{x} = 18$; $\Sigma f d^2 = 47$; $\Sigma f = 13$

$$\text{Variance, } s^2 = \frac{\Sigma f d^2}{\Sigma f} = \frac{47}{13} = 3.61$$

4. Standard deviation. Standard deviation is defined as the positive square root of the arithmetic mean of the squares of deviation of the observations from the arithmetic mean. It may also be defined as the square root of the variance. It is also known as ''root mean square deviation'' and generally denoted by small Greek letter σ (called Sigma) and also as $\mathbf{S^D}$. It may be calculated by the following formulae :

(i) $S^D = \sqrt{\frac{\Sigma d^2}{n-1}}$ used when data not grouped into frequency.

OR

(ii) $S^D = \sqrt{\frac{\Sigma f d^2}{\Sigma f}}$ used when data grouped into frequency distribution.

However, both these are illustrated below:

Illustration 1

The measurements of post-clitellar region of 10 earthworms show following observations, for which S^D is to be calculated:

Earthworm no.	1	2	3	4	5	6	7	8	9	10
Length (cm)	10.8	7.5	8.6	8.8	9.7	9.1	8.3	9.8	9.4	7.0

In this case S^D can be calculated by using the formula $\sqrt{\Sigma d^2 / n - 1}$; where Σd^2 = summation of squared deviation and n = number of observations.

Tabulation of data and calculation :

Length in cm (x)	$x - \bar{x}$ (d)	d^2
10.8	1.9	3.61
7.5	–1.4	1.96
8.6	–0.3	0.09
8.8	–0.1	0.01
9.7	0.8	0.64
9.1	0.2	0.04
8.3	–0.6	0.36
9.8	0.9	0.81
9.4	0.5	0.25
7.0	–1.9	3.61
$\Sigma = 89$ $\bar{x} = 8.9$		$\Sigma d^2 = 11.38$

$$\text{Mean} = \frac{\Sigma x}{n} = \frac{89}{10} = 8.9$$

$$\Sigma d^2 = 11.38$$

$$S^D = \sqrt{\frac{\Sigma d^2}{n-1}}$$

$$= \sqrt{\frac{11.38}{10-1}} = \sqrt{\frac{11.38}{9}} = \sqrt{1.26}$$

$$= 1.12$$

Hence, $S^D = 1.12$ cm

Illustration 2

Standard deviation from the following data is to be calcualted :

Items	15 – 17,	18 – 20,	21 – 23,	24 – 26,	27 – 29,	30 – 32,	33 – 35,	36 – 38,	39 – 41
Frequency	5	6	8	12	22	18	15	9	5

In this case S^D can be calculated by using the formula $\sqrt{\Sigma f d^2/\Sigma f}$; where $\Sigma f d^2$ = summation squared deviations multiplied by their respective frequencies and Σf = summation of frequencies.

Tabulation and calculation :

Item	Midpoint m	Frequency f	fm	$m - \bar{x}$ (d)	d^2	$f d^2$
15 – 17	16	5	80	–13	169	845
18 – 20	19	6	114	–10	100	600
21 – 23	22	8	176	–7	49	392
24 – 26	25	12	300	–4	16	192
27 – 29	28	22	616	–1	1	22
30 – 32	31	18	558	+2	4	72
33 – 35	34	15	510	+5	25	375
36 – 38	37	9	333	+8	64	576
39 – 41	40	5	200	+11	121	605
		$\Sigma f = 100$	$\Sigma fm = 2887$ $\bar{x} = 29$			$\Sigma fd^2 = 3679$

$$\text{Mean} = \frac{\Sigma fm}{n} = \frac{2887}{100}$$

$$= 28.87 \text{ ; say, } 29$$

$$S^D = \sqrt{\frac{\Sigma f d^2}{\Sigma f}} = \sqrt{\frac{3679}{100}} = \sqrt{36.79}$$

$$= 6.065$$

Standard error. Standard error is also referred to as **standard error of mean.** (SEM). It is a quality which can be calculated directly from standard deviation by using the following formula :

$$\text{SEM} = \frac{\text{Standard deviation}}{\sqrt{n}} = \frac{S^D}{\sqrt{n}}$$

Where, n represents the number of replicates or observations.

Coefficient of Variation

Whenever we compute the S^D of two samples or populations having different means, the S^D alone does not give us correct comparative idea about the extent of variations in two samples. If this S^D is expressed as percentage of the mean, then it will provide us a comparative idea of the extent of variation and will be useful for us in comparing the different samples. So, this unit is called **coefficient of variation** (c.v.) and can be calculated by using the following formula :

$$\text{Coefficient of variation} = \frac{\text{Standard deviation}}{\text{Mean}} \times 100$$

$$\left(\text{or C.V.} = \frac{\sigma}{\bar{x}} \times 100\right)$$

Illustration

From the following raw data one is to calculate the mean number of seeds per fruit in each bunch and which of the two bunches has the lower range of values or which of the two has more consistent result ?

Fruit No.	– 1	2	3	4	5	6	7	8	9	10
No. of seeds, bunch 1 (x_1)	– 7	8	6	9	6	8	7	8	5	6
No. of seeds, bunch 2 (x_2)	– 4	7	10	8	11	10	6	9	5	10

Tabulation of data and calculation :

	Bunch–1			Bunch–2		
Fruit No.	x_1	$x_1 - \bar{x}_1 = d$	d^2	x_2	$x_2 - \bar{x}_2 = d$	d^2
1	7	0	0	4	–4	16
2	8	+1	1	7	–1	1
3	6	–1	1	10	+2	4
4	9	+2	4	8	0	0
5	6	–1	1	11	+3	9
6	8	+1	1	10	+2	4
7	7	0	0	6	–2	4
8	8	+1	1	9	+1	1
9	5	–2	4	5	–3	9
10	6	–1	1	10	+2	4
n = 10	$\Sigma x_1 = 70$ $\bar{x}_1 = 7$		$\Sigma d^2 = 14$	$\Sigma x_2 = 80$ $\bar{x}_2 = 8$		$\Sigma d^2 = 52$

$n_1 = 10$
$\Sigma x_1 = 70$
$\bar{x}_1 = 7$

$$S^D = \sqrt{\frac{\Sigma d^2}{n-1}} = \sqrt{\frac{14}{10-1}} = \sqrt{\frac{14}{9}}$$
$$= \sqrt{1.56} = 1.25$$

$$C.V. = \frac{S^D}{Mean} \times 100$$
$$= \frac{1.25}{7} \times 100$$
$$= 17.86\%$$

$n_1 = 10$
$\Sigma x_2 = 80$
$\bar{x}_2 = 8$

$$S^D = \sqrt{\frac{\Sigma d^2}{n-1}} = \sqrt{\frac{52}{10-1}} = \sqrt{\frac{52}{9}}$$
$$= \sqrt{5.8} = 2.4$$

$$C.V. = \frac{S^D}{Mean} \times 100$$
$$= \frac{2.4}{8} \times 100$$
$$= 30\%$$

From the above two comparative results, it is evident that the fruits in second bunch have relatively more number of seeds per fruit because its mean is greater than the mean of first bunch. The number of seeds per fruit in second bunch is higher or in other words the variation in seed number in the fruits of second bunch is more, *i.e.*, the range is higher compared to the first bunch.

TESTS OF SIGNIFICANCE

Biological experiments covering different spheres often involve data work. All such data could be statistically analysed and can be said whether significant or insignificant. In fact, the test of significance concerns with finding out, between two items showing variation, whether the variation is significant or not. For this purpose, various tests of significances are applied, such as Z-test, t-test, chi-square test, etc. At present t-test and chi-square tests of significance have been illustrated.

t-test. In calculating t – test, the first step is to formulate a hypothesis referred to as **null hypothesis (H_O)** that there is no significant difference between two values or figures. For every null hypothesis, there is an alternate hypothesis called $\mathbf{H_1}$ which reveals that there is a significant difference.

The objective of t-test is to find out that which of the hypothesis is correct. If null hypothesis is rejected, it is concluded that difference is significant. For making decisions in biological experiments/problems, 5 per cent probability (**P = 0.05**) is accepted which means that there are only 5 per cent chances that the observed differences are due to variation. Therefore, if the value of significance level comes less than 5 per cent, the hypothesis is accepted and if value exceeds more than 5 per cent, the hypothesis is rejected. t is, however, calculated by the following formulae :

$$t = \frac{\bar{x}\ Mo}{S^D/\sqrt{n}} \quad \text{(For one sample), and}$$

$$t = \frac{\bar{x}_1 - \bar{x}_2}{\sqrt{\frac{S_1^{D^2}}{n_1 - 1} + \frac{S_2^{D^2}}{n_2 - 1}}} \quad \text{(For two samples)}$$

where; M_O = Hypothetical population mean (given)

S^D = Standard deviation

n, n_1, n_2 = Sample size (number of values/observations)

$x, \bar{x}_1, \bar{x}_2$ = Means of sample data

Illustration

Ten plants of *Pisum sativum* are chosen from a population at random whose heights are cm, 55 cm, 57 cm, 61 cm, 64 cm, 65 cm, 67 cm, 68 cm, 70 cm and 71 cm. Population mean 60 c From these data t is to be calculated to know whether difference in variation is significant or not

Calculation

Prepare a table as given below to find out sample mean and standard deviation.

S. No.	Hieght (cm) X	d (= X – 60)	d^2
1	52	– 8	64
2	55	– 5	25
3	57	– 3	9
4	61	1	1
5	64	4	16
6	65	5	25
7	67	7	49
8	68	8	64
9	70	10	100
10	71	11	121
Total	630	30	$\Sigma d^2 = 474$

Here given mean = 60

$$\text{Now } \overline{X} = \frac{\Sigma X}{N} = \frac{630}{10} = 63$$

$$\Sigma (X - \overline{X})^2 = \Sigma d^2 - \frac{(\Sigma d)^2}{N}$$

$$= 474 - \frac{(30)^2}{10}$$

$$= 384$$

Hence,

$$S^D = \sqrt{\frac{\Sigma (X - \overline{X})^2}{N-1}}$$

$$= \sqrt{\frac{384}{9}} = 6.53$$

$$\text{S.E. of } \overline{X} = \frac{S^D}{\sqrt{N}} = \frac{6.53}{\sqrt{10}} = 2.066$$

$H_O = 60$ (given)

$$t = \frac{\overline{X} - H_O}{S^D/\sqrt{n}} = \frac{63–60}{2.065} = \frac{3}{2.065} = 1.45$$

Degree of fredom = n – 1 = 9

Conclusion. In this case, the observed value of t is 1.45 which is less than 2.262 (tabu value of t at 5% probability and 9 degree of freedom; see table t for its value). It proves that variation in the height of plants is insignificant. Therefore, the hypothesis that the mean height plants is 60 cm is acceptable. H_o is correct.

Similarly t test for two samples can be calculated.

Chi-square test (Test for Goodness of fit). It is a test most commonly used in case genetical experiments to know how far the observed data agree with the expected one.

Uses of Chi-square test. This test is used for the following determinations :

1. It is used to determine if a certain distribution differs from some predetermined (parametr theoretical distribution.

2. It is used in testing hypothesis concerning the significance of the differences of responses of two or more groups to the effect of treatment of one or other.

3. It is used in goodness of fit.

Chi-square χ^2 is calculated by the following formulae ;

1. $\chi^2 = \sum \frac{(O - E)^2}{E}$ (used for **one sample**)

   where, O = Observed frequency in a class.
   E = Expected frequency in a class.
   Σ = summation of over all classes.

2. $\chi^2 = \sum \frac{(O_1 - E_1)^2}{E_1} + \frac{(O_2 + E_2)^2}{E_2}$ (used for **two samples**)

   Or, it is also written as for the number of observation (N), as under :

   $\chi^2 = \sum^{k} \frac{(O_i - E_i)^2}{E_i}$ Where, i = 1, and degree of freedom = k – 1

If the value of χ^2 is greater than the value of chi square at specified degree of freedom (k–1) at **P = 0.05** (given in χ^2 table), the difference will be significant and the data rejects the theory or fact (null hypothesis). And if the value of χ^2 of sample is lower than its value in the given table, then the data accepts the theory and the variation in data is significant. Such data indicate **Goodness of fit**.

To know whether the data is highly significant, the value of χ^2 is compared at **P = 0.01** (1 per cent probability).

Properties of Chi-square. Its following properties are to be noted :

1. In case where observed data and hypothetical data show perfect agreement, the value of χ^2 is zero. With the increase in the value of χ^2, the variation from accepted theory increases and goes on to infinity.

2. In its application, all values are taken into account and they are capable of being added together to give a total value of χ^2 based on total number of degree of freedom, *i.e.*,

$$\chi^2 (t) = \chi_1^2 + \chi_2^2 + \chi_3^2 \ldots\ldots + \chi_n^2$$

$$v_T = v_1 + v_2 + v_3 \ldots\ldots + v_n$$

χ_T^2 is also distributed as χ^2 with v_T degree of freedom.

3. It can be used where degree of freedom in the data is more than 30.

4. It should be used where number of observations are sufficiently large. The theoretical frequency should be atleast 10, but is no case less than 5.

Degree of freedom (d.f.). It is calculated as n–1. If there is only one constraint that the total of all frequencies is constant, *e.g.*, in Mendelian dihybrid cross, the total number of class is 4; the **d.f** will be n–1 = 4–1 = **3c** (three degree of freedom).

Illustrations

1. In a cross between black and white coat coloured mice, the F_2 individuals segregated into 787 black and 277 white coat coloured individuals. If you have to test that these results agree with the expected ratio of 3 : 1. Then apply chi- square test as worked out below :

Total no. of individuals = 787 + 277 – 1064

Applying ratio 3 : 1, expected value of class will be in

$$\text{Black} = 798 \left(1064 \times \frac{3}{4} = 798\right)$$

White = 266 $\left(1064 \times \frac{1}{4} = 266\right)$

The degree of freedom is n – 1 = 1 (2–1 = 1)

Now putting the values in formula ;

$$\chi^2 = \sum \frac{(O_1 - E_1)^2}{E_1} + \frac{(O_2 - E_2)^2}{E_2}$$

$$= \sum \frac{(787 - 798)^2}{798} + \frac{(277 - 266)^2}{266}$$

$$= \sum \frac{(11)^2}{798} + \frac{(11)^2}{266}$$

$$= \sum 0.15 + 0.45$$

$$= 0.6$$

Thus, the value of χ^2 of sample is 0.6 which is much lower than the value of χ^2 at **1 d.f.** at **P = 0.05** which is 3.841 (tabular value of χ^2). The data has a good fit to the ratio 3 : 1 or data accepts the hypotesis of **Goodness of fit**. (Value of $\chi^2_{(1)} = 3.841$).

2. In a cross between black and splashed white Andalusian fowls 140 fowls were obtained bearing black plume colour, 230 with blue plume colour and 130 with splashed white plume colour. If you have to test that these results agree with the expected ratio of incomplete dominance of 1 : 2 : 1. Then apply this test as worked out below :

Prepare the following table after initial calculation of χ^2 in the manner given in illustration 1.

Types	Frequencies		O–E	$(O - E)^2/E$
	Observed	Expected		
Black	140	125	15	1.8
Blue	230	250	– 20	1.6
White splashed	130	125	5	0.2
Total	500	500	Zero	3.6

Thus, the value of χ^2 of the observation is 3.6. The value of χ^2 at **2 d.f.** at **P = 0.05** is 5.991 ($\chi^2_{(2)} = 5.991$), the tabular value of χ^2, is much higher than χ^2 calculated from the observed data. Therefore, the data accepts the hypothesis of incomplete dominance and its ratio 1 : 2 : 1. The variation in observed data is by chance.

APPENDIX I

Table 1. Table of Critical Values of t

	Level of significance for one-tailed test					
df	.10	.05	.025	.01	.005	.0005
	Level of significance for two-tailed test					
	.20	.10	.05	.02	.01	.001
1.	3.078	6.314	12.706	31.821	63.657	636.619
2.	1.886	2.920	4.303	6.965	9.925	31.598
3.	1.638	2.353	3.182	4.541	5.841	12.941
4.	1.533	2.132	2.776	3.747	4.604	8.610
5.	1.476	2.015	2.571	3.365	4.032	6.859
6.	1.440	0.943	2.447	32.143	3.707	5.959
7.	1.415	1.895	2.365	2.998	3.499	5.405
8.	1.397	1.860	2.306	2.896	3.355	5.041
9.	1.383	1.833	2.262	2.821	3.250	4.781
10.	1.372	1.812	2.228	2.764	3.169	4.587
11.	1.363	1.796	2.201	2.718	3.106	4.437
12.	1.356	1.782	2.179	2.681	3.055	4.318
13.	1.350	1.771	2.160	2.650	3.012	4.221
14.	1.345	1.761	2.145	2.624	2.977	4.140
15.	1.341	1.753	2.131	2.602	2.947	4.073
16.	1.337	1.746	2.120	2.583	2.921	4.015
17.	1.333	1.740	2.110	2.567	2.898	3.965
18.	1.330	1.734	2.101	2.552	2.878	3.922
19.	1.328	1.720	2.093	2.539	2.861	3.883
20.	1.325	1.725	2.086	2.528	2.845	3.850
21.	1.323	1.721	2.080	2.518	2.831	3.819
22.	1.321	1.717	2.074	2.508	2.819	3.792
23.	1.319	1.714	2.069	2.500	2.807	3.767
24.	1.318	1.711	2.064	2.492	2.797	3.745
25.	1.316	1.708	2.606	2.485	2.787	3.725
26.	1.315	1.706	2.056	2.479	2.779	3.707
27.	1.314	1.703	2.052	2.473	2.771	3.690
28.	1.313	1.701	2.048	2.467	2.763	3.674
29.	1.311	1.699	2.045	2.462	2.756	3.659
30.	1.310	1.697	2.042	2.457	2.750	3.646
40.	1.303	1.684	2.021	2.423	2.704	3.551
60.	1.296	1.671	2.000	2.390	2.660	3.460
120.	1.289	1.658	1.980	2.358	2.617	3.373
∞	1.282	1.645	1.960	2.326	2.576	3.291

APPENDIX II

Table 2 : Table of Critical Values of Chi Square

df	Probability under H_0 that $x^2 \geq$ chi square													
	.99	.98	.95	.90	.80	.70	.50	.30	.20	.10	.05	.02	.01	.001
1.	.00016	.00063	.0039	.016	.064	.15	.46	1.07	1.64	2.71	3.84	5.41	6.64	10.83
2.	.02	.04	.10	.21	.45	.71	1.39	2.41	3.22	4.60	5.99	7.82	9.21	13.82
3.	.12	.18	.35	.58	1.00	1.42	2.37	3.66	4.64	6.25	7.82	9.84	11.34	16.27
4.	.30	.43	.71	1.06	1.65	2.20	3.36	4.88	5.99	7.78	9.49	11.67	13.28	18.46
5.	.55	.75	1.14	1.61	2.34	3.00	4.35	6.06	7.29	9.24	11.07	13.39	15.09	20.52
6.	.87	1.13	1.64	2.20	3.07	3.83	5.35	7.23	8.56	10.64	12.59	15.03	16.81	22.46
7.	1.24	1.56	2.17	2.83	3.82	4.67	6.35	8.38	9.80	12.02	14.07	16.62	18.48	24.32
8.	1.65	2.03	2.73	3.49	4.59	5.53	7.34	9.52	11.03	13.36	15.51	18.17	20.09	26.12
9.	2.09	2.53	3.32	4.17	5.38	6.39	8.34	10.66	12.24	14.68	16.92	19.68	21.67	27.88
10.	2.56	3.06	3.94	4.86	6.18	7.27	9.34	11.78	13.44	15.99	18.31	21.16	23.21	29.59
11.	3.05	3.61	4.58	5.58	6.99	8.15	10.34	12.90	14.63	17.28	19.68	22.62	24.72	31.26
12.	3.57	4.18	5.23	6.30	7.81	9.03	11.34	14.01	15.81	18.55	21.03	24.05	26.22	32.91
13.	4.11	4.76	5.89	7.04	8.63	9.93	12.34	15.12	16.98	19.81	22.36	25.47	27.69	34.53
14.	4.66	5.37	6.57	7.79	9.47	10.82	13.34	16.22	18.15	21.06	23.68	26.87	29.14	36.12
15.	5.23	5.98	7.26	8.55	10.31	11.72	14.34	17.32	19.31	22.31	25.00	28.26	30.58	37.70
16.	5.81	6.61	7.96	9.31	11.15	12.62	15.34	18.42	20.46	23.54	26.30	29.63	32.00	39.29
17.	6.41	7.26	8.67	10.08	12.00	13.53	16.34	19.51	21.62	24.77	27.59	31.00	33.41	40.75
18.	7.02	7.91	9.39	10.86	12.86	14.44	17.34	20.60	22.76	25.99	28.87	32.35	34.80	42.31
19.	7.63	8.57	10.12	11.65	13.72	15.35	18.34	21.69	23.90	27.20	30.14	33.69	36.19	43.82
20.	8.26	9.24	10.85	12.44	14.58	16.27	19.34	22.78	25.04	28.41	31.41	35.02	37.57	45.32
21.	8.90	9.92	11.59	13.24	15.44	17.18	20.34	23.86	26.17	29.62	32.67	36.34	38.93	46.80
22.	9.54	10.60	12.34	14.04	16.31	18.10	21.24	24.94	27.30	30.81	33.92	37.66	40.29	48.27
23.	10.20	11.29	13.09	14.85	17.19	19.02	22.34	26.02	28.43	32.01	35.17	38.97	41.64	49.73
24.	10.86	11.99	13.85	15.66	18.06	19.94	23.34	27.10	29.55	33.20	36.42	40.27	42.98	51.18
25.	11.52	12.70	14.61	16.47	18.94	20.87	24.34	28.17	30.68	34.38	37.65	41.57	44.31	52.62
26.	12.20	13.41	15.38	17.29	19.82	21.79	25.34	29.25	31.80	39.56	39.88	42.86	45.64	54.05
27.	12.88	14.12	16.15	18.11	20.70	22.72	26.34	30.32	32.91	36.74	40.11	44.14	46.96	55.48
28.	13.56	14.85	16.93	18.94	21.59	23.65	27.34	31.39	34.03	37.92	41.34	45.42	48.28	56.89
29.	14.26	15.57	17.71	19.77	22.48	24.58	28.34	32.46	35.14	39.09	42.56	46.69	49.59	58.30
30.	14.95	16.31	18.49	20.60	23.36	25.51	29.34	33.53	36.25	40.26	43.77	47.96	50.89	59.70

Zoology

For Degree Students

B.Sc. (Hons.) Semester - II

V.K. AGARWAL

ISBN: 9789352534111 | Pages: 952

Zoology

For Degree Students

B.Sc. (Hons.) Semester - III

V.K. AGARWAL

ISBN: 9789352533114 | Pages: 976

Zoology

For Degree Students

B.Sc. (Hons.) Semester - IV

V.K. AGARWAL

ISBN: 9789352534104 | Pages: 752

Cell Biology

P.S. VERMA
V.K. AGARWAL

ISBN: 9789385676147 | Pages: 1208

Environmental Biology

P.S. VERMA
V.K. AGARWAL

ISBN: 9788121908597 | Pages: 600

General and Applied Ichthyology

S.K. GUPTA
P.C. GUPTA

ISBN: 9788121926294 | Pages: 1160